Undergraduate Texts in Mathematics

Undergraduate Texts in Mathematics

(continued after Index)

John Stillwell

Mathematics
and Its History

With 163 Illustrations

Springer-Verlag
New York Berlin Heidelberg
London Paris Tokyo

John Stillwell
Department of Mathematics
Monash University
Clayton, Victoria 3168
Australia

Cover illustration of Euler's elastica was taken from *Isis*, **20** (1933), plates between pp. 72 and 73. Used with permission of *Isis*, Madison, Wisconsin.

Mathematics Subject Classification: 01-01

Library of Congress Cataloging-in-Publication Data
Stillwell, John.
  Mathematics and its history/John Stillwell.
    p. cm.—(Undergraduate texts in mathematics)
  Bibliography: p.
1. Mathematics—History. I. Title. II. Series.
QA21.S84 1989
510′.9—dc19

Printed on acid-free paper

Typeset by Asco Trade Typesetting Ltd., Hong Kong.
Printed and bound by R. R. Donnelley, Harrisonburg, Virginia.
Printed in the United States of America.

9 8 7 6 5 4 3 2 1

ISBN 0-387-96981-0 Springer-Verlag New York Berlin Heidelberg
ISBN 3-540-96981-0 Springer-Verlag Berlin Heidelberg New York

To Elaine, Michael, and Robert

Preface

One of the disappointments experienced by most mathematics students is that they never get a course in mathematics. They get courses in calculus, algebra, topology, and so on, but the division of labor in teaching seems to prevent these different topics from being combined into a whole. In fact, some of the most important and natural questions are stifled because they fall on the wrong side of topic boundary lines. Algebraists do not discuss the fundamental theorem of algebra because "that's analysis" and analysts do not discuss Riemann surfaces because "that's topology," for example. Thus if students are to feel they really know mathematics by the time they graduate, there is a need to unify the subject.

This book aims to give a unified view of undergraduate mathematics by approaching the subject through its history. Since readers should have had some mathematical experience, certain basics are assumed and the mathematics is not developed as formally as in a standard text. On the other hand, the mathematics is pursued more thoroughly than in most general histories of mathematics, as mathematics is our main goal and history only the means of approaching it. Readers are assumed to know basic calculus, algebra, and geometry, to understand the language of set theory, and to have met some more advanced topics such as group theory, topology, and differential equations. I have tried to pick out the dominant themes of this body of mathematics and to weave them together as strongly as possible by tracing their historical development.

In doing so, I have also tried to tie up some traditional loose ends. For example, undergraduates can solve quadratic equations. Why not cubics? They can integrate $1/\sqrt{1-x^2}$ but are told not to worry about $1/\sqrt{1-x^4}$. Why? Pursuing the history of these questions turns out to be very fruitful, leading to a deeper understanding of complex analysis and algebraic geometry, among other things. Thus I hope that the book will be not only a bird's eye view of undergraduate mathematics but also a glimpse of wider horizons.

Some historians of mathematics may object to my anachronistic use of modern notation and (fairly) modern interpretations of classical mathematics. This has certain risks, such as making the mathematics look simpler than it really was in its time, but the risk of obscuring ideas by cumbersome, unfamiliar notation is greater, in my opinion. Indeed, it is practically a truism that mathematical ideas generally arise before there is notation or language to express them clearly, and that ideas are implicit before they become explicit. Thus the historian, who is presumably trying to be both clear and explicit, often has no choice but to be anachronistic when tracing the origins of ideas.

Mathematicians may object to my choice of topics, since a book of this size is necessarily incomplete. My preference has been for topics with elementary roots and strong interconnections. The major themes are the concepts of number and space: their initial separation in Greek mathematics, their union in the geometry of Fermat and Descartes, and the fruits of this union in calculus and algebraic geometry. Certain important topics of today, such as Lie groups and functional analysis, are omitted on the grounds of their comparative remoteness from elementary roots. Others, such as probability theory, are mentioned only briefly, as most of their development seems to have occurred outside the mainstream. For any other omissions or slights I can only plead personal taste and a desire to keep the book within the bounds of a one- or two-semester course.

The book has grown from notes for a course given to senior undergraduates at Monash University over the past few years. The course was of half-semester length and a little over half the book was covered (Chapters 1–11 one year and Chapters 5–15 another year). Naturally I will be delighted if other universities decide to base a course on the book. There is plenty of scope for custom course design by varying the periods or topics discussed. However, the book should serve equally well as general reading for the student or professional mathematician.

Biographical notes have been inserted at the end of each chapter, partly to add human interest but also to help trace the transmission of ideas from one mathematician to another. These notes have been distilled mainly from secondary sources, the *Dictionary of Scientific Biography* (*DSB*) normally being used in addition to the sources cited explicitly. I have followed the *DSB*'s practice of describing the subject's mother by her maiden name. References are cited in the name [year] form, for example, Newton [1687] refers to the *Principia*, and the references are collected at the end of the book.

The manuscript has been read carefully and critically by John Crossley, Jeremy Gray, George Odifreddi, and Abe Shenitzer. Their comments resulted in innumerable improvements, and any flaws still remaining may be due to my failure to follow all their advice. To them, and to Anne-Marie Vandenberg for her usual excellent typing, I offer my sincere thanks.

Victoria, Australia   JOHN STILLWELL

Contents

CHAPTER 1

The Theorem of Pythagoras

1.1. Arithmetic and Geometry

If there is one theorem that is familiar to all mathematically educated people, it is surely the theorem of Pythagoras. It will probably be recalled as a property of right-angled triangles: the square of the hypotenuse equals the sum of the squares of the other two sides (Fig. 1.1). The "sum" is of course the sum of areas and the area of a square of side l is l^2, which is why we call it "l squared." Thus Pythagoras' theorem can also be expressed by the equation

$$a^2 + b^2 = c^2 \qquad (1)$$

where a, b, c are the lengths shown in Figure 1.1.

Conversely, a solution of (1) by positive numbers a, b, c can be realized by a right-angled triangle with sides a, b and hypotenuse c. It is clear that we can draw perpendicular sides a, b for any given positive numbers a, b, and then the hypotenuse c must be a solution of (1) to satisfy Pythagoras' theorem. This converse view of Pythagoras' theorem becomes interesting when we notice that (1) has some very simple solutions. For example

$$(a, b, c) = (3, 4, 5) \qquad (3^2 + 4^2 = 9 + 16 = 25 = 5^2)$$

$$(a, b, c) = (5, 12, 13) \qquad (5^2 + 12^2 = 25 + 144 = 169 = 13^2)$$

It is thought that in ancient times such solutions may have been used for the construction of right angles. For example, by stretching a closed rope with 12 equally spaced knots one can obtain a $(3, 4, 5)$ triangle with right angle between the sides 3, 4, as seen in Figure 1.2.

Whether or not this is a practical method for constructing right angles, the very existence of a geometrical interpretation of a purely arithmetical fact like

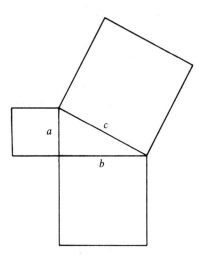

Figure 1.1

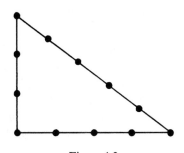

Figure 1.2

$$3^2 + 4^2 = 5^2$$

is quite wonderful. At first sight, arithmetic and geometry seem to be completely unrelated realms. Arithmetic is based on counting, the epitome of a *discrete* (or *digital*) process. The facts of arithmetic can be clearly understood as outcomes of certain counting processes, and one does not expect them to have any meaning beyond this. Geometry, on the other hand, involves *continuous*, rather than discrete objects, such as lines, curves, and surfaces. Continuous objects cannot be built from simpler elements by discrete processes, and one expects to *see* geometrical facts rather than arrive at them by calculation.

   Pythagoras' theorem was the first hint of a hidden, deeper relationship between arithmetic and geometry, and it has continued to hold a key position between these two realms throughout the history of mathematics. This has sometimes been a position of cooperation and sometimes one of conflict, as followed by the discovery that $\sqrt{2}$ is irrational (see Section 1.5). It is often the case that new ideas emerge from such areas of tension, resolving the conflict and allowing previously irreconcilable ideas to interact fruitfully. The tension between arithmetic and geometry is, without doubt, the most profound in mathematics, and it has led to the most profound theorems. Since Pythagoras' theorem is the first of these, and the most influential, it is a fitting subject for our first chapter.

1.2. Pythagorean Triples

Pythagoras lived around 500 B.C. (see 1.7), but the story of Pythagoras' theorem begins long before that, at least as far back as 1800 B.C. in Babylonia. The evidence is a clay tablet, known as Plimpton 322, which systematically lists a large number of integer pairs (a, c) for which there is an integer b satisfying

$$a^2 + b^2 = c^2 \tag{1}$$

A translation of this tablet, together with its interpretation and historical background, was first published by Neugebauer and Sachs [1945] (for a more recent discussion, see van der Waerden [1983], p. 2). Integer triples (a, b, c) satisfying (1)—for example, (3, 4, 5), (5, 12, 13), (8, 15, 17)—are now known as *Pythagorean triples*. Presumably the Babylonians were interested in them because of their interpretation as sides of right-angled triangles, though this is not known for certain. At any rate, the problem of finding Pythagorean triples was considered interesting in other ancient civilizations that are known to have possessed Pythagoras' theorem. Van der Waerden [1983] gives examples from China (between 200 B.C. and A.D. 220) and India (between 500 and 200 B.C.). The most complete understanding of the problem in ancient times was achieved in Greek mathematics, between Euclid (c. 300 B.C.) and Diophantus (c. A.D. 250).

We now know that the general formula for generating Pythagorean triples is

$$a = (p^2 - q^2)r, \qquad b = 2pqr, \qquad c = (p^2 + q^2)r$$

It is easy to see that $a^2 + b^2 = c^2$ when a, b, c are given by these formulas, and of course a, b, c will be integers if p, q, r are. Even though the Babylonians did not have the advantage of our algebraic notation, it is plausible that this formula, or the special case

$$a = p^2 - q^2, \qquad b = 2pq, \qquad c = p^2 + q^2$$

(which gives all solutions a, b, c without common factor) was the basis for triples they listed. Less general formulas have been attributed to Pythagoras himself (c. 500 B.C.) and Plato (see Heath [1921], pp. 80–81, Vol. 1); a solution equivalent to the general formula is given in Euclid's *Elements*, Book X (lemma following Prop. 28). As far as we know, this is the first statement of the general solution and the first proof that it is general. Euclid's proof is essentially arithmetical, as one would expect since the problem seems to belong to arithmetic.

However, there is a far more striking solution, which uses the geometric interpretation of Pythagorean triples. This emerges from the work of Diophantus, and it is described in the next section.

EXERCISE

1. If (a, b, c) is a Pythagorean triple and a, b, c have no common factor, show that one of a, b is even and the other is odd.

1.3. Rational Points on the Circle

We know from 1.1 that a Pythagorean triple (a, b, c) can be realized by a triangle with sides a, b and hypotenuse c. This in turn yields a triangle with fractional (or *rational*) sides $x = a/c$, $y = b/c$ and hypotenuse 1. All such triangles can be fitted inside the circle of radius 1 as shown in Figure 1.3. The sides x and y become what we now call the *coordinates* of the point P on the circle. The Greeks did not use this language; however, they could derive the relationship between x and y we call the *equation of the circle*. Since

$$a^2 + b^2 = c^2 \tag{1}$$

we have

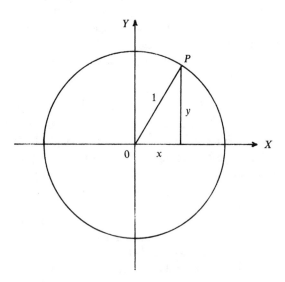

Figure 1.3

$$\left(\frac{a}{c}\right)^2 + \left(\frac{b}{c}\right)^2 = 1$$

so the relationship between $x = a/c$ and $y = b/c$ is

$$x^2 + y^2 = 1 \tag{2}$$

Consequently, finding integer solutions of (1) is equivalent to finding rational solutions of (2), or finding *rational points* on the curve (2).

Such problems are now called *Diophantine*, after Diophantus, who was the first to deal with them seriously and successfully. *Diophantine equations* have acquired the more special connotation of equations for which integer solutions are sought, although Diophantus himself sought only rational solutions. (There is an interesting open problem which turns on this distinction. Matiasevich [1970] proved that there is no algorithm for deciding which polynomial equations have integer solutions. It is not known whether there is an algorithm for deciding which polynomial equations have *rational* solutions.)

Most of the problems solved by Diophantus involve quadratic or cubic equations, usually with one obvious trivial solution. Diophantus used the obvious solution as a steppingstone to the nonobvious, but no account of his method survived. It was ultimately reconstructed by Fermat and Newton in the seventeenth century, and its general form will be considered later. At present, we need it only for the equation $x^2 + y^2 = 1$, which is an ideal showcase for the method in its simplest form.

A trivial solution of this equation is $x = -1$, $y = 0$, which is the point Q on the unit circle (Fig. 1.4). After a moment's thought, one realizes that a line through Q, with rational slope t,

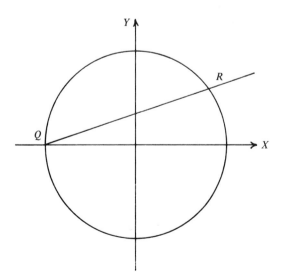

Figure 1.4

$$y = t(x + 1) \tag{1}$$

will meet the circle at a second rational point R. This is because substitution of $y = t(x + 1)$ in $x^2 + y^2 = 1$ gives a quadratic equation with rational coefficients and one rational solution ($x = -1$), hence the second solution must also be a rational value of x. But then the y value of this point will also be rational, since t and x will be rational in (1). Conversely, the line through Q and any other rational point R on the circle will have a rational slope. Thus by letting t run through all rational values, we find all rational points $R \neq Q$ on the unit circle.

What are these points? We find them by solving the equations just discussed. Substituting $y = t(x + 1)$ in $x^2 + y^2 = 1$ gives

$$x^2 + t^2(x + 1)^2 = 1$$

or

$$x^2(1 + t^2) + 2t^2x + (t^2 - 1) = 0$$

This quadratic equation in x has solutions $-1, (1 - t^2)/(1 + t^2)$. The nontrivial solution $x = (1 - t^2)/(1 + t^2)$, substituted in (1), gives $y = 2t/(1 + t^2)$.

EXERCISES

1. Derive from this the formula for Pythagorean triples given in 1.2.

2. Show that when one rational solution is known Diophantus' method gives all the rational solutions of any quadratic equation $p(x, y) = 0$ with rational coefficients.

3. Show that any right-angled triangle may be approximated arbitrarily closely by one with rational sides.

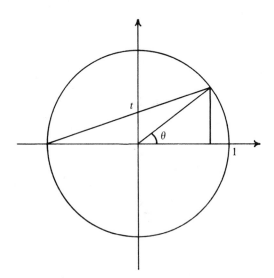

Figure 1.5

4. Use Figure 1.5 to show $t = \tan \theta/2$ and

$$\cos \theta = \frac{1 - t^2}{1 + t^2}, \quad \sin \theta = \frac{2t}{1 + t^2}$$

1.4. Right-angled Triangles

It is high time we looked at Pythagoras' theorem from the traditional point of view, as a theorem about right-angled triangles; however, we shall be rather brief about its proof. It is not known how the theorem was first proved, but probably it was by simple manipulations of area, perhaps suggested by rearrangement of floor tiles. Just how easy it can be to prove Pythagoras' theorem is shown by Figure 1.6 (given by Heath [1925] in his edition of Euclid's *Elements*, Vol. 1, p. 354). Each large square contains four copies of the given right-angled triangle. Subtracting these four triangles from the large square leaves, on the one hand, the squares on the two sides of the triangle (Fig. 1.6, *left*). On the other hand (*right*), it also leaves the square on the hypotenuse. This proof, like the hundreds of others that have been given for Pythagoras' theorem, rests on certain geometric assumptions. It is in fact possible to transcend geometric assumptions by using numbers as the foundation for geometry, and Pythagoras' theorem then becomes true almost

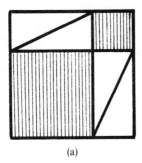
(a)

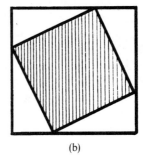
(b)

Figure 1.6

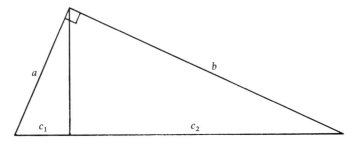

Figure 1.7

by definition, as an immediate consequence of the definition of distance (see Section 1.6).

To the Greeks, however, it did not seem possible to build geometry on the basis of numbers, due to a conflict between their notions of number and length. In the next section we shall see how this conflict arose.

EXERCISE

1. Show that the three triangles in Figure 1.7 are similar, and hence give another proof of Pythagoras' theorem by equating ratios of corresponding sides (Euclid's *Elements*, Book VI, Proposition 31).

1.5. Irrational Numbers

As mentioned, the Babylonians, although probably aware of the geometric meaning of Pythagoras' theorem, devoted most of their attention to the whole number triples it had brought to light, the Pythagorean triples. Pythagoras and his followers were even more devoted to whole numbers. It was they who discovered the role of numbers in musical harmony: dividing a vibrating string in two raises its pitch by an octave, dividing in three raises the pitch another fifth, and so on. This great discovery, the first clue that the physical world might have an underlying mathematical structure, inspired them to seek numerical patterns, which to them meant *whole number* patterns, everywhere. Imagine their consternation when they found that Pythagoras' theorem led to quantities that were not numerically comparable. They found lengths which were *incommensurable*, that is, not measurable as integer multiples of the same unit. The ratio between such lengths is therefore not a ratio of whole numbers, hence in the Greek view not a ratio at all, or *irrational*.

The incommensurable lengths discovered by the Pythagoreans were the side and diagonal of the unit square. It follows immediately from Pythagoras' theorem that

$$(\text{diagonal})^2 = 1 + 1 = 2$$

Hence if the diagonal and side are in the ratio m/n (where m and n can be assumed to have no common factor), we have

$$\frac{m^2}{n^2} = 2$$

whence

$$m^2 = 2n^2$$

The Pythagoreans were interested in odd and even numbers, so they probably observed that the latter equation, which says m^2 is even, also implies that *m* is even. Thus if

$$m = 2p$$

then

$$2n^2 = m^2 = 4p^2$$

hence

$$n^2 = 2p^2$$

which similarly implies *n* is even, contrary to the hypothesis that *m* and *n* have no common factor. (This proof appears in Aristotle's *Prior Analytics*. An alternative, more geometric proof will be mentioned in 3.4.)

This discovery had profound consequences. Legend has it that the first Pythagorean to make the result public was drowned at sea (see Heath [1921], pp. 65, 154, Vol. 1). It led to a split between the theories of number and space which was not healed until the nineteenth century (if then, some mathematicians would add). The Pythagoreans could not accept $\sqrt{2}$ as a number, but no one could deny that it was the diagonal of the unit square. Consequently, geometrical quantities had to be treated separately from numbers or, rather, without mentioning any numbers except rationals. Greek geometers thus developed ingenious techniques for precise handling of arbitrary lengths in terms of rationals, known as the *theory of proportions* and the *method of exhaustion*.

When these techniques were reconsidered in the nineteenth century by Dedekind, he realized that they provided an arithmetical interpretation of irrational quantities after all (Chapter 4). It was then possible, as Hilbert [1899] showed, to reconcile the apparent conflict between arithmetic and geometry. The key role of Pythagoras' theorem in this reconciliation is described in the next section.

1.6. The Definition of Distance

The numerical interpretation of irrationals gave each length a numerical measure and hence made it possible to give coordinates *x*, *y* to each point *P* in the plane. The simplest way is to take a pair of perpendicular lines (*axes*) *OX, OY* and let *x*, *y* be the lengths of the perpendiculars from *P* to *OX* and *OY* respectively (Fig. 1.8). Geometric relations entered by *P* are then reflected by arithmetical relations between *x* and *y*. This opens up the possibility of *analytic geometry*, whose development will be discussed in Chapter 6. Here we want only to see how coordinates give a precise meaning to the basic geometric notion of *distance*.

We have already said that the perpendicular distances from *P* to the axes are the numbers *x*, *y*. The distance between points on the same perpendicular

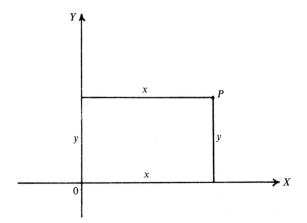

Figure 1.8

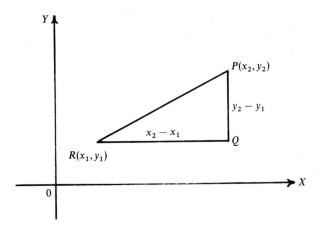

Figure 1.9

to an axis should therefore be defined as the difference between the appropriate coordinates. In Figure 1.9 this is $x_2 - x_1$ for RQ and $y_2 - y_1$ for PQ. But then Pythagoras' theorem tells us that the distance PR is given by

$$PR^2 = RQ^2 + PQ^2$$
$$= (x_2 - x_1)^2 + (y_2 - y_1)^2$$

That is,

$$PR = \sqrt{(x_2 - x_1)^2 + (y_2 - y_1)^2} \qquad (1)$$

Since this construction applies to arbitrary points P, R in the plane, we now have a general formula for the distance between two points.

We derived this formula as a consequence of geometric assumptions, in particular Pythagoras' theorem. Although this makes geometry amenable to arithmetical calculation—a very useful situation, to be sure—it does not say that geometry *is* arithmetic. In the early days of analytic geometry, the latter was a very heretical view (see Section 6.6). Eventually, however, Hilbert [1899] realized it could be made a fact by taking (1) as a *definition* of distance. Of course, all other geometric concepts have to be defined in terms of numbers too, but this boils down to defining a point, which is simply an *ordered pair* (x, y) of numbers. Equation (1) then gives the distance between the points (x_1, y_1) and (x_2, y_2).

When geometry is reconstructed in this way, all geometric facts become facts about numbers (though they do not necessarily become easier to see). In particular, Pythagoras' theorem becomes true by definition since it has been built into the definition of distance. This is not to say that Pythagoras' theorem ultimately is trivial. Rather, it shows that Pythagoras' theorem is precisely what is needed to interpret arithmetical facts as geometry.

I mentioned these comparatively recent developments only to bring Pythagoras' theorem up to date and to give a precise statement of its power to transform arithmetic into geometry. In ancient Greek times, geometry was based much more on seeing than on calculation. We shall see how the Greeks managed to build geometry on the basis of visually evident facts in the next chapter.

1.7. Biographical Notes: Pythagoras

Very little is known for certain about Pythagoras, although he figures in many legends. No documents have survived from the period in which he lived, so we have to rely on stories that were passed down for several centuries before being recorded. It appears that he was born on Samos, a Greek island near the coast of what is now Turkey, around 580 B.C. He traveled to the nearby mainland town of Miletus, where he learned mathematics from Thales (c. 624–547 B.C.), traditionally regarded as the founder of Greek mathematics. Pythagoras also traveled to Egypt and Babylon, where he presumably picked up additional mathematical ideas. Around 540 B.C. he settled in Croton, a Greek colony in what is now southern Italy.

There he founded a school whose members later became known as Pythagoreans. The motto of the school was "All is number," and the Pythagoreans tried to bring the realms of science, religion, and philosophy all under the rule of number. The very word *mathematics* ("that which is learned") is said to be a Pythagorean invention. The school imposed a strict code of conduct on its members, which included secrecy, vegetarianism, and a curious taboo on the eating of beans. The code of secrecy meant that mathematical results were considered to be the property of the school, and their individual discoverers were not identified to outsiders. Because of this, we do not know who discovered Pythagoras' theorem, the irrationality of $\sqrt{2}$, or other arithmetical results that will be mentioned in Chapter 3.

As mentioned in Section 1.5, the most notable scientific success of the Pythagorean school was the explanation of musical harmony in terms of whole number ratios. This success inspired the search for a numerical law governing the motions of planets, a "harmony of the spheres." Such a law probably cannot be expressed in terms that the Pythagoreans would have accepted; nevertheless, it seems reasonable to view the expansion of the number concept to meet the needs of geometry (and hence mechanics) as a natural extension of the Pythagorean program. In this sense, Newton's law of gravitation (Section 12.2) expresses the harmony that the Pythagoreans were looking for. Even in the strict sense, Pythagoreanism is very much alive today. With the digital computer, digital watches, digital audio and video coding everything, at least approximately, into sequences of whole numbers, we are closer than ever to a world in which "all is number."

Whether the complete rule of number is wise remains to be seen. It is said that when the Pythagoreans tried to extend their influence into politics they met with popular resistance. Pythagoras fled, but he was murdered in nearby Metapontum in 497 B.C.

CHAPTER 2

Greek Geometry

2.1. The Deductive Method

> He was 40 years old before he looked on Geometry; which happened accidentally. Being in a Gentleman's Library, Euclid's Elements lay open, and 'twas the 47 El. libri I. He read the Proposition. *By* G———sayd he (he would now and then sweare an emphaticall Oath by way of emphasis) *this is impossible*! So he reads the Demonstration of it, which referred him back to such a Proposition; which proposition he read. That referred him back to another, which he also read ... that at last he was demonstratively convinced of that trueth. This made him in love with Geometry.

This quotation about the philosopher Thomas Hobbes (1588–1679), from Aubrey's *Brief Lives*, beautifully captures the force of Greece's most important contribution to mathematics, the deductive method. (The proposition mentioned, incidentally, is Pythagoras' theorem.)

We have already seen that significant results were *known* before the period of classical Greece, but the Greeks were the first to construct mathematics by deduction from previously established results, resting ultimately on the most evident possible statements, called *axioms*. Thales (624–547 B.C.) is thought to be the originator of this method (see Heath [1921], p. 128), and by 300 B.C. it had become so sophisticated that Euclid's *Elements* set the standard for mathematical rigor down to the nineteenth century. The *Elements* were in fact too subtle for most mathematicians, let alone their students, so that in time they were boiled down to the simplest and driest propositions about straight lines, triangles, and circles. This part of the *Elements* is based on the following axioms (in the translation of Heath [1925], p. 154), which Euclid called *postulates* and *common notions*.

Postulates

Let the following be postulated
1. To draw a straight line from any point to any point.
2. To produce a finite straight line continuously in a straight line.
3. To describe a circle with any centre and distance.
4. That all right angles are equal to one another.
5. That, if a straight line falling on two straight lines make the interior angles on the same side less that two right angles, the two straight lines, if produced indefinitely, meet on that side on which are the angles less than the two right angles.

Common Notions

1. Things which are equal to the same thing are also equal to one another.
2. If equals be added to equals, the wholes are equal.
3. If equals be subtracted from equals, the remainders are equal.
4. Things which coincide with one another are equal to one another.
5. The whole is greater than the part.

It appears that Euclid's intention was to deduce geometric propositions from visually evident statements (the postulates) using evident principles of logic (the common notions). Actually, he often made unconscious use of visually plausible assumptions that are not among his postulates. His very first proposition used the unstated assumption that two circles meet if the center of one is on the circumference of the other (Heath [1925], p. 242). Nevertheless, such flaws were not noticed until the nineteenth century, and they were rectified by Hilbert [1899]. By themselves, they probably would not have been enough to end the *Elements* run of 22 centuries as a leading textbook. The *Elements* was overthrown by more serious mathematical upheavals in the nineteenth century. The so-called noneuclidean geometries, using alternatives to Euclid's fifth postulate, developed to the point where the old axioms could no longer be considered self-evident (see Chapter 17). At the same time, the concept of number matured to the point where irrational numbers became acceptable, and indeed preferable to intuitive geometric concepts, in view of the doubts about what the self-evident truths of geometry really were.

The outcome was a more adaptable language for geometry in which "points," "lines," and so on, could be defined, usually in terms of numbers, so as to suit the type of geometry under investigation. Such a development was long overdue, because even in Euclid's time the Greeks were investigating curves more complicated than circles, which did not fit conveniently in Euclid's system. Descartes [1637] introduced the coordinate method, which gives a single framework for handling both Euclid's geometry and higher curves (see Chapter 6), but it was not at first realized that coordinates allowed geometry to be entirely rebuilt on numerical foundations.

The comparatively trivial step (for us) of passing to axioms about numbers from axioms about points had to wait until the nineteenth century, when geometric axioms lost authority and number theoretic axioms gained it. We shall say more about these developments later (and of problems with the authority of axioms in general, which arose in the twentieth century). For the remainder of this chapter we shall look at some important nonelementary topics in Greek geometry, using the coordinate framework where convenient.

2.2. The Regular Polyhedra

Greek geometry is virtually complete as far as the elementary properties of plane figures are concerned. It is fair to say that only a handful of interesting elementary propositions about triangles and circles have been discovered since Euclid's time. Solid geometry is much more challenging, even today, so it is understandable that it was left in a less complete state by the Greeks. Nevertheless, they made some very impressive discoveries and managed to complete one of the most beautiful chapters in solid geometry, the enumeration of the regular polyhedra. The five possible regular polyhedra are shown in Figure 2.1.

Each polyhedron is bounded by a number of congruent polygonal faces, the same number of faces meet at each vertex, and in each face all the sides and angles are equal, hence the term *regular polyhedron*. A regular polyhedron is a spatial figure analogous to a regular polygon in the plane. But whereas there are regular polygons with any number $n \geqslant 3$ of sides, there are only five regular polyhedra.

This fact is easily proved and may go back to the Pythagoreans (see, e.g., Heath [1921], p. 159). One considers the possible polygons that can occur as faces, their angles, and the numbers of them that can occur at a vertex. For a 3-gon (triangle) the angle is $\pi/3$, so three, four, or five can occur at a vertex, but six cannot, as this would give a total angle 2π and the vertex would be flat. For a 4-gon, the angle is $\pi/2$, so three can occur at a vertex, but not four. For

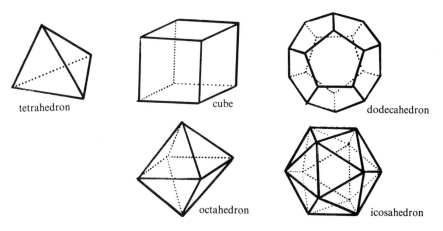

tetrahedron   cube   dodecahedron

octahedron   icosahedron

Figure 2.1

a 5-gon the angle is $3\pi/5$, so three can occur at a vertex, but not four. For a 6-gon the angle is $2\pi/3$, so not even three can occur at a vertex. But at least three faces must meet at each vertex of a polyhedron, so 6-gons (and, similarly, 7-gons, 8-gons, ..., n-gons) cannot occur as faces of a regular polyhedron. This leaves only the five possibilities just listed, which correspond to the five known regular polyhedra.

But do we really know that these five exist? There is no difficulty with the tetrahedron, cube, or octahedron, but it is not so clear that, say, 20 equilateral triangles will fit together to form a closed surface. Euclid found this problem difficult enough to be placed near the end of the *Elements*, and few of his readers ever mastered his solution. A beautiful direct construction was given by Luca Pacioli, a friend of Leonardo da Vinci's, in his book *De divina proportione* [1509]. Pacioli's construction uses three copies of the *golden rectangle*, with sides 1 and $(1 + \sqrt{5})/2$, interlocked as in Figure 2.2. The 12 vertices define 20 triangles such as ABC, and it suffices to show that these are equilateral, that is, $AB = 1$. This is a straightforward exercise in Pythagoras' theorem (Exercise 1).

The regular polyhedra will make another important appearance in connection with yet another nineteenth-century development, the theory of finite groups and Galois theory. Before the regular polyhedra made this triumphant comeback, they also took part in a famous fiasco: Kepler's theory of planetary distances (Kepler [1596]). Kepler's theory is summarized by his famous diagram (Fig. 2.3) of the five polyhedra, nested in such a way as to produce six spheres of radii proportional to the distances of the six planets then known.

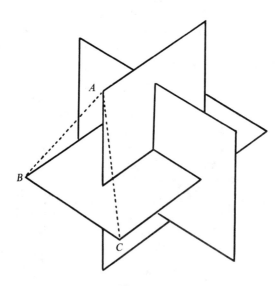

Figure 2.2

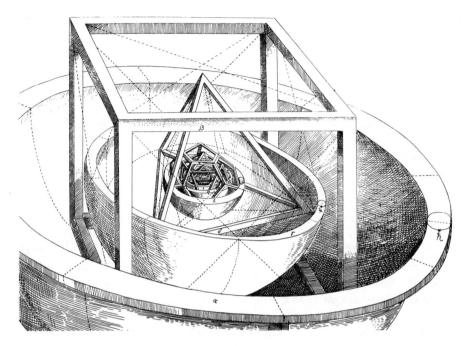

Figure 2.3

Unfortunately, although mathematics could not permit any more regular polyhedra, nature could permit more planets, and Kepler's theory was ruined when Uranus was discovered in 1781.

EXERCISES

1. Verify Pacioli's construction of the icosahedron. *Hint*: It may help to use the fact that $x = (1 + \sqrt{5})/2$ satisfies $x^2 - x - 1 = 0$.

2. Deduce the radius of a sphere in which an icosahedron with edges of length 1 is inscribed.

3. (Apollonius) If a dodecahedron D and an icosahedron I are inscribed in the same sphere, show that

$$\frac{\text{surface area } D}{\text{surface area } I} = \frac{\text{volume } D}{\text{volume } I}$$

4. Show that the angle α between the faces of a regular tetrahedron satisfies $\cos \alpha = 1/3$.

5. Show that there are only three regular tessellations of the plane by regular polygons. (A tessellation is a covering of the plane by polygons that meet only along the edges.)

2.3. Ruler and Compass Constructions

Greek geometers prided themselves on their logical purity; nevertheless, they were guided by intuition about physical space. One aspect of Greek geometry that was peculiarly influenced by physical considerations was the theory of constructions. Much of the elementary geometry of straight lines and circles can be viewed as the theory of constructions by ruler and compasses. The very subject matter, lines and circles, reflects the instruments used to draw them. And many of the elementary problems of geometry—for example, to bisect a line segment or angle, construct a perpendicular, or draw a circle through three given points—can be solved by ruler and compass constructions.

When coordinates are introduced, it is not difficult to show that the points which can be constructed from points $P_1, \ldots, P_n$ have coordinates in the field generated by the coordinates of $P_1, \ldots, P_n$ by the square root operation (see, e.g., Moise [1963]). Square roots arise, of course, because of Pythagoras' theorem: if the points (a, b) and (c, d) have been constructed, then so has the distance $\sqrt{(c - a)^2 + (d - b)^2}$ between them (Section 1.6 and Fig. 2.4). Conversely, it is possible to construct $\sqrt{l}$ for any given length l (exercise 2).

Looked at from this point of view, ruler and compass constructions look very special and unlikely to yield numbers such as $\sqrt[3]{2}$, for example. However, the Greeks tried very hard to solve just this problem, which was known as *duplication of the cube.* (So-called because to double the volume of a cube one must multiply the side by $\sqrt[3]{2}$.) Other notorious problems were *trisection of the angle* and *squaring the circle.* The latter problem was to construct a square equal in area to a given circle or to construct the number π, which amounts to the same thing. They never seem to have given up these goals, though the possibility of a negative solution was admitted and solutions by less elementary means were tolerated. We shall see some of these in the next sections.

The impossibility of solving these problems by ruler and compass construc-

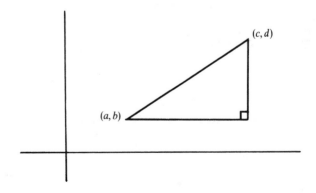

Figure 2.4

tions was not proved until the nineteenth century. For the duplication of the cube and trisection of the angle, impossibility was shown by Wantzel [1837]. Wantzel seldom receives credit for settling these problems, which had baffled the best mathematicians for 2000 years, perhaps because his methods were superseded by the more powerful theory of Galois.

The impossibility of squaring the circle was proved by Lindemann [1882], in a very strong way. Not only is π undefinable by rational operations and square roots; it is also *transcendental*, that is, not the root of any polynomial equation with rational coefficients. Like Wantzel's work, this was a rare example of a major result being proved by a minor mathematician. In Lindemann's case the explanation is perhaps that a major step had already been taken when Hermite [1873] proved the transcendence of e. Accessible proofs of both these results can be found in Klein [1924]. Lindemann's subsequent career was mathematically undistinguished, even embarrassing. In response to skeptics who thought his success with π had been a fluke, he took aim at the most famous unsolved problem in mathematics, "Fermat's last theorem" (see Chapter 10 for the origin of this problem). His efforts fizzled out in a series of inconclusive papers, each one correcting an error in the one before. Fritsch [1984] has written an interesting biographical article on Lindemann.

There is one ruler and compass problem that is still open: which regular n-gons are constructible? Gauss discovered in 1796 that the 17-gon is constructible and then showed (Gauss [1801], Art. 366) that a regular n-gon is constructible if and only if $n = 2^m p_1 p_2 \ldots p_k$ where each p_i is a prime of the form $2^{2^h} + 1$. The proof of necessity was actually completed by Wantzel [1837]. However, it is still not explicitly known what these primes are, or even whether there are infinitely many of them. The only ones known are for $h = 0, 1, 2, 3, 4$.

EXERCISES

1. Show, using similar triangles, that if lengths l_1 and l_2 are constructible, then so are $l_1 l_2$ and l_1/l_2.

2. Use Figure 2.5 to explain why $\sqrt{l}$ is constructible from l.

3. Show that the diagonal of a regular pentagon of side 1 is $(1 + \sqrt{5})/2$, and deduce that the regular pentagon is constructible.

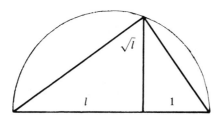

Figure 2.5

2.4. Conic Sections

Conic sections are the curves obtained by intersecting a circular cone by
a plane: ellipses (including circles), parabolas, and hyperbolas (Fig. 2.6).
Nowadays we know the conic sections better in terms of their equations in
cartesian coordinates:

$$\frac{x^2}{a^2} + \frac{y^2}{b^2} = 1 \qquad \text{(ellipse)}$$

$$y = ax^2 \qquad \text{(parabola)}$$

$$\frac{x^2}{a^2} - \frac{y^2}{b^2} = 1 \qquad \text{(hyperbola)}$$

More generally, any second-degree equation represents a conic section or
a pair of straight lines, a result that was proved by Descartes [1637].

The invention of conic sections is attributed to Menaechmus (fourth century
B.C.), a contemporary of Alexander the Great. Alexander is said to have asked
Menaechmus for a crash course in geometry, but Menaechmus refused, saying,
"There is no royal road to geometry." Menaechmus used conic sections to
give a very simple solution of the problem of duplication of the cube. In
analytic notation, this can be described as finding the intersection of the
parabola $y = \frac{1}{2}x^2$ with the hyperbola $xy = 1$. This yields

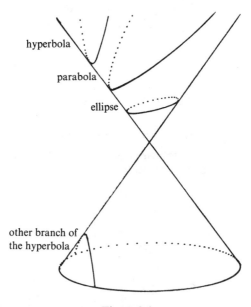

hyperbola

parabola

ellipse

other branch of
the hyperbola

Figure 2.6

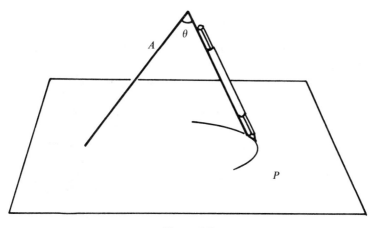

Figure 2.7

$$x\frac{1}{2}x^2 = 1 \qquad \text{or} \qquad x^3 = 2$$

so $\sqrt[3]{2}$ is constructed as the x coordinate of the intersection.

Although the Greeks accepted this as a "construction" for duplicating the cube, they apparently never discussed instruments for actually drawing conic sections. This is very puzzling since a natural generalization of the pair of compasses immediately suggests itself (Fig. 2.7). The arm A is set at a fixed position relative to a plane P, while the other arm rotates about it at a fixed angle θ, generating a cone with A as its axis of symmetry. The pencil, which is free to slide in a sleeve on this second arm, traces the section of the cone lying in the plane P. According to Coolidge [1945], p. 149, this instrument for drawing conic sections was first described as late as A.D. 1000 by the Arab mathematician al-Kuji. Yet nearly all the *theoretical* facts one could wish to know about conic sections had already been worked out by Apollonius (c. 250–200 B.C.)!

The theory and practice of conic sections finally met when Kepler [1609] discovered the orbits of the planets to be ellipses, and Newton [1687] explained this fact by his law of gravitation. This wonderful vindication of the theory of conic sections has often been described in terms of basic research receiving its long overdue reward, but perhaps one can also see it as a rebuke to Greek disdain for applications. (Kepler would not have been sure which it was. To the end of his days he was proudest of his theory explaining the distances of the planets in terms of the five regular polyhedra (Section 2.2). The fascinating and paradoxical character of Kepler has been warmly described in two excellent books, Koestler [1959] and Banville [1981].)

2.5. Higher Degree Curves

The Greek lacked a systematic theory of higher degree curves, because they lacked a systematic algebra. They could find what amounted to cartesian equations of individual curves (or "symptoms," as they called them; see van der Waerden [1954], p. 241), but they did not consider equations in general or notice any of their properties relevant to the study of curves, for example, the degree. Nevertheless, they studied many interesting special curves, which Descartes and his followers cut their teeth on when algebraic geometry finally emerged in the seventeenth century. An excellent, well illustrated, account of these early investigations may be found in Brieskorn and Knörrer [1981], Ch. 1.

In this section we must confine ourselves to brief remarks on a few examples.

The Cissoid of Diocles (c. 100 B.C.)

This curve is defined using an auxiliary circle, which for convenience we take to be the unit circle, and vertical lines through x and $-x$. It is the set of all points P seen in Figure 2.8. The portion shown results from varying x between

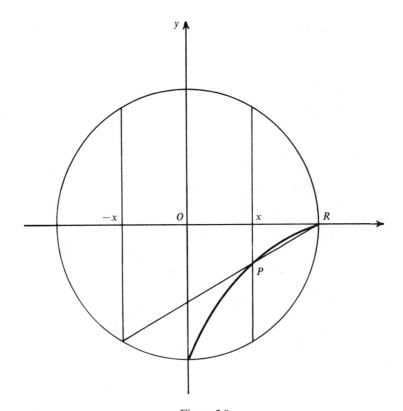

Figure 2.8

0 and 1. It is a cubic curve with cartesian equation

$$y^2(1 + x) = (1 - x)^3$$

This equation shows that if (x, y) is a point on the curve, then so is $(x, -y)$. Hence one gets the complete picture of it by reflecting the portion shown in Figure 2.8 in the x axis. The result is a sharp point at R, a *cusp*, a phenomenon that first arises with cubic curves. Diocles showed that the cissoid could be used to duplicate the cube, which is plausible (though still not obvious!) once one knows this curve is cubic.

The Spiric Sections of Perseus (c. 150 B.C.)

Apart from the sphere, cylinder, and cone—whose sections are all conic sections—one of the few surfaces studied by the Greeks was the *torus*. This surface, generated by rotating a circle about an axis outside the circle, but in the same plane, was called a *spira* by the Greeks—hence the name spiric sections for the sections by planes parallel to the axis. These sections, which were first studied by Perseus, have four qualitatively distinct forms (see Fig. 2.9, which is adapted from Brieskorn and Knörrer [1981], p. 20).

These forms—convex ovals, "squeezed" ovals, the figure 8, and pairs of

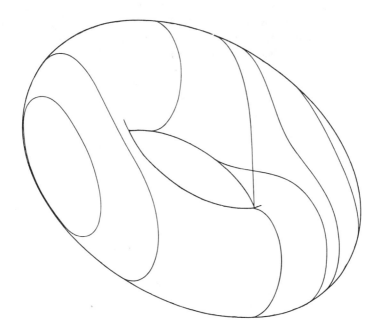

Figure 2.9

ovals—were rediscovered in the seventeenth century when analytic geometers looked at curves of degree 4, of which the spiric sections are examples. For suitable choice of torus, the figure 8 curve becomes the *lemniscate of Bernoulli* and the convex ovals become *Cassini ovals*. Cassini (1625–1712) was a distinguished astronomer but an opponent of Newton's theory of gravitation. He rejected Kepler's ellipses and instead proposed Cassini ovals as orbits for the planets.

The Epicycles of Ptolemy (A.D. 140)

These curves are known from a famous astronomical work, the *Almagest* of Claudius Ptolemy. Ptolemy himself attributes the idea to Apollonius. It seems almost certain that this is the Apollonius who mastered conic sections, which is ironic, because epicycles were his candidates for the planetary orbits, destined to be defeated by those very same conic sections.

An epicycle, in its simplest form, is the path traced by a point on a circle that rolls on another circle (Fig. 2.10). More complicated epicycles can be defined by having a third circle roll on the second, and so on. These curves were introduced by the Greeks to try to reconcile the complicated movements of the planets relative to the fixed stars, with a geometry based on the circle. In principle, this is possible! Lagrange [1772] showed that *any* motion along the celestial equator can be approximated arbitrarily closely by epicyclic motion, and a more modern version of the result may be found in Sternberg [1969]. But Ptolemy's mistake was to accept the apparent complexity of the motions of the planets as a fact in the first place. As we now know, the motion

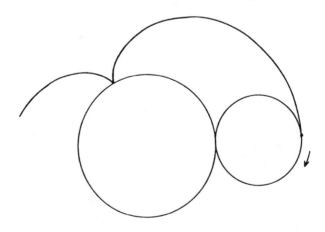

Figure 2.10

becomes simple when one considers motion relative to the sun rather than the earth and allows orbits to be ellipses.

Epicycles still have a role to play in engineering, and their mathematical properties are interesting. Some of them are closed curves and turn out to be algebraic, i.e., of the form $p(x, y) = 0$ for polynomial p. Others, such as those that result from rolling circles whose radii have an irrational ratio, lie densely in a certain region of the plane and hence cannot be algebraic; an algebraic curve $p(x, y) = 0$ can meet a straight line $y = mx + c$ in only a finite number of points, corresponding to roots of the polynomial equation $p(x, mx + c) = 0$, and the dense epicycles meet some lines infinitely often.

EXERCISE

1. Derive the equation of the cissoid.

2.6. Biographical Notes: Euclid

Even less is known about Euclid than about Pythagoras. We know only that he flourished around 300 B.C. and taught in Alexandria, the Greek city in Egypt founded by Alexander the Great in 322 B.C. Two stories are told about him. The first—the same that is told about Menaechmus and Alexander—has Euclid telling King Ptolemy I "there is no royal road to geometry." The second concerns a student who asked the perennial question: "What shall I gain from learning mathematics?" Euclid called his slave and said: "Give him a coin if he must profit from what he learns."

The most important fact of Euclid's life was undoubtedly his writing of the *Elements*, though we do not know how much of the mathematics in it was actually his own work. Certainly the elementary geometry of triangles and circles was known before Euclid's time. Some of the most sophisticated parts of the *Elements*, too, are due to earlier mathematicians. The theory of irrationals in Book V is due to Eudoxus (c. 400–347 B.C.), as is the "method of exhaustion" of Book XII (see Chapter 4). The theory of regular polyhedra in Book XIII is due, at least partly, to Theaetetus (c. 415–369 B.C.). But whatever Euclid's "research" contribution may have been, it was dwarfed by his contribution to the organization and dissemination of mathematical knowledge. For 2000 years the *Elements* was not only the core of mathematical education but at the heart of Western culture. The most glowing tributes to the *Elements* do not in fact come from mathematicians but from philosophers, politicians, and others. We saw Hobbes' response to Euclid in Section 2.1. Here are some others:

> He studied and nearly mastered the six books of Euclid since he was a member of Congress. He began a course of rigid mental discipline with the intent to

improve his faculties, especially his powers of logic and language. Hence his fondness for Euclid, which he carried with him on the circuit till he could demonstrate with ease all the six books.

—Abraham Lincoln (writing of himself),
Short Autobiography

At the age of eleven, I began Euclid. ... This was one of the great events of my life, as dazzling as first love. I had not imagined there was anything so delicious in the world.

—Bertrand Russell, *Autobiography*

Perhaps the low cultural status of mathematics today, not to mention the mathematical ignorance of politicians and philosophers, reflects the lack of an *Elements* suitable for the modern world.

CHAPTER 3

Greek Number Theory

3.1. The Role of Number Theory

In the first chapter we saw that number theory has been important in the history of mathematics for at least as long as geometry, and from a foundational point of view it may be more important. Despite this, number theory has never submitted to a systematic treatment like that undergone by elementary geometry in Euclid's *Elements*. At all stages in its development, number theory has had glaring gaps because of the intractability of elementary problems. Most of the really old unsolved problems in mathematics, in fact, are simple questions about the natural numbers 1, 2, 3, ... The nonexistence of a general method for solving Diophantine equations (Section 1.3) and the problem of identifying the primes of the form $2^{2^h} + 1$ (Section 2.3) has been noted. Other unsolved number theory problems will be mentioned in the sections that follow.

As a consequence, the role of number theory in the history of mathematics has been quite different from that of geometry. Geometry has played a stabilizing and unifying role, to the point of retarding further development at times and creating the popular impression that mathematics is a static subject. For those able to understand it, number theory has been a spur to progress and change. Only a minority of mathematicians have contributed to advances in number theory, but they include some of the greats—for example, Diophantus, Fermat, Euler, Lagrange, and Gauss. This book stresses those advances in number theory that sprang from its deep connections with other parts of mathematics, particularly geometry, since these were the most significant for mathematics as a whole. Nevertheless, there are topics in number theory that are too interesting to ignore, even though they seem (at present) to be outside the mainstream. We discuss a few of them in the next section.

3.2. Polygonal, Prime, and Perfect Numbers

The *polygonal numbers*, which were studied by the Pythagoreans, result from a naive transfer of geometric ideas to number theory. From Figure 3.1 it is an easy exercise to calculate an expression for the mth n-agonal number as the sum of a certain arithmetic series (exercise 1) and to show, for example, that a square is the sum of two triangular numbers. Apart from Diophantus' work, which contains impressive results on sums of squares, Greek results on polygonal numbers were of this elementary type.

On the whole, the Greeks seem to have been mistaken in attaching much importance to polygonal numbers. There are no major theorems about them, except perhaps the following two. The first is the theorem conjectured by Bachet [1621] (in his edition of Diophantus' works) that every positive integer is the sum of four integer squares. This was proved by Lagrange [1770]. A generalization, which Fermat [1670] stated without proof, is that every posi-

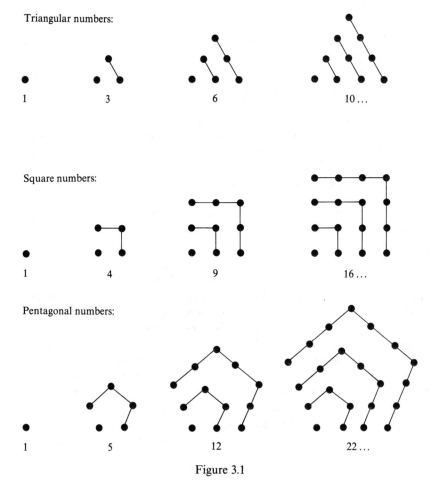

Figure 3.1

tive integer is the sum of n n-agonal numbers. This was proved by Cauchy [1813], though the proof is a bit of a letdown because all but four of the numbers can be 0 or 1. A short proof of Cauchy's theorem has been given by Nathanson [1987]. The other remarkable theorem about polygonal numbers is the formula

$$\prod_{n=1}^{\infty} (1 - x^n) = 1 + \sum_{k=1}^{\infty} (-1)^k (x^{(3k^2-k)/2} + x^{(3k^2+k)/2})$$

proved by Euler [1750] and known as Euler's pentagonal number theorem, since the exponents $(3k^2 - k)/2$ are the pentagonal numbers. (For a proof see Hall [1967], p. 33.)

(The four square theorem and the pentagonal number theorem were both absorbed around 1830 into Jacobi's theory of theta functions, a much larger theory.)

The *prime numbers* were also considered within the geometric framework, as the numbers with no rectangular representation. A prime number, having no factors apart from itself and 1, has only a "linear" representation. Of course this is no more than a restatement of the definition of prime, and most theorems about prime numbers require much more powerful ideas; however, the Greeks did come up with one gem. This is the proof that there are infinitely many primes, in Book IX of Euclid's *Elements*.

Given any finite collection of primes $p_1, p_2, \ldots, p_n$, we can find another by considering

$$p = p_1 p_2 \ldots p_n + 1$$

This number is not divisible by $p_1, p_2, \ldots, p_n$ (each leaves remainder 1). Hence either p itself is a prime, and $p > p_1, p_2, \ldots, p_n$, or else it has a prime factor $\neq p_1, p_2, \ldots, p_n$.

A *perfect number* is one that equals the sum of its factors (including 1 but excluding itself). For example, $6 = 1 + 2 + 3$ is a perfect number, as is $28 = 1 + 2 + 4 + 7 + 14$. Although this concept goes back to the Pythagoreans, only two noteworthy theorems about perfect numbers are known. Euclid concludes Book IX of the *Elements* by proving that if $2^n - 1$ is prime, then $2^{n-1}(2^n - 1)$ is perfect (exercise 3). These perfect numbers are of course even, and Euler [1849] (a posthumous publication) proved that every even perfect number is of Euclid's form. Euler's surprisingly simple proof may be found in Burton [1985], p. 504. It is not known whether there are any odd perfect numbers; this may be the oldest open problem in mathematics.

In view of Euler's theorem, the existence of even perfect numbers depends on the existence of primes of the form $2^n - 1$. These are known as Mersenne primes, after Marin Mersenne (1588–1648), who first drew attention to the problem of recognizing primes of this form. It is not known whether there are infinitely many Mersenne primes, though larger and larger ones seem to be found quite regularly. In recent years each new world record prime has been a Mersenne prime, giving a corresponding world record perfect number.

1. Show that the kth pentagonal number is $(3k^2 - k)/2$.

2. Show that each square is the sum of two consective triangular numbers.

3. Show that if $2^n - 1$ is prime, then $2^{n-1}(2^n - 1)$ is perfect.

3.3. The Euclidean Algorithm

This algorithm is named after Euclid because its earliest known appearance is in Book VII of the *Elements*. However, in the opinion of many historians (e.g., Heath [1921], p. 399) the algorithm and some of its consequences were probably known earlier. At the very least, Euclid deserves credit for a masterly presentation of the fundamentals of number theory, based on this algorithm.

The euclidean algorithm is used to find the highest common factor (hcf) of two positive integers a, b. The first step is to construct the pair (a_1, b_1) where

$$a_1 = \max(a, b) - \min(a, b)$$
$$b_1 = \min(a, b)$$

and then one simply repeats this operation of subtracting the smaller number from the larger. That is, if the pair constructed at step i is (a_i, b_i), then the pair constructed at step $i + 1$ is

$$a_{i+1} = \max(a_i, b_i) - \min(a_i, b_i)$$
$$b_{i+1} = \min(a_i, b_i)$$

The algorithm terminates at the first stage when $a_{i+1} = b_{i+1}$, and this common value is $\mathrm{hcf}(a, b)$. This is because taking differences preserves any common factors, hence when $a_{i+1} = b_{i+1}$ we have

$$\mathrm{hcf}(a, b) = \mathrm{hcf}(a_1, b_1) = \cdots = \mathrm{hcf}(a_{i+1}, b_{i+1}) = a_{i+1} = b_{i+1}$$

The very simplicity of the algorithm makes it easy to draw some important consequences. Euclid of course did not use our notation but nevertheless he had essentially the following results.

1. If $\mathrm{hcf}(a, b) = 1$, then there are integers m, n such that $ma + nb = 1$. The equations

$$a_1 = \max(a, b) - \min(a, b)$$
$$b_1 = \min(a, b)$$
$$\vdots$$
$$a_{i+1} = \max(a_i, b_i) - \min(a_i, b_i)$$
$$b_{i+1} = \min(a_i, b_i)$$

show successively that a_1, b_1 are integral linear combinations, $ma + nb$, of a and b, hence so are a_2, b_2, hence so are a_3, b_3, ..., and finally this is true of $a_{i+1} = b_{i+1}$. But $a_{i+1} = b_{i+1} = 1$ since $\text{hcf}(a, b) = 1$, hence $1 = ma + nb$ for some integers m, n.

2. If p is a prime number that divides ab, then p divides a or b.

To see this, suppose p does *not* divide a. Then since p has no factors of its own, we have $\text{hcf}(p, a) = 1$. Hence by the previous result we get integers m, n such that

$$ma + np = 1$$

Multiplying each side by b gives

$$mab + nbp = b$$

By hypothesis, p divides ab, hence p divides *both* terms on the left-hand side, and therefore p divides the right-hand side b.

3. Each positive integer has a unique factorization into primes (Fundamental Theorem of Arithmetic).

Suppose on the contrary that some integer n has two different prime factorizations:

$$n = p_1 p_2 \cdots p_j = q_1 q_2 \cdots q_k$$

By dividing out common factors, if necessary, we can assume there is a p_i that is not among the q's. But this is a contradiction to the previous result, because p_i divides $n = q_1 q_2 \cdots q_k$, yet it does not divide any of $q_1, q_2, \ldots, q_k$ individually, since these are prime numbers $\neq p_i$.

EXERCISES

1. If $\text{hcf}(a, b) = d$, show that there are integers m, n such that $ma + nb = d$.

2. (Solution of linear Diophantine equations) Give an algorithm that can be used to decide, given integers a, b, c, whether there are integers m, n such that

$$ma + nb = c$$

and to find such m, n if they exist.

3.4. Pell's Equation

The Diophantine equation $x^2 - Dy^2 = 1$, where D is a nonsquare integer, is known as Pell's equation because Euler mistakenly attributed a solution of it to the seventeenth-century English mathematician Pell (it should have been attributed to Brouncker). Pell's equation is probably the best known Diophantine equation after the equation $a^2 + b^2 = c^2$ for Pythagorean triples, and in some ways it is more important. Solution of Pell's equation is the main step in the solution of the general quadratic Diophantine equation in two variables (see, e.g., Gelfond [1960]) and also a key tool in proving the theorem of Matiasevich mentioned in Section 1.3 that there is no algorithm for solving all Diophantine equations (see, e.g., Davis [1973]). In view of this, it is fitting that Pell's equation should make its first appearance in the foundations of Greek mathematics, and it is impressive to see how well the Greeks understood it.

The simplest instance of Pell's equation,

$$x^2 - 2y^2 = 1$$

was studied by the Pythagoreans in connection with $\sqrt{2}$. If x, y are large solutions to this equation, then $x/y \simeq \sqrt{2}$ and in fact the Pythagoreans found a way of generating larger and larger solutions by means of the recurrence relations

$$x_{n+1} = x_n + 2y_n$$

$$y_{n+1} = x_n + y_n$$

A short calculation shows that

$$x_{n+1}^2 - 2y_{n+1}^2 = -(x_n^2 - 2y_n^2)$$

hence if (x_n, y_n) satisfies $x^2 - 2y^2 = \pm 1$, then (x_{n+1}, y_{n+1}) satisfies $x^2 - 2y^2 = \mp 1$. Starting with the trivial solution $(x_0, y_0) = (1, 0)$ of $x^2 - 2y^2 = 1$, we get successively larger solutions $(x_2, y_2), (x_4, y_4), \ldots$.

But how might these recurrence relations have been discovered in the first place? Van der Waerden [1976] and Fowler [1980, 1982] suggest that the key is the euclidean algorithm applied to line segments, an operation the Greeks called *anthyphairesis*. Given any two lengths a, b, one can define the sequence $(a_1, b_1), (a_2, b_2), \ldots$, as in Section 3.2 by repeated subtraction of the smaller length from the larger. If a, b are integer multiples of some unit, then the process terminates as in Section 3.3, but if b/a is irrational, it continues forever. We can well imagine that the Pythagoreans would have been interested in anthyphairesis applied to $a = 1, b = \sqrt{2}$. Here is what happens. We represent a, b by sides of a rectangle, and each subtraction of the smaller number from the larger is represented by cutting off the square on the shorter side (Fig. 3.2). We notice that the rectangle remaining after step 2, with sides $\sqrt{2} - 1$ and $2 - \sqrt{2} = \sqrt{2}(\sqrt{2} - 1)$, is the same shape as the original, though the long side is now vertical instead of horizontal. It follows that similar steps will recur forever, which is another proof that $\sqrt{2}$ is irrational, incidentally. Our present interest, however, is in the relation between successive similar rectangles. If we let the long and short sides of successive similar rectangles be x_{n+1}, y_{n+1} and x_n, y_n, we can derive a recurrence relation for x_{n+1}, y_{n+1} from Figure 3.3:

$$x_{n+1} = x_n + 2y_n$$

$$y_{n+1} = x_n + y_n$$

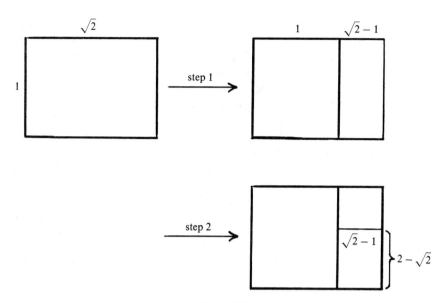

Figure 3.2

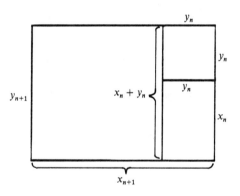

Figure 3.3

exactly the relations of the Pythagoreans! The difference is that our x_n, y_n are not integers, and they satisfy $x^2 - 2y^2 = 0$ not $x^2 - 2y^2 = 1$. Nevertheless, one feels that Figure 3.3 gives the most natural interpretation of these relations. The fact that the same relations generate solutions of the equations $x^2 - 2y^2 = 1$ was possibly discovered as a result of wishing that the euclidean algorithm terminated with $x_1 = y_1 = 1$. If the Pythagoreans started with $x_1 = y_1 = 1$ and applied the recurrence relations, then they could have found that (x_n, y_n) satisfies $x^2 - 2y^2 = (-1)^n$, as we did earlier.

Many other instances of the Pell equation $x^2 - Dy^2 = 1$ occur in Greek mathematics, and these can be understood in a similar way by applying anthyphairesis to the rectangle with sides 1, $\sqrt{D}$. In the seventh century A.D. the Indian mathematician Brahmagupta gave a recurrence relation for generating solutions of $x^2 - Dy^2 = 1$. The Indians called the euclidean algorithm on lengths 1, $\sqrt{D}$ the "pulverizer" because it breaks them down to smaller and smaller pieces. To obtain a recurrence one has to know that a rectangle proportional to the original eventually recurs, a fact that was rigorously proved only in 1768 by Lagrange. The later European work on Pell's equations, which began in the seventeenth century with Brouncker and others, was based on the continued fraction for $\sqrt{D}$, though this amounts to the same thing as anthyphairesis (see exercises). For a condensed but detailed history of Pell's equation, see Dickson [1920], pp. 341–400.

An interesting aspect of the theory is the very irregular relationship between D and the number of steps of anthyphairesis before a rectangle proportional to the original recurs. If the number of steps is large, the smallest nontrivial solution of $x^2 - Dy^2 = 1$ is enormous. A famous example is the so-called *cattle problem* of Archimedes (287–212 B.C.). This problem leads to the equation

$$x^2 - 4{,}729{,}494y^2 = 1$$

the smallest solution of which was found by Amthor [1880] to have 206,545 digits!

EXERCISES

1. The continued fraction of a real number $\alpha > 0$ is written

$$\alpha = n_1 + \cfrac{1}{n_2 + \cfrac{1}{n_3 + \cfrac{1}{n_4 \cdots}}}$$

where $n_1, n_2, n_3, n_4, \ldots$, are integers obtained by the following operations. Let $n_1 =$ integer part of α. Then $\alpha - n_1 < 1$ and $\alpha_1 = 1/(\alpha - n_1) > 1$, so we can take $n_2 =$ integer part of α_1. Then $\alpha_1 - n_2 < 1$ and $\alpha_2 = 1/(\alpha_1 - n_2) > 1$, so we can take $n_3 =$ integer part of α_2, and so on. Interpret the operations of detaching the integer part and inverting the remainder in terms of anthyphairesis.

2. Show

$$\sqrt{2} = 1 + \cfrac{1}{2 + \cfrac{1}{2 + \cfrac{1}{2 \cdots}}}$$

3. Find the continued fraction for $\sqrt{3}$.

3.5. Diophantus' Chord and Tangent Constructions

In Section 1.3 we used a method of Diophantus to find all rational points on the circle. If $p(x, y) = 0$ is any quadratic equation in x and y with rational coefficients, and if the equation has one rational solution $x = r_1$, $y = s_1$, then we can find any rational solution by drawing a rational line $y = mx + c$ through the point (r_1, s_1) and finding its other intersection with the curve $p(x, y) = 0$. The two intersections with the curve, $x = r_1, r_2$, say, are given by the roots r_1, r_2 of the equation

$$p(x, mx + c) = 0$$

This means $p(x, mx + c) = k(x - r_1)(x - r_2)$, and since all coefficients on the left-hand side are rational and r_1 is rational, then k and r_2 must also be rational. The y value when $x = r_2$, $y = s_2 = mr_2 + c$, is rational since m and c are, hence (r_2, s_2) is another rational point on $p(x, y) = 0$. Conversely, any line through two rational points is rational, hence all rational points are found in this way.

Now if $p(x, y) = 0$ is a curve of degree 3, its intersections with a line $y = mx + c$ are given by the roots of the cubic equation $p(x, mx + c) = 0$. If we know two rational points on the curve, then the line through them will be rational, and its third intersection with the curve will also be rational, by an argument like the preceding one. This fact becomes more useful when one realizes that the two known rational points can be taken to coincide, in which case the line is the tangent through the known rational point. Thus from one rational solution we can generate another by the tangent construction, and from two we can construct a third by taking the chord between the two.

Diophantus found rational solutions to cubic equations in what seems to have been essentially this way. The surviving works of Diophantus reveal little of his methods, but a plausible reconstruction—an algebraic version of the tangent and chord constructions—has been given by Bashmakova [1981]. Probably the first to understand Diophantus' methods was Fermat, in the seventeenth century, and the first to give the tangent and chord interpretation was Newton [late 1670s].

In contrast to the quadratic case, we have no choice in the slope of the rational line for cubics. Thus it is by no means clear that this method will give us *all* rational points on a cubic. A remarkable theorem, conjectured by Poincaré [1901] and proved by Mordell [1922], says that all rational points can be generated by tangent and chord constructions applied to finitely many points. However, it is still not known whether there is an algorithm for finding a finite set of such rational generators on each cubic curve.

EXERCISES

1. Explain the solution $x = 21/4$, $y = 71/8$ to $x^3 - 3x^2 + 3x + 1 = y^2$ given by Diophantus (Heath [1910], p. 242) by construction of the tangent through the obvious rational point on this curve.

2. Derive the following result of Viète [1593], p. 145, by the tangent construction. Given a rational point (a, b) show that another rational point on the curve $x^3 - y^3 = a^3 - b^3$ is

$$x = a\frac{a^3 - 2b^3}{a^3 + b^3}, \qquad y = b\frac{2a^3 - b^3}{a^3 + b^3}$$

3.6. Biographical Notes: Diophantus

Diophantus lived in Alexandria during the period when Greek mathematics, along with the rest of Western civilization, was generally in decline. The catastrophes that engulfed the West with the fall of Rome and the rise of Islam, culminating in the burning of the library in Alexandria in A.D. 640, buried almost all details of Diophantus' life. His dates can be placed with certainty only betwen A.D. 150 and 350, since he mentions Hypsicles (known to be c. 150) and is mentioned by Theon of Alexandria (c. 350). One other scrap of evidence, a letter of Michael Psellus (eleventh century), suggests A.D. 250 as the most likely time when Diophantus flourished. Apart from this, the only clue to Diophantus' life is a conundrum in the *Greek Anthology* (c. A.D. 600):

> God granted him to be a boy for the sixth part of his life, and adding a twelfth part to this, He clothed his cheeks with down. He lit him the light of wedlock after a seventh part, and five years after his marriage He granted him a son. Alas! late-born wretched child; after attaining the measure of half his father's life, chill Fate took him. After consoling his grief by this science of numbers for four years he ended his life. (Cohen and Drabkin [1958], p. 27)

If this information is correct, then Diophantus married at 33 and had a son who died at 42, four years before Diophantus himself died at 84.

Diophantus' work went almost unnoticed for many centuries, and only parts of it survive. The first stirrings of interest in Diophantus occurred in the Middle Ages, but much of the credit for the eventual revival of Diophantus belongs to Rafael Bombelli (1526–1572) and Wilhelm Holtzmann (known as Xylander, 1532–1576). Bombelli discovered a copy of Diophantus' *Arithmetic* in the Vatican library and published 143 problems from it in his *Algebra* [1572]. Xylander published the first Latin translation of the *Arithmetic* in 1575. The most famous edition of the *Arithmetic* was that of Bachet de Meziriac [1621]. Bachet glimpsed the possibility of general principles behind the special problems of the *Arithmetic* and, in his commentary on the book, alerted his contemporaries to the challenge of properly understanding Diophantus and carrying his ideas further. It was Fermat who took up this challenge and made the first significant advances in number theory since the classical era (see Chapter 10).

CHAPTER 4

Infinity in Greek Mathematics

4.1. Fear of Infinity

Reasoning about infinity is one of the characteristic features of mathematics as well as its main source of conflict. We saw, in Chapter 1, the conflict that arose from the discovery of irrationals, and in this chapter we shall see that the rejection of irrational numbers by the Greeks was just part of a general rejection of infinite processes. In fact, until the late nineteenth century most mathematicians were reluctant to accept infinity as more than "potential." The infinitude of a process, collection, or magnitude was understood as the possibility of its indefinite continuation, and no more—certainly not the possibility of eventual completion. For example, the natural numbers 1, 2, 3, ..., can be accepted as a potential infinity—generated from 1 by the process of adding 1—without accepting that there is a completed totality $\{1, 2, 3, \dots\}$. The same applies to any sequence x_1, x_2, x_3, ... (of rational numbers, say), where x_{n+1} is obtained from x_n by a definite rule.

And yet a beguiling possibility arises when x_n tends to a limit x. If x is something we already accept, for geometric reasons, say, then it is very tempting to view x as somehow the "completion" of the sequence x_1, x_2, x_3, ... It seems that the Greeks were afraid to draw such conclusions. According to tradition, they were frightened off by the paradoxes of Zeno, around 450 B.C.

We know of Zeno's arguments only through Aristotle, who quotes them in his *Physics* in order to refute them, and it is not clear what Zeno himself wished to achieve. Was there, for example, a tendency toward speculation about infinity that he disapproved of? His arguments are so extreme they could almost be parodies of loose arguments about infinity he heard among his contemporaries. Consider his first paradox, the *dichotomy*:

> There is no motion because that which is moved must arrive at the middle (of
> its course) before it arrives at the end. (Aristotle, *Physics*, Book VI, Ch. 9)

The full argument presumably is that before getting anywhere one must first
get $\frac{1}{2}$-way, and before that $\frac{1}{4}$-way, and before that $\frac{1}{8}$-way, ad infinitum. The
completion of this infinite sequence of steps no longer seems impossible to
most mathematicians, since it represents nothing more than an infinite set of
points within a finite interval. It must have frightened the Greeks though,
because in all their proofs they were very careful to avoid completed infinities
and limits.

The first mathematical processes we would recognize as infinite were probably devised by the Pythagoreans, for example, the recurrence relations

$$x_{n+1} = x_n + 2y_n$$

$$y_{n+1} = x_n + y_n$$

for generating integer solutions of the equations $x^2 - 2y^2 = \pm 1$. We saw in
Section 3.4 why it is likely that these relations arose from an attempt to
understand $\sqrt{2}$, and it is easy for us to see that $x_n/y_n \to \sqrt{2}$ as $n \to \infty$.

However, it is unlikely that the Pythagoreans would have viewed $\sqrt{2}$ as
a "limit" or seen the sequence as a meaningful object at all. The most we can
say is that, by stating a recurrence, the Pythagoreans *implied* a sequence with
limit $\sqrt{2}$, but only a much later generation of mathematicians could accept
the infinite sequence as such and appreciate its importance in defining the
limit.

In a problem where we would find it natural to reach a solution α by a
limiting process, the Greeks would instead eliminate any solution *but* α. They
would show that any number $< \alpha$ was too small and any number $> \alpha$ too large
to be the solution. In the following sections we shall study some examples of
this style of proof and see how it ultimately bore fruit in the foundations
of mathematics. As a method of finding solutions to problems, however, it
was sterile: how does one guess the number α in the first place? When
mathematicians returned to problems of finding limits in the seventeenth
century, they found no use for the rigorous methods of the Greeks. The
dubious seventeenth-century methods of infinitesimals were criticized by the
Zeno of the time, Bishop Berkeley, but little was done to meet his objections
until much later, since infinitesimals did not seem to lead to incorrect results.
It was Dedekind, Weierstrass, and others in the nineteenth century who
eventually restored Greek standards of rigor.

The story of rigor lost and rigor regained took an amazing turn when
a previously unknown manuscript of Archimedes, *The Method*, was discovered in 1906. In it he reveals that his deepest results were found using
dubious infinitary arguments, and only later proved rigorously. Because, as
he says, "It is of course easier to supply the proof when we have previously
acquired some knowledge of the questions by the method, than it is to find it
without any previous knowledge."

The importance of this statement goes beyond its revelation that infinity can be used to discover results that are not initially accessible to logic. Archimedes was probably the first mathematician candid enough to explain that there is a difference between the way theorems are discovered and the way they are proved.

4.2. Eudoxus' Theory of Proportions

The theory of proportions is credited to Eudoxus (c. 400–350 B.C.) and is expounded in Book V of Euclid's *Elements*. The purpose of the theory is to enable lengths (and other geometric quantities) to be treated as precisely as numbers, while only admitting the use of rational numbers. We saw the motivation for this in Section 1.5: the Greeks could not accept irrational numbers, but they accepted irrational geometric quantities such as the diagonal of the unit square. To simplify the exposition of the theory, let us call lengths *rational* if they are rational multiples of a fixed length.

Eudoxus' idea was to say that a length λ is determined by those rational lengths less than it and those greater than it. To be precise, he says $\lambda_1 = \lambda_2$ if any rational length $< \lambda_1$ is also $< \lambda_2$, and vice versa. Likewise $\lambda_1 < \lambda_2$ if there is a rational length $> \lambda_1$ but $< \lambda_2$. This definition uses the rationals to give an infinitely sharp notion of length while avoiding any overt use of infinity. Of course the infinite set of rational lengths $< \lambda$ is present in spirit, but Eudoxus avoids mentioning it by speaking of an arbitrary rational length $< \lambda$.

The theory of proportions was so successful that it delayed the development of a theory of real numbers for 2000 years. This was ironic, because the theory of proportions can be used to define irrational numbers just as well as lengths. It was understandable though, because the common irrational lengths, such as the diagonal of the unit square, arise from constructions that are intuitively clear and finite from the geometric point of view. Any *arithmetic* approach to $\sqrt{2}$, whether by sequences, decimals, or continued fractions, is infinite and therefore less intuitive. Until the nineteenth century this seemed a good reason for considering geometry to be a better foundation for mathematics than arithmetic. Then the problems of geometry came to a head, and mathematicians began to fear geometric intuition as much as they had previously feared infinity. There was a purge of geometric reasoning from the textbooks and industrious reconstruction of mathematics on the basis of numbers and sets of numbers. Set theory will be discussed further in Chapter 20. Suffice to say, for the moment, that set theory depends on the acceptance of completed infinities.

The beauty of the theory of proportion was its adaptability to this new climate. Instead of rational lengths, take rational numbers. Instead of comparing existing irrational lengths by means of rational lengths, construct irrational numbers from scratch using sets of rationals! The length $\sqrt{2}$ is determined by the two sets of positive rationals

$$L_{\sqrt{2}} = \{r|r^2 < 2\}, \qquad U_{\sqrt{2}} = \{r|r^2 > 2\}$$

Dedekind [1872] decided to let $\sqrt{2}$ be this pair of sets! In general, let any partition of the positive rationals into sets L, U such that any member of L is less than any member of U be a positive real number. This idea, now known as a *Dedekind cut*, is more than just a twist of Eudoxus; it gives a complete and uniform construction of all real numbers, or points on the line, using just the rationals. In short, this is an explanation of the *continuous* in terms of the *discrete*, finally resolving the fundamental conflict in Greek mathematics. Dedekind was understandably pleased with his achievement. He wrote (Dedekind [1872/1901], p. 2):

> The statement is so frequently made that the differential calculus deals with continuous magnitude, and yet an explanation of this continuity is nowhere given .... It then only remained to discover its true origin in the elements of arithmetic and thus at the same time secure a real definition of the essence of continuity. I succeeded Nov. 24 1858.

EXERCISES

1. Define $\alpha + \beta$ and $\alpha \times \beta$ in terms of the Dedekind cuts for α, β.

2. (Dedekind) Show that $\sqrt{2} \times \sqrt{3} = \sqrt{6}$.

3. Show, using Dedekind cuts, that any bounded set of real numbers has a least upper bound.

4.3. The Method of Exhaustion

The method of exhaustion, also credited to Eudoxus, is a generalization of his theory of proportions. Just as an irrational length is determined by the rational lengths on either side of it, more general unknown quantities become determined by arbitrarily close approximations using known figures. Examples given by Eudoxus (and expounded in Book XII of Euclid's *Elements*) are an approximation of the circle by inner and outer polygons (Fig. 4.1) and an approximation of a pyramid by stacks of prisms (Fig. 4.2, which shows the most obvious approximation, not the cunning one actually used by Euclid). In both cases the approximating figures are known quantities, on the basis of the theory of proportions and the theorem that area of triangle = $\frac{1}{2}$ base × height.

The polygonal approximations are used to show that the area of any circle is proportional to the square on its radius, as follows. Suppose $P_1 \subset P_2 \subset P_3 \subset \cdots$, are the inner polygons and $Q_1 \supset Q_2 \supset Q_3 \supset \cdots$, are the outer polygons. Each polygon is obtained from its predecessor by bisecting the arcs between its vertices, as shown in Figure 4.1. It can then be shown, by

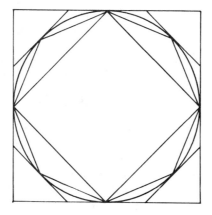

Figure 4.1

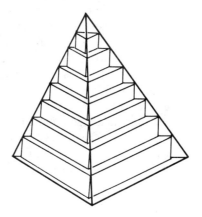

Figure 4.2

elementary geometry, that the area difference $Q_i - P_i$ can be made arbitrarily small, and hence P_i approximates the area C of the circle arbitrarily closely.

On the other hand, elementary geometry also shows that the area P_i is proportional to the square, R^2, of the radius. Writing the area as $P_i(R)$ and using the theory of proportions to handle ratios of areas, we have

$$P_i(R) : P_i(R') = R^2 : R'^2 \tag{1}$$

Now let $C(R)$ denote the area of the circle of radius R, and suppose

$$C(R) : C(R') < R^2 : R'^2 \tag{2}$$

By choosing a P_i that approximates C sufficiently closely we also get

$$P_i(R) : P_i(R') < R^2 : R'^2$$

which contradicts (1). Hence the $<$ sign in (2) is incorrect, and we can similarly show that $>$ is incorrect. Thus the only possibility is

$$C(R) : C(R') = R^2 : R'^2$$

that is, the area of a circle is proportional to the square of its radius.

Notice that "exhaustion" does not mean using an infinite sequence of steps to show that area is proportional to the square of the radius. Rather, one shows that any *disproportionality* can be refuted in a *finite* number of steps (by going to a suitable P_i). This is typical of the way in which exhaustion arguments avoid mention of limits and infinity.

In the case of the pyramid, one uses elementary geometry again to show that stacks of prisms approximate the pyramid arbitrarily closely. Then exhaustion shows that the volume of a pyramid, like that of a prism, is proportional to base × height (see exercises below). Finally, there is a clever argument to show that the constant of proportionality is $\frac{1}{3}$. We can restrict to the case of triangular pyramids (since any pyramid can be cut into these), and Figure 4.3 shows how a triangular prism is cut into three triangular pyramids. Any two of these pyramids can be seen to have equal base and height—although which face is taken to be the base depends on which pyramids are being compared—hence all three are equal in volume. Each is therefore one-third of the prism, that is, $\frac{1}{3}$ base × height.

It is interesting that Euclid does *not* need the method of exhaustion in the theory of area for polygons. All this can be done by dissection arguments such as that showing area of triangle $= \frac{1}{2}$ base × height (Fig. 4.4). In fact, it was shown by W. Bolyai [1832] that any polygons P, Q of equal area can be cut

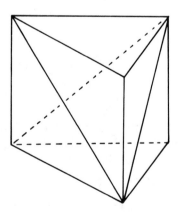

Figure 4.3

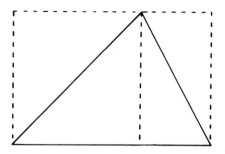

Figure 4.4

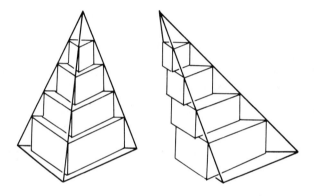

Figure 4.5

into polygonal pieces $P_1, \ldots, P_n$ and $Q_1, \ldots, Q_n$ such that P_i is congruent to Q_i. Thus we can *define* polygons to be equal in area if they possess dissections into such correspondingly congruent pieces. In Hilbert's famous list of mathematical problems (Hilbert [1900]), the third was to decide whether an analogous definition was possible for polyhedra. Dehn [1900] showed that it was not; in fact, a tetrahedron and a cube of equal volume cannot be dissected into corresponding congruent polyhedral pieces. Hence infinite processes of some kind, such as the method of exhaustion, are needed to define equality of volume. A readable account of Dehn's theorem and related results may be found in Boltianskii [1978].

EXERCISES

1. Show that any two pyramids with the same base and height can be approximated arbitrarily closely by the same prisms, differently stacked (Fig. 4.5).

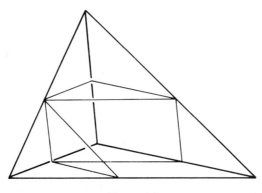

Figure 4.6

2. Deduce that pyramids of the same base and height have equal volume. (It then follows, by a previous argument, that pyramid $= \frac{1}{3}$ prism with same base and height, without having to show the volume of the pyramid is proportional to base × height. This is Legendre's proof; see Heath [1925], Book XII, Proposition 5.)

3. (Euclid) Figure 4.6 shows a dissection of a triangular pyramid into two pyramids, similar to the original but half its height, and two prisms of equal volume. Show that the two prisms occupy more than half the volume of the original pyramid. (Hence, by iterating the construction within the subpyramids, one can approximate the pyramid arbitrarily closely by prisms.)

4.4. The Area of a Parabolic Segment

The method of exhaustion was brought to full maturity by Archimedes (287–212 B.C.). Among his most famous results were the volume and surface area of the sphere and the area of a parabolic segment. As mentioned in Section 4.1, Archimedes first discovered these results by nonrigorous methods, later confirming them by the method of exhaustion. Perhaps the most interesting and natural of his exhaustion proofs is the one for the area of the parabolic segment. The segment is exhausted by polygons similarly to Eudoxus' exhaustion of the circle, but the area is obtained outright and not merely in proportion to another figure.

To simplify the construction slightly we assume that the segment is cut off by a chord perpendicular to the axis of the symmetry of the parabola. Archimedes divides the parabolic segment into triangles $\Delta_1, \Delta_2, \Delta_3, \ldots$, as shown in Figure 4.7 (labeled by their subscripts). The middle vertex of each triangle lies on the parabola halfway between the other two (measured horizontally). These triangles clearly exhaust the parabolic segment, and so it remains to compute their area. Quite surprisingly, this turns into a geometric series.

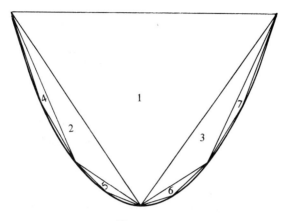

Figure 4.7

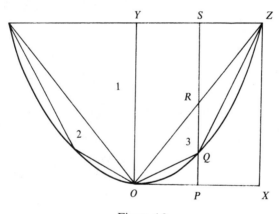

Figure 4.8

We shall briefly indicate how this comes about by studying Δ_3 (Fig. 4.8). Since $OP = \frac{1}{2}OX$, $PQ = \frac{1}{4}PS$ by definition of the parabola. On the other hand, $SR = \frac{1}{2}PS$, hence $QR = \frac{1}{4}PS$. Now Δ_3 is the sum of the triangles RQZ and OQR, which have the same base RQ and "height" $OP = PX$, hence equal area. We have just seen that RQZ has half the base of SRZ and it has the same height, hence (calling figures equal when they have the same area)

$$\Delta_3 = SRZ$$

$$= \frac{1}{4}OYZ$$

$$= \frac{1}{8}\Delta_1.$$

By symmetry, $\Delta_2 = \Delta_3$, so $\Delta_2 + \Delta_3 = \frac{1}{4}\Delta_1$.

A similar argument shows

$$\Delta_4 + \Delta_5 + \Delta_6 + \Delta_7 = \frac{1}{16}\Delta_1$$

and so on, each new chain of triangles having one-fourth the area of the chain before. Consequently,

$$\text{area of parabolic segment} = \Delta_1\left(1 + \frac{1}{4} + \left(\frac{1}{4}\right)^2 + \cdots\right)$$

$$= \frac{4}{3}\Delta_1$$

Of course, Archimedes does not use the infinite series but uses exhaustion, showing that any area $< \frac{4}{3}\Delta_1$ can be exceeded by taking sufficiently many of the triangles Δ_i. The sum of the *finite* geometric series needed for this was known from Euclid's *Elements*, Book IX, where Euclid used it for the theorem about perfect numbers (Section 3.2).

4.5. Biographical Notes: Archimedes

Archimedes is one of the few ancient mathematicians whose life is known in any detail, thanks to the attention he received from classical authors such as Plutarch, Livy, and Cicero and his participation in the historically significant siege of Syracuse in 212 B.C. He was born in Syracuse (a Greek city in what is now Sicily) around 287 B.C. and did most of his important work there, though he may have studied for a time in Alexandria. He seems to have been related to the ruler of Syracuse, King Hieron II, or at least on good terms with him. There are many stories of mechanisms invented by Archimedes for the benefit of Hieron: compound pulleys for moving ships, ballistic devices for the defense of Syracuse, and a model planetarium.

The most famous story about Archimedes is the one told by Vitruvius (*De architectura*, Book IX, Ch. 3) which has Archimedes leaping from his bath with a shout of "Eureka!" when he realized that weighing a crown immersed in water would give a means of testing whether it was pure gold. Historians doubt the authenticity of this story, but it does at least recognize Archimedes' understanding of hydrostatics.

In ancient times Archimedes' reputation rested on his mechanical inventions, which no doubt were more understandable to most people than his pure mathematics. However, it can also be argued that his theoretical mechanics (including the law of the lever, centers of mass equilibrium, and hydrostatic pressure) was his most original contribution to science. Before Archimedes there was no mathematical theory of mechanics at all, only the thoroughly incorrect mechanics of Aristotle. In pure mathematics, Archimedes did not make any comparable *conceptual* advances, except perhaps in his *Method*, which uses his ideas from statics as a means of discovering results on areas and volumes. The concepts that Archimedes needed for proofs in geometry—the theory of proportions and the method of exhaustion—had already been supplied by Eudoxus, and it was Archimedes' phenomenal insight and technique that lifted him head and shoulders above his contemporaries.

The story of Archimedes' death has often been told, though with varying details. He was killed by a Roman soldier when Syracuse fell to the Romans under Marcellus in 212 B.C. Probably he was doing mathematics at the time of his death, but whether he enraged a soldier by saying "Stand away from my diagram!" is conjectural. This story has come down to us from Tzetzes (*Chiliad*, Book II). Other versions of the death of Archimedes are given in Plutarch's *Marcellus* (Ch. XIX). Plutarch also tells us that Archimedes asked that his gravestone be inscribed with a figure and description of his favorite result, the relation between the volumes of the sphere and the cylinder. (He showed that the volume of the sphere is two-thirds that of the enveloping cylinder. See Heath [1897], p. 43.) A century and a half later, Cicero (*Tuscular Disputations*, Book V) reported finding the gravestone when he was quaestor in Sicily in 75 B.C. The grave had been neglected, but the figure of sphere and cylinder was still recognizable.

CHAPTER 5

Polynomial Equations

5.1. Algebra

The word "algebra" comes from the Arabic word *al-jabr* meaning "restoring."
It passed into mathematics through the book *Al-jabr w'al mûqabala* (Science
of restoring and opposition) of al-Khwārizmī [830], a work on the solution
of equations. In this context, "restoring" meant adding equal terms to both
sides and "opposition" meant setting the two sides equal. For many centuries,
al-jabr more commonly meant the resetting of broken bones, and the surgical
meaning accompanied the mathematical one when "al-jabr" became "algebra"
in Spanish, Italian, and English. Even today the surgical meaning is included
in the *Oxford English Dictionary*. Al-Khwārizmī's own name has given us the
word "algorithm," so his work has had a lasting impact on mathematics, even
though its content was quite elementary.

His algebra went no further than the solution of quadratic equations,
which had already been understood by the Babylonians, presented from the
geometric viewpoint by Euclid, and reduced to a formula by Brahmagupta
[628] (see Section 5.3). Brahmagupta's work, the high point of Indian mathematics to that time, was more advanced than al-Khwārizmī's in several
respects—notation, admission of negative numbers, and the treatment of
Diophantine equations—even though it predated al-Khwārizmī and was very
likely known to him. Indian mathematics had spread to the Arab world with
the general promotion of culture by the eighth-century caliphs of Baghdad,
and Arab mathematicians acknowledged the Indian origin of certain ideas,
for instance, decimal numerals. Why then did al-Khwārizmī's work rather
than Brahmagupta's become the definitive "algebra"?

Perhaps this is a case (like "Pell's equation," to mention another pertinent
example) where a mathematical term caught on for accidental reasons. How-

ever, it may also have been that the time was ripe for the idea of algebra to be cultivated, and the simple algebra of al-Khwārizmī served this purpose better than those of his more sophisticated predecessors. In Indian mathematics, algebra was inseparable from number theory and elementary arithmetic. In Greek mathematics, algebra was hidden by geometry. Other possible sources of algebra, Babylonia and China, were lost or cut off from the West until it was too late for them to be influential. Arabic mathematics developed at the right time and place to absorb both the geometry of the West and the algebra of the East and to recognize algebra as a separate field with its own methods. The concept of algebra that emerged—the theory of polynomial equations— proved its worth by holding firm for 1000 years. Only in the nineteenth century did algebra grow beyond the bounds of the theory of equations, and this was a time when most fields of mathematics were outgrowing their established habitats.

The early algebraic methods seemed only superficially different from geom- etric methods, as we shall see in the case of quadratic equations in Section 5.3. Algebraic methods for solving equations became distinct from, and superior to, the geometric only with the advent of new manipulative techniques and efficient notation in the sixteenth century (Section 5.5). Algebra did not break away from geometry, however, but actually gave geometry a new lease on life, thanks to the development of analytic geometry by Fermat and Descartes around 1630. This recombination of algebra and geometry at a higher level will be discussed in Chapter 6. It led to the modern field of algebraic geometry.

The story of algebraic geometry unfolds along with the story of polynomial equations, becoming entwined with many other mathematical threads in the process. We shall study several of the decisive early events in this story. One we have already seen is Diophantus' chord and tangent methods for finding rational solutions of equations (Section 3.5). Another relevant event, though not in fact historically connected with Western mathematics, was the method of elimination developed by Chinese mathematicians between the early Christian era and the Middle Ages. Since this method predates any comparable method in the West, and concerns equations of the lowest degree, it is logical to discuss it first.

5.2. Linear Equations and Elimination

The Chinese discovered a method for solving linear equations in any number of unknowns during the Han dynasty (206 B.C.–A.D. 220). It appears in the famous book *Jiuzhang suanshu* (Nine chapters of mathematical art), which was written during this period, and survives today in a third-century version with a commentary by Liu Hui. The method was essentially what we call "Gaussian elimination," systematically eliminating terms in a system

$$a_{11}x_1 + a_{12}x_2 + \cdots + a_{1n}x_n = b_1$$
$$\vdots \qquad\qquad\qquad \vdots$$
$$a_{n1}x_1 + a_{n2}x_2 + \cdots + a_{nn}x_n = b_n$$

by subtracting a suitable multiple of each equation from the one below it until a triangular system is obtained:

$$a'_{11}x_1 + a'_{12}x_2 + \cdots + a'_{1n}x_n = b'_1$$
$$a'_{22}x_2 + \cdots + a'_{2n}x_n = b'_2$$
$$\ddots \qquad\qquad \vdots$$
$$a'_{nn}x_n = b'_n$$

then solving for x_n, x_{n-1}, ..., x_1 in turn by successive substitutions. This type of calculation was particularly suited to a Chinese device called the counting board, which held the array of coefficients and facilitated manipulations similar to those we perform with matrices. For further details, see Li Yan and Du Shiran [1987].

Around the twelfth century Chinese mathematicians discovered that elimination could be adapted to simultaneous polynomial equations in two or more variables. For example, one can eliminate y between a pair of equations

$$a_0(x)y^m + a_1(x)y^{m-1} + \cdots + a_m(x) = 0 \qquad (1)$$

$$b_0(x)y^m + b_1(x)y^{m-1} + \cdots + b_m(x) = 0 \qquad (2)$$

where the $a_i(x)$, $b_j(x)$ are polynomials in x. The y^m term can be eliminated by forming the equation $b_0(x) \times (1) - a_0(x) \times (2)$, say,

$$c_0(x)y^{m-1} + c_1(x)y^{m-2} + \cdots + c_{m-1} = 0 \qquad (3)$$

We can form a second equation of degree $m - 1$ in y by multiplying (3) by y, then again eliminating y^m between (3) $\times$ y and (1), giving, say,

$$d_0(x)y^{m-1} + d_1(x)y^{m-2} + \cdots + d_{m-1} = 0 \qquad (4)$$

The problem is now reduced to eliminating y between the equations (3) and (4), which are of lower degree in y than (1) and (2). Thus one can continue inductively until an equation in x alone is obtained. This method was extended to four variables in the work of Zhū Shijié [1303] entitled *Siyuan yujian* (Jade mirror of four unknowns).

As we shall see in Chapter 6, the two-variable polynomial problem arose in the West in the seventeenth century, in the context of finding intersections of curves. This led first to a rediscovery of the method of elimination for polynomials; only later was this method based on an understanding of linear equations. The well-known Cramer's rule for the solution of linear equations was named after its appearance in a book on algebraic curves (Cramer [1750]).

5.3. Quadratic Equations

As early as 2000 B.C., the Babylonians could solve pairs of simultaneous equations of the form

$$x + y = p$$

$$xy = q$$

which are equivalent to the quadratic equations

$$x^2 + q = px$$

The original pair was solved by a method that gave the two roots of the quadratic:

$$x, y = \frac{p}{2} \pm \sqrt{\left(\frac{p}{2}\right)^2 - q}$$

when both were positive (the Babylonians did not admit negative numbers). The steps in the method were:

(i) Form $\dfrac{x + y}{2}$

(ii) Form $\left(\dfrac{x + y}{2}\right)^2$

(iii) Form $\left(\dfrac{x + y}{2}\right)^2 - xy$

(iv) Form $\sqrt{\left(\dfrac{x + y}{2}\right)^2 - xy} = \dfrac{x - y}{2}$

(v) Find x, y by inspection of the values in (i), (iv)

(See Boyer [1968], p. 34, for an actual example.) Of course these steps were not expressed in symbols but only applied to specific numbers. Nevertheless, a general method is implicit in the many specific cases solved.

An explicit general method, expressed as a formula in words, was given by Brahmagupta [628]:

> To the absolute number multiplied by four times the [coefficient of the] square, add the square of the [coefficient of the] middle term; the square root of the same, less the [coefficient of the] middle term, being divided by twice the [coefficient of the] square is the value. (Colebrook [1817], p. 346])

This is the solution

$$x = \frac{\sqrt{4ac + b^2} - b}{2a}$$

of the equation

$$ax^2 + bx = c$$

yet one wonders whether Brahmagupta understood it quite this way when, a few lines later, he gives another rule which is trivially equivalent to the first when expressed in our notation:

$$x = \frac{\sqrt{ac + (b/2)^2} - (b/2)}{a}$$

The methods of the Babylonians and Brahmagupta clearly give correct solutions, but their basis is not clear. The meaning of square roots, for example, was not questioned as it was by the Greeks. A rigorous basis for the solution of quadratic equations can be found in Euclid's *Elements*, Book VI. Proposition 28 can be interpreted as a solution of the general quadratic equation in the case where there is a positive root, as Heath ([1925], Vol. 2, p. 263) explains. However, the algebraic interpretation is far from obvious even when one specializes the proposition, which is about parallelograms, to one about rectangles. It seems unlikely that Euclid was aware of the algebra, or he would have expressed it by much simpler geometry.

The transition from geometry to algebra can be observed in al-Khwārizmī's solution of a quadratic equation ([830], pp. 13–15). The solution is still expressed in geometric language, but now the geometry is a direct embodiment of the algebra. It is really the standard algebraic solution, but with "squares" and "products" understood literally as geometric squares and rectangles. To solve $x^2 + 10x = 39$, represent x^2 by a square of side x, and $10x$ by two $5 \times x$ rectangles as in Figure 5.1. The extra square of area 25 "completes the square" of side $x + 5$ to one of area $25 + 39$, since 39 is the given value of $x^2 + 10x$. Thus the big square has area 64, hence its side $x + 5 = 8$. This gives the solution $x = 3$.

Euclid and al-Khwārizmī did not admit negative lengths, so the solution $x = -13$ to $x^2 + 10x = 39$ does not appear. This is quite natural, since geometry admits only one square with area 64. Avoidance of negative coef-

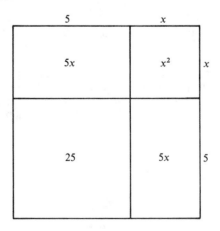

Figure 5.1

ficients, however, causes some unnatural algebraic complications. There is not one general quadratic equation, but three, corresponding to the different ways of distributing positive terms between the two sides: $x^2 + ax = b$, $x^2 = ax + b$, $x^2 + b = ax$.

5.4. Quadratic Irrationals

The roots of quadratic equations with rational coefficients are numbers of the form $a + \sqrt{b}$, where a, b are rational. Euclid took the theory of irrationals further in Book X of the *Elements* with a very detailed study of numbers of the form $\sqrt{\sqrt{a} \pm \sqrt{b}}$, where a, b are rational. Book X is the longest book in the *Elements* and it is not clear why Euclid devoted so much space to this topic: perhaps because some of it is needed for the study of regular polyhedra in Book XIII (cf. exercise 2.2.2), perhaps simply because it was Euclid's favorite topic, or perhaps it was one in which he had some original contributions to show off. It is said that Apollonius took the theory of irrationals further, but unfortunately his work on the subject is lost.

After this, there seems to have been no progress in the theory of irrationals until the Renaissance, except for a remarkable isolated result by Fibonacci [1225]. Fibonacci showed that the roots of $x^3 + 2x^2 + 10x = 20$ are not any of Euclid's irrationals. This is *not* a proof, as some historians have thought, that the roots cannot be constructed by ruler and compasses. Fibonacci did not rule out *all* expressions built from rationals and square roots; nevertheless, it was the first step into the world of irrationals beyond Euclid.

At this point it is worth asking how difficult it is to show that a specific number, say, $\sqrt[3]{2}$, cannot be constructed from rational numbers by square roots. The answer will depend on how well the reader manages the following exercises. The manipulation required would certainly not have been beyond the sixteenth-century algebraists. The subtle part is finding a suitable classification of expressions according to complexity—extending Euclid's classification to expressions in which radical signs are nested to arbitrary depth—and using induction on the level of complexity. This type of thinking did not emerge until the 1820s, hence the relatively late proof that $\sqrt[3]{2}$ is not constructible by ruler and compasses (Wantzel [1837]).

EXERCISES

1. Show that $\sqrt[3]{2}$ is irrational.

2. Let

$$F_0 = \{\text{rational numbers}\} \quad \text{and} \quad F_{k+1} = \{a + b\sqrt{c_k} \mid a, b \in F_k\} \quad \text{for some } c_k \in F_k.$$

Show that each F_k is a field, that is,

$$x, y \in F_k \Rightarrow x \pm y, \; xy, \; \frac{x}{y} (y \neq 0) \in F_k$$

3. Show that if $a, b, c \in F_k$ but $\sqrt{c} \notin F_k$, then $a + b\sqrt{c} = 0 \Leftrightarrow a = b = 0$ (for $k = 0$ this is in the *Elements*, Book X, Prop. 79).

4. Suppose $\sqrt[3]{2} = a + b\sqrt{c}$ where $a, b, c \in F_k$, but that $\sqrt[3]{2} \notin F_k$. (We know $\sqrt[3]{2} \notin F_0$ by exercise 1). Cube both sides and deduce that

$$2 = a^3 + 3ab^2 c \qquad \text{and} \quad 0 = 3a^2 b + b^3 c$$

5. Deduce that $\sqrt[3]{2} = a - b\sqrt{c}$ also, which is a contradiction.

5.5. The Solution of the Cubic

> In our own days Scipione del Ferro of Bologna has solved the case of the cube and first power equal to a constant, a very elegant and admirable accomplishment. Since this art surpasses all human subtlety and the perspicuity of mortal talent and is a truly celestial gift and a very clear test of the capacity of men's minds, whoever applies himself to it will believe that there is nothing that he cannot understand. In emulation of him, my friend Niccolò Tartaglia of Brescia, wanting not to be outdone, solved the same case when he got into a contest with his [Scipione's] pupil, Antonio Maria Fior, and, moved by my many entreaties, gave it to me ... having received Tartaglia's solution and seeking a proof of it, I came to understand that there were a great many other things that could also be had. Pursuing this thought and with increased confidence, I discovered these others, partly by myself and partly through Lodovico Ferrari, formerly my pupil. (Cardano [1545/1968], p. 8)

The solution of cubic equations in the early sixteenth century was the first clear advance in mathematics since the time of the Greeks. It revealed the power of algebra which the Greeks had not been able to harness, power that was soon to clear a new path to geometry, which was virtually a royal road (analytic geometry and calculus). Cardano's elation at the discovery is completely understandable. Even in the twentieth century, personally discovering the solution of the cubic equation has been the inspiration for at least one distinguished mathematical career (see Kac [1984]).

As far as the history of the original discovery goes, we do not know much more than Cardano tells us. Scipione del Ferro died in 1526, so the first solution was known before then. Tartaglia discovered his solution on February 12, 1535, probably independently, because he solved all problems in the contest with del Ferro's pupil Fior, whereas Fior did not. Cardano has been accused by almost everyone, from Tartaglia on, of stealing Tartaglia's solution, but his own account seems to distribute credit quite fairly. For more background, see the introduction and preface to Cardano [1545/1968] and Crossley [1988].

Cardano presents his algebra in the geometric style of al-Khwārizmī (whom he describes as the originator of algebra at the beginning of the book), with the case distinctions that result from avoidance of negative coefficients. By ignoring these complications, his solution can be described as follows. The cubic equation $x^3 + ax^2 + bx + c = 0$ is first transformed into one with no quadratic term by a linear change of variable, namely, $x = y - a/3$. One then has, say,

$$y^3 = py + q$$

By setting $y = u + v$, the left-hand side becomes

$$(u^3 + v^3) + 3uv(u + v) = 3uvy + (u^3 + v^3)$$

which will equal the previous right-hand side if

$$3uv = p$$

$$u^3 + v^3 = q$$

Eliminating v gives a quadratic in u^3,

$$u^3 + \left(\frac{p}{3u}\right)^3 = q$$

with roots

$$\frac{q}{2} \pm \sqrt{\left(\frac{q}{2}\right)^2 - \left(\frac{p}{3}\right)^3}$$

By symmetry, we obtain the same values for v^3. And since $u^3 + v^3 = q$, if one of the roots is taken to be u^3, the other is v^3. Without loss of generality we can take

$$u^3 = \frac{q}{2} + \sqrt{\left(\frac{q}{2}\right)^2 - \left(\frac{p}{3}\right)^3}$$

$$v^3 = \frac{q}{2} - \sqrt{\left(\frac{q}{2}\right)^2 - \left(\frac{p}{3}\right)^3}$$

and hence

$$y = u + v = \sqrt[3]{\frac{q}{2} + \sqrt{\left(\frac{q}{2}\right)^2 - \left(\frac{p}{3}\right)^3}} + \sqrt[3]{\frac{q}{2} - \sqrt{\left(\frac{q}{2}\right)^2 - \left(\frac{p}{3}\right)^3}}$$

5.6. Angle Division

Another important contributor to algebra in the sixteenth century was Viète (1540–1603). He helped emancipate algebra from the geometric style of proof by introducing letters for unknowns and using plus and minus signs to facilitate manipulation. Yet at the same time he strengthened its ties with geometry at a higher level by relating algebra to trigonometry. A case in point is his solution of the cubic by circular functions (Viète [1591], Ch. VI, Theorem 3), which shows that solving the cubic is equivalent to trisecting an arbitrary angle.

If we take the cubic in the form

$$x^3 + ax + b = 0$$

we can reduce to an equation

$$4y^3 - 3y = c$$

with just one parameter, by setting $x = ky$ and choosing k so that

$$\frac{k^3}{ak} = \frac{-4}{3} \quad \text{or} \quad k = \sqrt{\frac{-4a}{3}}$$

The point of the expression $4y^3 - 3y$ is that

$$4\cos^3\theta - 3\cos\theta = \cos 3\theta$$

hence by setting $y = \cos\theta$ we obtain

$$\cos 3\theta = c$$

If we are given c, then we can construct a triangle with angle $\cos^{-1} c = 3\theta$. Trisection of this angle gives us the solution $y = \cos\theta$ of the equation. Conversely, the problem of trisecting an angle with cosine c is equivalent to solving the cubic equation $4y^3 - 3y = c$.

(Of course, there is a problem with trigonometric interpretation when $|c| > 1$, which requires complex numbers for its resolution. Complex numbers are also involved in Cardano's formula, since the expression under the square root sign, $(q/2)^2 - (p/3)^3$, can be negative. It so happens that Viète's method requires complex numbers only when Cardano's does not, so between the two of them, complex numbers are avoided. Nevertheless, cubic equations are the birthplace of complex numbers, as we shall see when we study complex numbers in more detail later.)

Astonishingly, the problem of dividing an angle into any odd number of equal parts turns out to have an algebraic solution analogous to the algebraic solution of the cubic. Viète himself [1579] took the problem as far as finding expressions for $\cos n\theta$ and $\sin n\theta$ as polynomials in $\cos\theta$, $\sin\theta$, at least for certain values of n. Newton read Viète in 1663/4 and found the equation

$$y = nx - \frac{n(n^2 - 1)}{3!}x^3 + \frac{n(n^2 - 1)(n^2 - 3^2)}{5!}x^5 + \cdots$$

relating $y = \sin n\theta$ and $x = \sin\theta$ (see Newton [1676] in Turnbull [1960]). He asserted this result for arbitrary n, but we are interested in the case of odd integral n, when it reduces to a polynomial equation. The surprise is that Newton's equation then has a solution by nth roots analogous to the Cardano formula for cubics,

$$x = \frac{1}{2}\sqrt[n]{y + \sqrt{y^2 - 1}} + \frac{1}{2}\sqrt[n]{y - \sqrt{y^2 - 1}} \qquad (1)$$

although only for n of the form $4m + 1$. This formula appears out of the blue in de Moivre [1707]. (It also appears in the unpublished Leibniz [1675], though without the restriction on n. See Schneider [1968], pp. 224–228.) He does not explain how he found it, but it is comprehensible to us as

$$\sin\theta = \frac{1}{2}\sqrt[n]{\sin n\theta + i\cos n\theta} + \frac{1}{2}\sqrt[n]{\sin n\theta - i\cos n\theta} \qquad (2)$$

a consequence of *our* version of de Moivre's formula

$$(\cos\theta + i\sin\theta)^n = \cos n\theta + i\sin n\theta \qquad (3)$$

when $n = 4m + 1$ (See exercises 1 and 2.)

Viète himself came remarkably close to (3) in a posthumously published work [1615]. He observed that the products of $\sin\theta$, $\cos\theta$ which occur in $\cos n\theta$, $\sin n\theta$ are the alternate terms in the expansion of $(\cos\theta + \sin\theta)^n$, except for certain minus signs. He failed only to notice that the signs could be explained by inserting the coefficient i for $\sin\theta$. In any case, such an explanation would have seemed ad hoc in Viète's time, in the absence of other good reasons for accepting $i = \sqrt{-1}$. Similarly, de Moivre's [1707] formula would have seemed natural to his contemporaries, who were far more comfortable with Cardano's formula than they were with i. In Section 13.5 we shall see how the perception of de Moivre's formula changed with the development of complex numbers.

EXERCISES

1. Use (3) and $\sin\alpha = \cos(\pi/2 - \alpha)$, $\cos\alpha = \sin(\pi/2 - \alpha)$ to show that

$$(\sin\theta + i\cos\theta)^n = \begin{cases} \sin n\theta + i\cos n\theta & \text{when } n = 4m + 1 \\ -\sin n\theta - i\cos n\theta & \text{when } n = 4m + 3 \end{cases}$$

2. Deduce that (2) is correct for $n = 4m + 1$ and false for $n = 4m + 3$, and hence that (1) is a correct relation between $y = \sin n\theta$ and $x = \sin\theta$ only when $n = 4m + 1$.

3. Show that (1) is a correct relation between $y = \cos n\theta$ and $x = \cos\theta$ for *all* n (de Moivre [1730]).

5.7. Higher Degree Equations

The general equation of fourth degree

$$x^4 + ax^3 + bx^2 + cx + d = 0$$

was solved by Cardano's student Ferrari, and the solution was published in

Cardano [1545], p. 237. A linear transformation reduces the equation to the form

$$x^4 + px^2 + qx + r = 0$$

or

$$(x^2 + p)^2 = px^2 - qx + p^2 - r$$

Then for any y

$$(x^2 + p + y)^2 = (px^2 - qx + p^2 - r) + 2y(x^2 + p) + y^2$$
$$= (p + 2y)x^2 - qx + (p^2 - r + 2py + y^2)$$

The quadratic $Ax^2 + Bx + C$ on the right-hand side will be a square if $B^2 - 4AC = 0$, which is a cubic equation for y. We can therefore solve for y and take the square root of both sides of the equation for x, which then becomes quadratic and hence also solvable. The final result is a formula for x using just square and cube roots of rational functions of the coefficients.

This impressive bonus to the solution of cubic equations raised hopes that higher degree equations could also be solved by formulas built from the coefficients by rational operations and roots, and *solution by radicals*, as it was called, become a major goal of algebra for the next 250 years. However, all such efforts to solve the general equation of fifth degree (quintic) failed. The most that could be done was to reduce it to the form

$$x^5 - x - A = 0$$

with only one parameter. This was done by Bring [1786], and a sketch of his method may be seen in Pierpont [1895]. Bring's result appeared in a very obscure publication and went unnoticed for 50 years, or it might have rekindled hopes for the solution of the quintic by radicals. As it happened, Ruffini [1799] offered the first proof that this is impossible. Ruffini's proof was not completely convincing; however, he was vindicated when a satisfactory proof was given by Abel [1826], and again with the beautiful general theory of equations of Galois [1831].

A positive outcome of Bring's result was the nonalgebraic solution of the quintic by Hermite [1858]. The reduction to an equation with one parameter opened the way to a solution by transcendental functions, analogous to Viète's solution of the cubic by circular functions. The appropriate functions, the elliptic modular functions, had been discovered by Gauss, Abel, and Jacobi, and Galois [1831'] had hinted at their relation to quintic equations. This extraordinary convergence of mathematical ideas was the subject of Klein [1884].

In view of the difficulties with the quintic, there was naturally very little progress with the general equation of degree n. However, two simple but important contributions were made by Descartes [1637]. The first was the superscript notation for powers we now use: x^3, x^4, x^5, and so on. (Though not x^2, oddly enough. The square of x continued to be written xx until well into the next century.) The second was the theorem ([1637], p. 159) that a polynomial $p(x)$ which takes the value 0 when $x = a$ has a factor $(x - a)$. Since division of a polynomial $p(x)$ of degree n by $(x - a)$ leaves a polynomial of degree $n - 1$, Descartes' theorem raised the hope of factorizing each nth-degree polynomial into n linear factors. As we shall see in Chapter 13, this hope was fulfilled with the development of complex numbers.

EXERCISE

1. Show that $x^n - a^n$ has a factor $x - a$, whence so has $p(x) - p(a)$, and hence deduce Descartes' theorem.

5.8. Biographical Notes: Tartaglia, Cardano, and Viète

Little is known about Scipione del Ferro, the discoverer of the first solution to cubic equations, other than his dates (1465–1526) and the fact that he was a professor of arithmetic and geometry at Bologna from 1496. This has possibly resulted in Tartaglia and Cardano receiving more mathematical credit than they deserve. On the other hand, there is no denying that the personalities of Tartaglia (Fig. 5.2) and Cardano (Fig. 5.3), their contrasting lives, and their quarrel make a story that is fascinating in its own right.

Niccolò Tartaglia was born in Brescia in 1499 or 1500 and died in Venice in 1557. The name "Tartaglia" (meaning "stutterer") was actually a nickname; his real name is believed to have been Fontana. Tartaglia's childhood was scarred by poverty, following the death of his father, a mail courier, around 1506, and injuries suffered when Brescia was sacked by the French in 1512. Despite taking refuge in the cathedral, Tartaglia received five serious head wounds, including one to the mouth, which left him with his stutter. His life was saved only by the devoted nursing of his mother, who literally licked his wounds. Around the age of 14 he went to a teacher to learn the alphabet, but he ran out of money for his lessons by the letter K. This much is in Tartaglia's own sketch of his life ([1546], p. 69). After that, the story goes, he stole a copybook and taught himself to read and write, sometimes using tombstones as slates for want of paper.

By 1534 he had a family and, still short of money, he moved to Venice. There he gave public mathematics lessons in the church of San Zanipolo and published various scientific works. The famous disclosure of his method for solving cubic equations occurred on a visit to Cardano's house in Milan on March 25, 1539. When Cardano published it in 1545, Tartaglia angrily accused him of dishonesty. Tartaglia [1546], p. 120, claimed that Cardano had solemnly sworn never to publish the solution and to write it down only in cipher. Ferrari, who had been an 18-year-old servant of Cardano at the time,

Figure 5.2. Niccolò Tartaglia.

came to Cardano's defense, declaring that he had been present and there had been no promise of secrecy. In a series of 12 printed pamphlets, known as the *Cartelli* (reprinted by Masotti [1974]), Ferrari and Tartaglia traded insults and mathematical challenges; the two finally squared off in a public contest in the church of Santa Maria del Giardino, Milan, in 1548. It seems that Ferrari got the better of the exchange, as there was little subsequent improvement in Tartaglia's fortunes. He died alone, still impoverished, nine years later.

Apart from his solution of the cubic, Tartaglia is remembered for other contributions to science. It was he who discovered that a projectile should be fired at 45° to achieve maximum range [1546], p. 6. His conclusion was based on incorrect theory, however, as is clear from Tartaglia's diagrams

Figure 5.3. Girolamo Cardano.

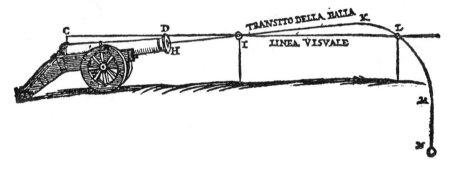

Figure 5.4

of trajectories (e.g., Fig. 5.4; [1546], p. 16]). Tartaglia's Italian translation of the *Elements* was the first printed translation of Euclid in a modern language, and he also published an Italian translation of some of Archimedes' works. For information on these, and Tartaglia's mechanics, see Rose [1975], pp. 151–154.

Girolamo Cardano, often described in English books by the anglicized name Jerome Cardan, was born in Pavia in 1501 and died in Rome in 1576. His father Fazio was a lawyer and physician who encouraged Girolamo's studies but otherwise seems to have treated him rather harshly, as did his mother, Chiara Micheri, whom Cardano described as "easily provoked, quick of memory and wit, and a fat, devout little woman." Cardano entered the University of Pavia in 1520 and completed a doctorate of medicine at Padua in 1526.

He married in 1531 and, after struggling until 1539 for acceptance, became a successful physician in Milan—so successful, in fact, that his fame spread all over Europe. He evidently had a remarkable skill in diagnosis, though his contributions to medical knowledge were slight in comparison with those of his contemporaries Andreas Vesalius and Ambroise Paré. Mathematics was one of his many interests outside his profession. Cardano also secured a niche in the history of cryptography for an encoding device known as the Cardano grille (see Kahn [1967], pp. 143–145) and in the history of probability, where he was the first to make calculations, though not always correctly (see David [1962], pp. 40–60, and Ore [1953], which contains a translation of Cardano's book on games of chance).

The violence and intrigue of Renaissance Italy soured Cardano's life just as much as Tartaglia's, though in a different way. An uncle died of poisoning, attempts were made to poison both Cardano and his father (so Cardano claimed), and in 1560 Cardano's oldest son was beheaded for the crime of poisoning his wife. Cardano, who believed his son's only fault was to marry the girl in the first place, never got over this calamity. He could no longer bear to live in Milan and moved to Bologna. There he suffered another blow when his protégé Ferrari died in 1565—poisoned by his sister, it was said. In 1570 Cardano was imprisoned by the Inquisition for heresy. After a few months he recanted, was released, and moved to Rome.

In the year before he died, Cardano wrote *The Book of My Life* [1575/1930], which is not so much autobiography as self-advertisement. It contains a few scenes from his childhood and returns again and again to the tragedy of his oldest son, but most of the book is devoted to boasting. There is a chapter of testimonials from patients, a chapter on important people who sought his services, a list of authors who cited his works, a list of his sayings he considered quotable, and a collection of tall stories that would have done Baron von Münchhausen proud. Admittedly, there is also a (very short) chapter called "Things in Which I Have Failed" and frequent warnings about the vanity of all earthly things, but Cardano invariably tramples all such outbreaks of humility in his rush to admire other facets of his excellent self.

On the quarrel with Tartaglia, *The Book of My Life* is almost silent. Among the authors who have cited him, Cardano lumps Tartaglia with those for whom he "cannot understand by what impertinence they have managed to get themselves into the ranks of the learned." Only at the end of the book does Cardano concede that "in mathematics I received a few suggestions, but very few, from brother Niccolò." Thus we are forced back to the *Cartelli* and Tartaglia's writings. The most accessible analysis of these works, with translations of relevant passages, is in Ore [1953], Ch. 4.

Figure 5.5. François Viète.

François Viète (Fig. 5.5) was born in 1540 in Fontenay-le-Comte, a town in what is now the Vendée department of France. His father, Etienne, was a lawyer and his mother, Marguerite Dupont, was well connected to ruling circles in France. Viète was educated by the Franciscans in Fontenay and at the University of Poitiers. He received his bachelor's degree in law in 1560 and then returned to Fontenay to commence practice. For the rest of his life he was engaged mainly in law or related judicial and court services, doing mathematics only during periods of leisure. His clients are said to have included Queen Mary of England and Eleanor of Austria, and from 1574 to 1584 he acted as an advisor and negotiator for King Henry III of France. At that stage he was banished through the efforts of political rivals, but he returned to court in 1589 when Henry III moved his seat of government from Paris to Tours. Following the assassination of Henry III in 1589, he served Henry IV until 1602. Viète died in 1603.

The most famous exploit of Viète's professional career was his deciphering of Spanish dispatches for Henry IV during the war against Spain. King Philip II of Spain, unable to believe that this was humanly possible, protested to the pope that the French were using black magic. The pope may well have been impressed, but not enough to believe that magic was involved, as the Vatican's own experts had broken one of Philip's codes 30 years earlier (see Kahn [1967], pp. 116–118).

An equally famous mathematical feat of Viète's, and equally magical to his contemporaries, was his solution of a 45th-degree equation posed by Adriaen van Roomen in 1593:

$$45x - 3795x^3 + 95{,}634x^5 - \cdots + 945x^{41} - 45x^{43} + x^{45} = N$$

Viète saw immediately that this equation resulted from the expansion of $\sin 45\theta$ in powers of $\sin\theta$, and was able to give 23 solutions (he did not recognize negative solutions). This was one contest, incidentally, that did not generate any bitterness—it led to a firm friendship between the two mathematicians.

CHAPTER 6

Analytic Geometry

6.1. Steps Toward Analytic Geometry

The basic idea of analytic geometry is the representation of curves by equations, but this is not the whole idea. If it was, then the Greeks would be considered the first analytic geometers. Menaechmus was perhaps the first to discover equations of curves, along with his discovery of the conic sections, and we have seen how he used equations to obtain $\sqrt[3]{2}$ as the intersection of a parabola and a hyperbola (Section 2.4). Apollonius' study of conics used equations obtained as by-products of geometric arguments.

What was lacking in Greek mathematics was both the inclination and the technique to manipulate equations to obtain information about curves. The Greeks used curves to study algebra rather than the other way around. Menaechmus' construction of $\sqrt[3]{2}$ is an excellent example of this: extraction of roots was not a given operation but one that had to be secured by geometric construction. Similarly, an equation was not an entity in its own right but a property of a curve which could be discovered after the curve had been constructed geometrically. This was a natural state of affairs as long as equations were written out in words. When, as in Apollonius, an equation takes half a page to write out, it is difficult to form a general concept of equation, function, or curve. Hence the lack of a general concept of curve in Greek mathematics—it was just too complicated to handle in their language.

In the Middle Ages the idea of coordinates emerged in a different way in the work of Oresme (c. 1323–1382). Coordinates had been used in astronomy and geography since Hipparchus (c. 150 B.C.); in fact, Oresme called his coordinates "longitude" and "latitude," but he seems to have been the first to use them to represent functions such as velocity as a function of time. Setting up the coordinate system *before* determining the curve was Oresme's step beyond the Greeks, but he too lacked the algebra to go further.

The step that finally made analytic geometry feasible was the solution of equations and the improvement of notation in the sixteenth century, which we discussed in the previous chapter. This step made it possible to consider equations, and hence curves, in some generality, and to have confidence in one's ability to manipulate them. As we shall see in the next section, the two founders of analytic geometry, Fermat and Descartes, were both strongly influenced by these developments.

For more details on the development of analytic geometry, the reader is referred to an excellent book by Boyer [1956].

6.2. Fermat and Descartes

There have been several occasions in the history of mathematics when an important discovery was made independently and almost simultaneously by two individuals: noneuclidean geometry by Bolyai and Lobachevsky, elliptic functions by Abel and Jacobi, the calculus by Newton and Leibniz, for example. To the extent that we can rationally explain these remarkable events, it must be on the basis of ideas already "in the air," of conditions becoming favorable for their crystallization. As I tried to show in the previous section, conditions were favorable for analytic geometry at the beginning of the seventeenth century. Thus it is not completely surprising that the subject was independently discovered by Fermat [1629] and Descartes [1637]. (Descartes' work *La Géométrie* may in fact have been started in the 1620s. In any case it is independent of Fermat, whose work was not published until 1679.)

It is a surprise to learn, however, that both Fermat and Descartes began with an analytic solution of the same classical geometric problem, the four-line problem of Apollonius, and that the main discovery of each was that second-degree equations correspond to conic sections. Up to this point Fermat was more systematic than Descartes, but that was as far as he went. He was content to leave his work in a "simple and crude" state, confident that it would grow in stature when nourished by new inventions.

Descartes, on the other hand, treated many higher degree curves and clearly understood the power of algebraic methods in geometry. He wanted to withhold this power from his contemporaries, however, particularly the rival mathematician Roberval, as he admitted in a letter to Mersenne (see Boyer [1956], p. 104). *La Géométrie* was written to boast about his discoveries, not to explain them. There is little systematic development and proofs are frequently omitted with a sarcastic remark such as, "I shall not stop to explain this in more detail, because I should deprive you of the pleasure of mastering it yourself" (p. 10). Descartes' conceit is so great that it is a pleasure to see him come a cropper occasionally, as on p. 91: "The ratios between straight and curved lines are not known, and I believe cannot be discovered by human minds." He was referring to the then unsolved problem of determining the length of curves, but he spoke too soon, for in 1657 Neil and Van Heuraet found the length of an arc of the semicubical parabola $y^2 = x^3$, and the calculus soon made such problems routine. (A full and interesting account of the story of arc length may be found in Hofmann [1974], Ch. 8.)

6.3. Algebraic Curves

I could give here several other ways of tracing and conceiving a series of curved lines, each curve more complex than any preceding one, but I think the best way to group together all such curves and then classify them in order is by recognizing the fact that all points of those curves which we may call "geometric," that is, those which admit of precise and exact measurement, must bear a definite relation to all points of a straight line, and that this relation must be expressed by means of a single equation. (Descartes [1637/1954], p. 48)

In this passage Descartes defines what we now call *algebraic curves*. The fact that he calls them "geometric" shows his attachment to the Greek idea that curves are the product of geometric constructions. He is using the notion of equation not to define curves directly but to restrict the notion of geometric construction more severely than the Greeks did, thereby restricting the concept of curve. As we saw in Section 2.5, the Greeks considered some constructions, such as rolling one circle on another, which are capable of producing transcendental curves. Descartes called such curves "mechanical" and found a way to exclude them by his restriction to curves "expressed by means of a single equation." It becomes clear in the lines following the preceding quotation that he means polynomial equations, since he gives a classification of equations by degree.

Although Descartes' rejection of transcendental curves was short-sighted, as the calculus soon provided techniques to handle them, it nevertheless proved fruitful to concentrate on algebraic curves. The notion of degree, in particular, was a useful measure of complexity. First-degree curves are the simplest possible, namely, straight lines; second-degree are the next simplest, conic sections. With third-degree curves one sees the new phenomena of inflections, double points, and cusps. Inflection and cusp are familiar from $y = x^3$ and $y^2 = x^3$, respectively; we also saw a cusp on the cissoid (Section 2.5). A classical example of a cubic with a double point is the *folium* (leaf) *of Descartes* (Descartes [1638]),

$$x^3 + y^3 = 3axy$$

The "leaf" is the closed portion to the right of the double point; Descartes misunderstood the rest of the curve through neglect of negative coordinates. The true shape of the folium was first given by Huygens [1692]. Figure 6.1 is Huygens' drawing, which also shows the asymptote to the curve.

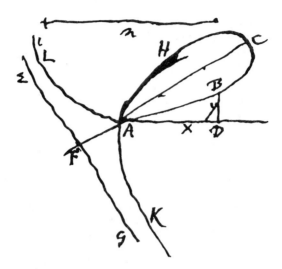

Figure 6.1

Figure 6.2

An excellent account of the early history of curves can be found in Brieskorn and Knörrer [1981], Ch. 1. Many individual curves, with diagrams, equations, and historical notes, can be found in Gomes Teixeira [1908, 1909, 1915]. The development of Descartes' concept of curve has been studied by Bos [1981].

EXERCISES

1. Show that the folium of Descartes has parametric equations

$$x = \frac{3at}{1 + t^3}, \qquad y = \frac{3at^2}{1 + t^3}$$

and use these equations to show that it is tangential to the axes at 0.

2. The *roses of Grandi* are given by the polar equations

$$r = a \sin n\theta, \qquad r = a \cos n\theta$$

where n is an integer. (Figure 6.2 shows some of these curves, as given by Grandi [1723]). Show that the roses of Grandi are algebraic.

6.4. Newton's Classification of Cubics

Since first- and second-degree curves are straight lines and conics, they were well understood before the advent of analytic geometry. Up to the end of the eighteenth century most mathematicians considered them not amenable to further clarification, and hence an unsuitable subject for the new methods. A famous example is the Greek-style treatment of planetary orbits in Newton's *Principia* [1687]. The classical attitude to low degree curves was summed up by d'Alembert in his article on geometry in the *Encylopédie* (1751):

> Algebraic calculation is not to be applied to the propositions of elementary geometry because it is not necessary to use this calculus to facilitate demonstrations, and it appears that there are no demonstrations which can really be facilitated by this calculus except for the solution of problems of second degree by the line and circle.

Thus the first new problem opened up by analytic geometry, and also the first considered properly to belong to the subject, was the investigation of cubic curves. These curves were classified, more or less completely, by Newton [1695] (see Ball [1890] for a commentary).

Newton [1667] began this work with the general cubic in x and y,

$$ay^3 + bxy^2 + cx^2y + dx^3 + ey^2 + fxy + gx^2 + hy + kx + l = 0$$

making a general transformation of axes, leading to an equation with 84 terms, then showing that the latter equation could be reduced to one of the forms

$$Axy^2 + By = Cx^3 + Dx^2 + Ex + F$$
$$xy = Ax^3 + Bx^2 + Cx + D$$
$$y^2 = Ax^3 + Bx^2 + Cx + D$$
$$y = Ax^3 + Bx^2 + Cx + D$$

Newton then divided the curves into species according to the roots of the right-hand side, obtaining 72 species (and overlooking 6). His paper does not contain detailed proofs; these were supplied by Stirling [1717], along with four of the species Newton had missed. Newton's classification was criticized by some later mathematicians, such as Euler, for lacking a general principle. A unifying principle was certainly desirable, to reduce the complexity of the classification. And such a principle was already implicit in one of Newton's

CUR   CUR

Fig. 71.

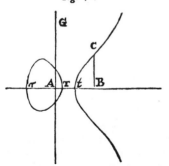

of the Form of a Bell, with an Oval at its Vertex. And this makes a *Sixty seventh Species.*

If two of the Roots are equal, a Parabola will be formed, either *Nodated* by touching an Oval,

Fig. 72.

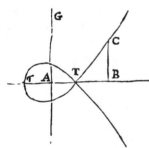

Fig. 73.

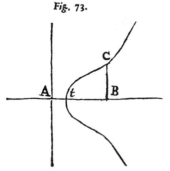

or *Punctate*, by having the Oval infinitely small. Which two *Species* are the *Sixty eighth* and *Sixty ninth.*

If three of the Roots are equal, the Parabola will be *Cuspidate* at the Vertex. And this is the

Fig. 75.

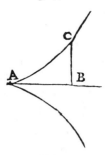

Neilian Parabola, commonly called Semi-cubical. Which makes the *Seventieth Species.*

If two of the Roots are impossible, there will (See *Fig. 73.*)

Fig. 73.

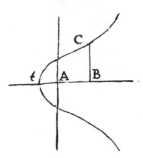

be a *Pure* Parabola of a Bell-like Form. And this makes the *Seventy first Species.*

Figure 6.3

passing remarks, section 29, "On the Genesis of Curves by Shadows." This principle, which will be explained in the next chapter, reduces cubics to the five types seen in Figure 6.3 (taken from an English translation of Newton's paper published in 1710; see Whiteside [1964]).

The reader may wonder where the most familiar cubic, $y = x^3$, appears among these five. The answer is that it is equivalent to the one with a cusp, in Newton's Figure 75. This will be explained in the next chapter.

6.5. Construction of Equations and Bézout's Theorem

In Sections 6.1, 6.2, and 6.3 the development of analytic geometry was outlined from the first observations of equations as properties of curves to the full realization that equations *defined* curves and that the concept of (polynomial) equation was the key to the concept of (algebraic) curve. With hindsight, we can say that Descartes' *La Géométrie* [1637] was the major step in the maturation of the subject, but the book does not conclusively establish what analytic geometry is. In fact, it is largely devoted to two transitional topics in the development of the subject: the sixteenth-century theory of equations and the now almost forgotten discipline called "construction of equations."

The paradigm construction of an equation was Menaechmus' construction of $\sqrt[3]{2}$ by intersecting a parabola and hyperbola. From a geometric point of view, one is using familiar curves (parabola and hyperbola) to construct a less familiar length ($\sqrt[3]{2}$). This becomes sharper when expressed algebraically: curves of degree 2 are being used to solve an equation of degree 3, $x^3 = 2$. In the 1620s Descartes discovered something more general: a method of solving any third- or fourth-degree equation by intersecting curves of degree 2, a parabola and a circle. His friend Beeckman [1628] reported in a note that "M. Descartes made so much of this invention that he confessed never to have found anything superior himself and even that nobody else had ever found anything better" (translation of Bos [1981], p. 330). Descartes was not as superior as he thought, since Fermat independently made the same discovery in an unpublished work [1629], strengthening the already extraordinary coincidence between his work and that of Descartes. However, Fermat apparently did not pursue the idea further, and Descartes did.

In *La Géométrie* Descartes found a particular cubic curve, the so-called cartesian parabola, whose intersections with a suitable circle yield the solution of any given fifth- or sixth-degree equation. Descartes concludes the book with this result, blithely telling the reader that

> it is only necessary to follow the same method to construct all problems, more and more complex, ad infinitum; for in the case of a mathematical progression, whenever the first two or three terms are given, it is easy to find the rest. (Descartes [1637/1954], p. 240)

In reality it was not easy, and efforts to find a satisfactory general construction for nth-degree equations petered out around 1750. The story of the rise and fall of this field of mathematics has been told by Bos [1981, 1984].

In their search for a general construction, mathematicians had casually assumed that a curve of degree m meets a curve of degree n in mn points. The first statement of this principle, which became known as Bézout's theorem, seems to have been made by Newton on May 30, 1665:

> For ye number of points in wch two lines may intersect can never bee greater y^n y^e rectangle of y^e numbers of their dimensions. And they always intersect in soe many points, excepting those wch are imaginarie onely. (Newton [1665'], p. 498)

Bézout's theorem leads one to expect that solutions of an equation $r(x) = 0$ of degree $k = m \cdot n$ might be obtainable from the intersections of a suitable degree m curve with a suitable degree n curve. In algebraic terms, one seeks equations

$$p(x, y) = 0 \tag{1}$$

$$q(x, y) = 0 \tag{2}$$

of degrees m, n respectively, from which elimination of y yields the given equation

$$r(x) = 0 \tag{3}$$

as "resultant." This is how mathematicians in the West first encountered the problem of elimination which the Chinese had solved some centuries earlier (Section 5.2).

However, apart from the fact that construction of equations was inverse to elimination, and much harder, Western mathematicians needed two additional facts about elimination itself. First, that elimination between equations of degrees m and n gave a resultant of degree mn; second, that an equation of degree mn has mn roots. The second statement, as mentioned in Section 5.7, becomes a fact only when complex numbers are admitted. The first becomes a fact only when "points at infinity" are admitted. If, for example, (1) and (2) are equations of parallel lines, then (3) is of "degree 0" and has *no* solutions. However, one can consider parallel lines to meet "at infinity," and the geometric framework for this idea, projective geometry, developed at about the same time as analytic geometry. Unfortunately, it was not realized until the nineteenth century that projective geometry and analytic geometry needed each other. Until then, projective geometry developed without coordinates, and all attempts to prove Bézout's theorem (notably by Maclaurin [1720], Euler [1748'], Cramer [1750], and Bézout [1779]) foundered for want of a proper method for counting points at infinity. As a result, Bézout's theorem, which turned out to be the main achievement of the theory of construction of equations, was not properly proved until long after the theory itself had been abandoned.

The origins of projective geometry, and the fruits of its merger with analytic geometry, will be discussed in Chapter 7.

6.6. The Arithmetization of Geometry

As previously emphasized, the early analytic geometers—Descartes in particular—did not accept that geometry could be *based* on numbers or algebra. Perhaps the first to take the idea of arithmetizing geometry seriously was Wallis (1616–1703). Wallis [1657], Chs. XXIII and XXV, gave the first arithmetic treatment of Euclid's Books II and V, and he had earlier given the first purely algebraic treatment of conic sections [1655]. He initially derived equations from the classical definitions by sections of the cone but then proceeded conversely to derive their properties from the equations, "without the embranglings of the cone," as he put it.

Wallis was ahead of his time. Thomas Hobbes, introduced at the beginning of Chapter 2, described Wallis' treatise on conics as a "scab of symbols" and denounced "the whole herd of them who apply their algebra of geometry" (Hobbes [1656], p. 316, and [1672], p. 447). The example and authority of Newton probably reinforced the opinion that algebra was inappropriate in the geometry of lines or conic sections; we saw in Section 6.4 how this remained the accepted view until at least 1750.

Algebra did not catch on in elementary geometry until it was taken up by Lagrange [1773] and supported by influential textbooks of Monge and Lacroix around 1800. But by the time elementary geometry had been brought into the theory of equations, higher geometry had broken out, depending more and more on calculus and the emerging theories of complex functions, abstract algebra, and topology, which bloomed in the nineteenth century. Higher geometry broke away to form differential geometry and algebraic geometry, leaving the elementary residue we call "analytic geometry" today.

Despite its lowly status, analytic geometry was given an important foundational role by Hilbert [1899]. Hilbert took Wallis' arithmetization to its logical conclusion by assuming only the real numbers and sets as given and constructing geometry from them.

Thus from the set $\mathbb{R}$ of reals, one constructs the *euclidean plane* as the set of ordered pairs (x, y) ("points") where $x, y \in \mathbb{R}$. A *straight line* is a set of points (x, y) in the plane such that $ax + by + c = 0$ for some constants a, b, c. Lines are *parallel* if their x and y coefficients are proportional. The *distance* between points (x_1, y_1) and (x_2, y_2) is defined to be $\sqrt{(x_2 - x_1)^2 + (y_2 - y_1)^2}$. As explained in Section 1.6, this definition is motivated by Pythagoras' theorem, which is the keystone in the bridge from arithmetic to geometry.

With these definitions, all axioms and propositions of euclidean geometry become provable propositions about equations. For example, the axiom that nonparallel lines have a point in common corresponds to the theorem that linear equations

$$a_1 x + b_1 y + c_1 = 0$$

$$a_2 x + b_2 y + c_2 = 0$$

have a solution when $a_1 b_2 - b_1 a_2 \neq 0$.

Hilbert did not believe, any more than Newton did, that numbers were the true subject matter of geometry. He strongly supported geometric intuition as a method of discovery, as the book Hilbert and Cohn-Vossen [1932] makes clear. The purpose of his arithmetization was to give a secure logical foundation to geometry after the nineteenth-century developments which discredited geometry and installed arithmetic as the ultimate authority in mathematics. This foundation is no longer quite as secure as it seemed in 1900, as we shall see in Chapter 20; nevertheless, it is still the most secure foundation we know.

6.7. Biographical Notes: Descartes

René Descartes (Figure 6.4) was born in La Haye (now called La Haye–Descartes) in the French province of Touraine in 1596 and died in Stockholm in 1650. His father, Joachim, was a councilor in the high court of Rennes in Brittany; his mother, Jeanne, was the daughter of a lieutenant general from Poitiers and the owner of property that was eventually to assure Descartes of financial independence. His mother died in 1597, and Descartes was raised by his maternal grandmother and a nurse. He does not seem to have been close to his father, brother, or sister, seldom mentioning them to others and writing to them only on matters of business.

Joachim Descartes was away from home for half the year because of his court duties, but he saw enough of René to observe his exceptional curiosity, calling him his "little philosopher." In 1606 he enrolled him in the Jesuit College of La Flèche, which had recently been founded by Henry IV in Anjou. The young Descartes was given special privileges at school, in recognition of his intellectual promise and delicate health. He was one of the few boys to have his own room, was permitted books forbidden to other students, and was allowed to stay in bed until late in the morning. Spending several morning hours in bed thinking and writing became his lifelong habit and, when he finally had to break it in the Swedish winter, the consequences were fatal.

The most dramatic event of his schooldays was the assassination of Henry IV in 1610. Since Henry IV was not only the founder of the school but also the most popular king in French history, his death was a profound shock. La Flèche became the venue for an elaborate funeral ceremony, the climax of which was the burial of the king's heart. Descartes was one of 24 students chosen to participate in the ceremony.

Figure 6.4. Descartes. (Louvre Museum)

He left La Flèche in 1614 and, after legal studies at Poitiers, which seem to have left no impression on him, went to Holland as an unpaid volunteer in the army of Prince Maurice of Nassau in 1618. This was not an unusual decision for a young Frenchman of means at the time, since the Dutch were fighting France's enemy, Spain, and Descartes seems to have joined the army to see the world, not because of any taste for barracks life or combat. As it happened, there was a lull in the war at the time, and Descartes had two years of virtual leisure to reflect on science and philosophy.

When in Breda, on November 10, 1618, he saw a mathematical problem posted on a wall. Since his Dutch was not yet fluent, he asked a bystander to translate for him. This was how Descartes met Isaac Beeckman, who became his first instructor in mathematics and a lifelong friend. The following November 10, Descartes was in Bavaria. He spent a day of intense thought in a heated room ("stove" he called it) and that night had a dream he later considered to be a revelation of the path he should follow in developing his philosophy. Whether the dream also revealed the path to analytic geometry, as some have conjectured, will probably never be known. Descartes' own description of the dream has been lost, and we have only a summary by his first biographer, Baillet ([1691], p. 85), which is not helpful. In any case, it seems a little ludicrous to award Descartes priority over Fermat on the basis of a dream. Could a counterclaim of priority be lodged if the dream of a teenaged Fermat came to light?

In 1628 Descartes moved to Holland, where he spent most of the rest of his life. He lived a simple but leisurely life and finally settled down to working out the ideas conceived nine years earlier. The relative isolation suited him, as he was hostile to other scientific giants of his time such as Galileo, Fermat, and Pascal and preferred to communicate with scholars who could understand him without challenging his superiority. One such was Marin Mersenne, who had been a senior student at La Flèche in Descartes' time and was his main scientific contact in France. Others were Princess Elizabeth of Bohemia and Queen Christina of Sweden, with both of whom Descartes had extensive correspondences.

A positive side to Descartes' intolerance of intellectual rivals was an apparently genuine interest in the affairs of his neighbors in Holland. He encouraged local youths who showed talent in mathematics, and he was known in the region as someone to turn to in times of trouble (see Vrooman [1970], pp. 194–196). The one serious love of his life was a servant girl named Helen, who bore him a daughter, Francine, in 1635. Admittedly, his interest in this case did not extend to marrying Helen, but the death of Francine from scarlet fever in 1640 caused him the greatest sorrow of his life.

In 1649 Descartes agreed to journey to Stockholm to become tutor to Queen Christina. This was the culmination of his correspondence with her and of negotiations through Descartes' friend Chanut, the French ambassador. The queen, who was noted for her physical as well as mental vigor, slept no more than five hours a night and rose at 4 a.m. Descartes had to arrive at 5 a.m. to give her lessons in philosophy. The program commenced on January 14, 1650, during the coldest winter for over 60 years. One can imagine the shock to Descartes' system of such early rising followed by a journey from the ambassador's residence to the palace. However, it was actually Chanut who succumbed to the cold first. On January 18 he came down with pneumonia, and Descartes apparently caught it from him. Chanut recovered but Descartes did not, and he died on February 11, 1650.

Descartes is, of course, as well known for his philosophy as his analytic geometry. The *Geometry* was originally an appendix to his main philosophical work, the *Discourse on Method*. The other appendices were the *Dioptrics*, a treatise on optics, and the *Meteorics*, the first attempt to give a scientific theory of the weather. In the *Dioptrics*, Descartes did not inform his readers that Ptolemy, al-Haytham, Kepler, and Snell had already discovered the main principles of optics; nevertheless, he presented the subject with greater clarity and thoroughness than before, undoubtedly advancing both the theory and practice of optical instrumentation. As for the *Meteorics*, we now know how premature it was to attempt a theory of the weather in 1637, so it is understandable that this treatise has more misses than hits. His big hit was a correct explanation of rainbows (except for the colors, whose explanation was completed by Newton), which Descartes was able to give on the basis of his optics. More typical, unfortunately, was his explanation of thunder: it was caused by clouds bumping together and not related to lightning. An excellent survey of Descartes' scientific work and philosophy, with a particularly detailed analysis of the *Geometry*, is given by Scott [1952].

CHAPTER 7

Projective Geometry

7.1. Perspective

Perspective may be simply described as the realistic representation of spatial scenes on a plane. This of course has been a concern of painters since ancient times, and some Roman artists seem to have achieved correct perspective by the first century B.C.; an impressive example is shown in Wright [1983], p. 38. However, this may have been a stroke of individual genius rather than the success of a theory, because the vast majority of ancient paintings show incorrect perspective. If indeed there was a classical theory of perspective, it was well and truly lost during the Dark Ages. Medieval artists made some charming attempts at perspective but always got it wrong, and errors persisted well into the fifteenth century. (Errors still survive in twentieth-century mathematics texts. Figure 7.1 shows a fifteenth-century artistic example from Wright [1983], p. 41 alongside a twentieth-century mathematical example from the exposé of Grünbaum [1985].)

The discovery of a method for correct perspective is usually attributed to the Florentine painter-architect Brunelleschi (1377–1446), around 1420. The first published method appears in the treatise *On Painting* by Alberti [1436]. The latter method, which came to be known as *Alberti's veil*, was to set up a piece of transparent cloth, stretched on a frame, in front of the scene to be painted. Then, viewing the scene with one eye, in a fixed position, one could trace the scene directly onto the veil. Figure 7.2 shows this method, with a peephole to maintain a fixed eye position, as depicted by Dürer [1525].

Alberti's veil was fine for painting actual scenes, but to paint an imaginary scene in perspective some theory was required. The basic principles used by Renaissance artists were the following:

 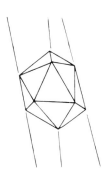

Figure 7.1

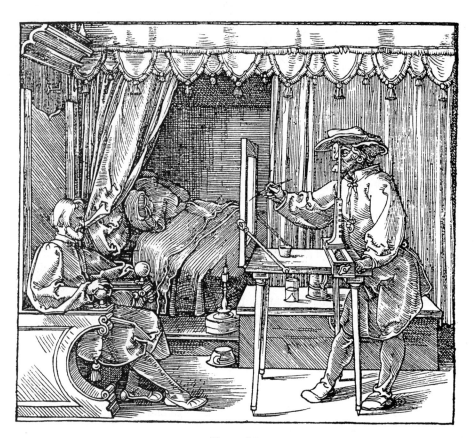

Figure 7.2

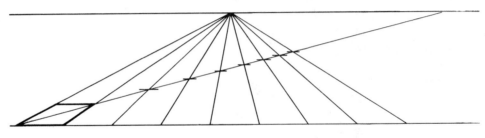

Figure 7.3

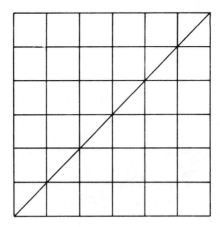

Figure 7.4

(i) A straight line in perspective remains straight.

(ii) Parallel lines either remain parallel or converge to a single point (their *vanishing point*).

These principles suffice to solve a problem artists frequently encountered: the perspective depiction of a square-tiled floor. Alberti [1436] solved the special case of this problem in which one set of floor lines is horizontal, that is, parallel to the horizon. His method, which became known as the *costruzione legittima*, is indicated in simplified form in Figure 7.3. The nonhorizontal floor lines are determined by spacing them equally along the base line (imagined to touch the floor) and letting them converge to a vanishing point on the horizon. The horizontal floor lines are then determined by choosing one of them arbitrarily, thus determining one tile in the floor, and then producing the diagonal of this tile to the horizon. The intersections of this diagonal with the nonhorizontal lines are the points through which the horizontal lines pass. This is certainly true on the actual floor (Fig. 7.4), hence it remains true in the perspective view.

The same principles suffice to generate a perspective view of a tiled floor given an arbitrarily situated tile (exercise 1).

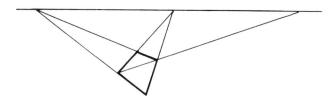

Figure 7.5

1. Use the lines shown in Figure 7.5 to determine all lines in the pavement generated by the given tile.

7.2. Anamorphosis

It is clear from the Alberti veil construction that a perspective view will not look absolutely correct except when seen from the viewpoint used by the artist. Experience shows, however, that distortion is not noticeable except from extreme viewing positions. Following the mastery of perspective by the Italian artists, an interesting variation developed, in which the picture looks right from only one, extreme, viewpoint. The first known example of this style, known as *anamorphosis*, is an undated drawing by Leonardo da Vinci from the *Codex Atlanticus* (compiled between 1483 and 1518). Figure 7.6 shows part of this drawing, a child's face which looks right when viewed with the eye near the right-hand edge of the page.

The idea was taken up by German artists around 1530. The most famous example occurs in Holbein's painting *The Two Ambassadors* (1533). A mysterious streak across the bottom of the picture becomes a skull when viewed from near the picture's edge. For excellent views of this picture and a history of anamorphosis, see Baltrušaitis [1977] and Wright [1983], pp. 146–156. The art of anamorphosis reached its technically most advanced form in France in the early seventeenth century. It seems no coincidence that this was also the time and place of the birth of projective geometry. In fact, the key figures in the two fields, Niceron and Desargues, were well aware of each others' work.

Niceron (1613–1646) was a student of Mersenne and, like him, a monk in the order of Minims. He executed some extraordinary anamorphic wall paintings, up to 55 meters long, and also explained the theory in *La perspective curieuse* (Niceron [1638]). Figure 7.7 is his illustration of anamorphosis of a chair (from Baltrušaitis [1977], p. 44). The anamorphosis, viewed normally, shows a chair like none ever seen, yet from a suitably extreme point one sees an ordinary chair in perspective. This example encapsulates an important

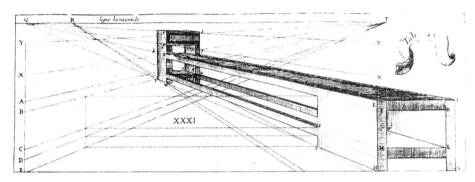

Figure 7.7

mathematical fact: *a perspective view of a perspective view is not in general a perspective view.* Iteration of perspective views gives what we now call a *projective* view, and Niceron's chair shows that projectivity is a broader concept than perspectivity. As a consequence, *projective geometry*, which studies the properties that are invariant under projection, is broader than the theory of perspective. Perspective itself did not develop into a mathematical theory, *descriptive geometry*, until the end of the eighteenth century.

7.3. Desargues' Projective Geometry

The mathematical setting in which one can understand Alberti's veil is the family of lines ("light rays") through a point (the "eye"), together with a plane *V* (the "veil") (Fig. 7.8). In this setting, the problems of perspective and anamorphosis were not very difficult, but the *concepts* were interesting and a challenge to traditional geometric thought. Contrary to Euclid, one had the following:

 (i) Points at infinity ("vanishing points") where parallels met.
(ii) Transformations which changed lengths and angles (projections).

The first to construct a mathematical theory incorporating these ideas was Desargues (1591–1661), although the idea of points at infinity had already been used by Kepler [1604], p. 93. Desargues' book *Brouillon project d'une*

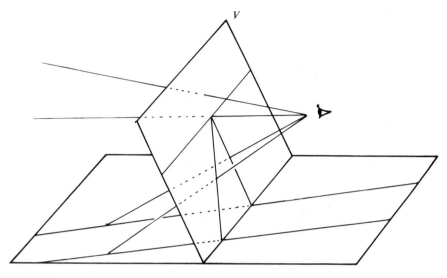

Figure 7.8

atteinte aux événemens des rencontres du cône avec un plan (Schematic sketch of what happens when a cone meets a plane) [1639] suffered an extreme case of delayed recognition, being completely lost for 200 years. Fortunately, his two most important theorems, the so-called Desargues' theorem and the invariance of the cross-ratio, were published in a book on perspective (Bosse [1648]). The text of Desargues [1639] and a portion of Bosse [1648] containing Desargues' theorem may be found in Taton [1951]. An English translation, with an extensive historical and mathematical analysis, is in Field and Gray [1987].

Kepler and Desargues both postulated one point at infinity on each line, closing the line to a "circle of infinite radius." All lines in a family of parallels share the same point at infinity. Nonparallel lines, having a finite point in common, do not have the same point at infinity. Thus any two distinct lines have exactly one point in common—a simpler axiom than Euclid's. Strangely enough, the line at infinity was only introduced into the theory by Poncelet [1822], even though it is the most obvious line in perspective drawing, the horizon. Desargues made extensive use of projections in the *Brouillon projet*; he was the first to use them to prove theorems about conic sections.

Desargues' theorem is a property of triangles in perspective illustrated by Figure 7.9. The theorem states that the points X, Y, Z at the intersections of corresponding sides lie in a line. This is obvious if the triangles are in space, since the line is the intersection of the planes containing them. The theorem in the plane is subtly but fundamentally different and requires a separate proof, as Desargues realized. In fact, Desargues' theorem was shown to play a key role in the foundations of projective geometry by Hilbert [1899].

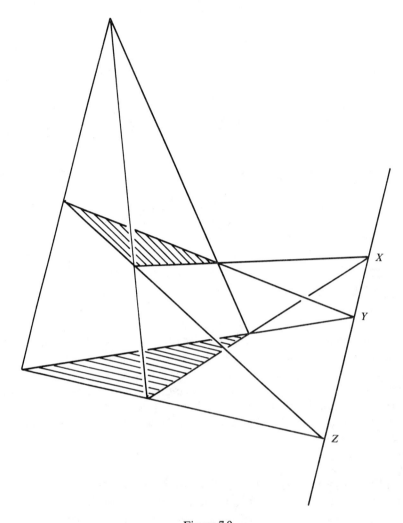

Figure 7.9

The invariance of the cross-ratio answers a natural question first raised by Alberti: since length and angle are not preserved by projection, what is? No property of three points on a line can be invariant because it is possible to project any three points on a line to any three others (exercise 1). At least four points are therefore needed, and the cross-ratio is in fact a projective invariant of four points. The cross-ratio $(ABCD)$ of points A, B, C, D on a line (in that order) is $\dfrac{CA}{CB} \bigg/ \dfrac{DA}{DB}$. Its invariance is most simply seen by reexpressing it in terms of angles using Figure 7.10. Let O be any point outside the line and consider the areas of the triangles OCA, OCB, ODA, and ODB. First compute them from bases on AB and height h, then recompute using OA and OB as bases

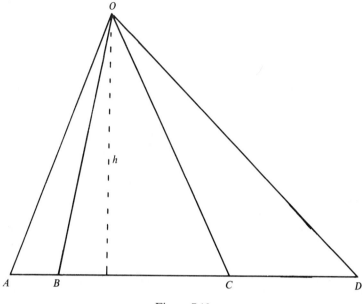

Figure 7.10

and heights expresed in terms of the sines of angles at O:

$$\tfrac{1}{2}h \cdot CA = \text{area } OCA = \tfrac{1}{2}OA \cdot OC \sin \underline{/COA}$$

$$\tfrac{1}{2}h \cdot CB = \text{area } OCB = \tfrac{1}{2}OB \cdot OC \sin \underline{/COB}$$

$$\tfrac{1}{2}h \cdot DA = \text{area } ODA = \tfrac{1}{2}OA \cdot OD \sin \underline{/DOA}$$

$$\tfrac{1}{2}h \cdot DB = \text{area } ODB = \tfrac{1}{2}OB \cdot OD \sin \underline{/DOB}$$

Substituting the values of CA, CB, DA, and DB from these equations we find (following Möbius [1827]) the cross-ratio in terms of angles at O:

$$\frac{CA}{CB}\bigg/\frac{DA}{DB} = \frac{\sin \underline{/COA}}{\sin \underline{/COB}}\bigg/\frac{\sin \underline{/DOA}}{\sin \underline{/DOB}}$$

Any four points A', B', C', D' in perspective with A, B, C, D from a point O have the same angles (Fig. 7.11), hence they will have the same cross-ratio. But then so will any four points A'', B'', C'', D'' projectively related to A, B, C, D since a projectivity is by definition the product of a sequence of perspectivities.

EXERCISE

1. Show that any three points on a line can be sent to any other three points on a line by a projection.

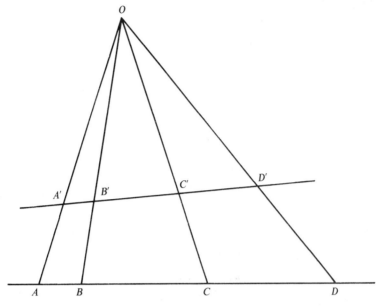

Figure 7.11

7.4. The Projective View of Curves

The problems of perspective drawing mainly involved the geometry of straight lines. There were, it is true, problems such as drawing ellipses to look like perspective views of circles, but artists were generally content to solve such problems by interpolating smooth-looking curves in a suitable straight-line framework. An example is the drawing of a chalice by Uccello (1397–1475) in Figure 7.12.

A mathematical theory of perspective for curves became possible in principle with the advent of analytic geometry. When a curve is specified by an equation $f(x, y) = 0$, the equation of any perspective view is obtainable by suitably transforming x and y. However, this transformational viewpoint, even though quite simple algebraically, emerged only with Möbius [1827]. The first works in projective geometry, by Desargues [1639] and Pascal [1640], used the language of classical geometry, even though the language of equations was available from Descartes [1637]. This was understandable, not only because the analytic method was so obscure in Descartes, but also because the advantages of the projective method could be more clearly seen when it was used in a classical setting. Desargues and Pascal confined themselves to straight lines and conic sections, showing how projective geometry could easily reach and surpass the results obtained by the Greeks. Moreover,

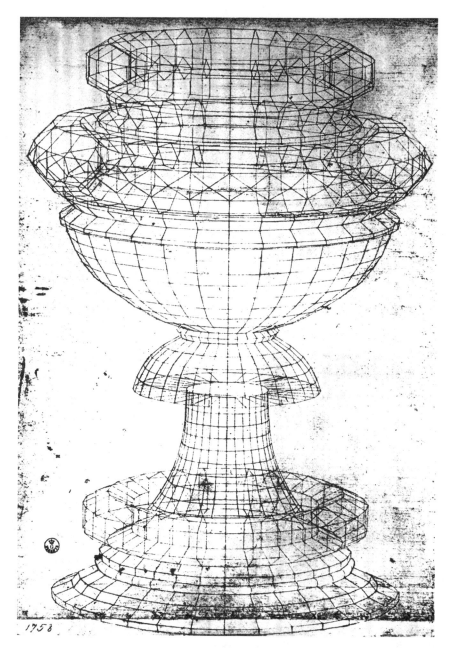

1758

Figure 7.12. Drawing of a chalice by Uccello. (Uffizi, Florence)

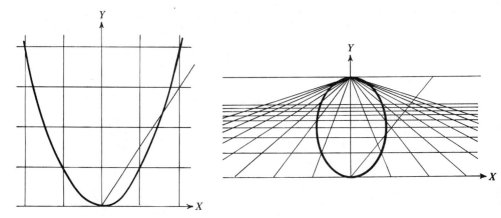

Figure 7.13

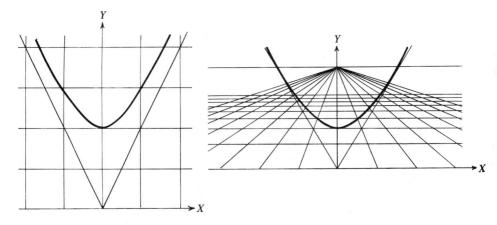

Figure 7.14

the projective viewpoint gave something else that would have been incomprehensible to the Greeks: a clear account of the behavior of curves at infinity.

For example, Desargues ([1639], in Taton [1951], p. 137) distinguished the ellipse, parabola, and hyperbola by their numbers of points at infinity, 0, 1, and 2, respectively. The points at infinity on the parabola and hyperbola can be seen quite plainly by tilting the ordinary views of them into perspective views (Figs. 7.13 and 7.14). The parabola has just one point at infinity because it crosses each ray through 0, except the y-axis, at just one finite point. As for the hyperbola, its two points at infinity are where it touches its asymptotes, as seen in Figure 7.14. The continuation of the hyperbola above the horizon

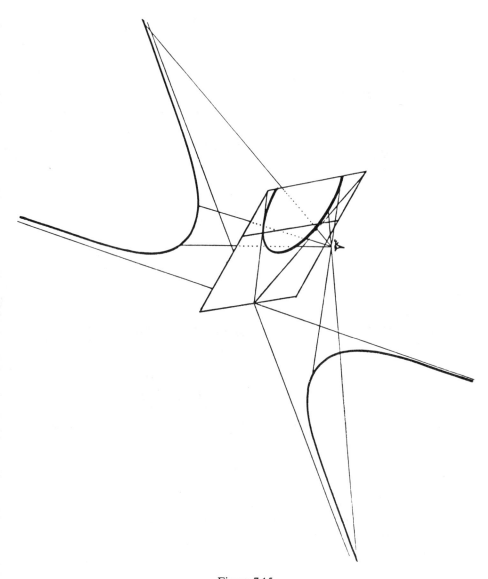

Figure 7.15

results from projecting the lower branch through the same center of projection (Fig. 7.15).

Projective geometry goes beyond describing the behavior of curves at infinity. The line at infinity is no different from any other line and can be deprived of its special status. Then all projective views of a curve are equally valid, and one can say, for example, that all conic sections are ellipses when

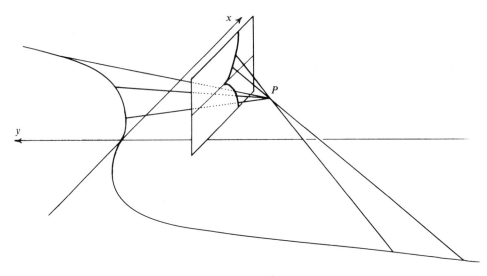

Figure 7.16

suitably viewed. This is no surprise if one remembers conic sections not as second-degree curves but as sections of the cone. Of course they all look the same from the vertex of the cone.

More surprisingly, a great simplification of cubic curves also occurs when they are viewed projectively. As mentioned in Section 6.4, Newton [1695] classified cubic curves into 72 types (and missed 6). However, in his Section 29, "On the Genesis of Curves by Shadows." Newton claimed that each cubic curve can be projected onto one of just five types. As mentioned in Section 6.4, this includes the result that $y = x^3$ can be projected onto $y^2 = x^3$. The proof of this is an easy calculation when coordinates are introduced (see exercise 7.5.3), but one already gets an inkling of it from the perspective view of $y = x^3$ (Fig. 7.16). The lower half of the cusp is the view of $y = x^3$ below the horizon; the upper half comes from projecting the view behind one's head through P to the picture plane in front.

Conversely, $y^2 = x^3$ has an inflection at infinity. Newton's projective classification comes about by studying the behavior at infinity of all cubics and observing that each has characteristics already possessed, not necessarily at infinity, by curves of the form

$$y^2 = Ax^3 + Bx^2 + Cx + D$$

Newton had already divided these into five types in his analytic classification (they are the five shown in Fig. 6.3). Newton's result was improved only in the nineteenth century when projective classification over the complex numbers reduced the number of types of cubics to just three. We shall discuss this later in connection with the development of complex numbers (Section 15.5).

7.5. Homogeneous Coordinates and the Projective Plane

The way in which projective geometry allows infinity to be put on the same footing as the finite points of the plane is intuitively clear when one thinks of the horizon in a picture, which is a line like any other. However, the most convenient way to formalize the idea is to introduce coordinates. This did not happen in Desargues' time, perhaps because of the resistance to coordinates in elementary geometry that was then prevalent (see Sections 6.4 and 6.5). Suitable coordinates, now known as *homogeneous coordinates*, were invented by Möbius [1827] and Plücker [1830]. Homogeneous coordinates give a natural extension of the cartesian plane $\mathbb{R}^2$ by points at infinity by assigning new coordinates to the points already present and creating new points with the coordinates left over.

The homogeneous coordinates of a point $(X, Y) \in \mathbb{R}^2$ are all real triples (Xz, Yz, z) with $z \neq 0$, that is, all real triples (x, y, z) with $x/z = X$, $y/z = Y$. If, following Klein [1925], we take X, Y to be the x, y coordinates in the plane $z = 1$, then these triples are just the coordinates of points $\neq 0$ on the line in $\mathbb{R}^3$ from 0 to (X, Y) (Fig. 7.17). Thus homogeneous coordinates give a one-to-one correspondence between points $(X, Y) \in \mathbb{R}^2$ and nonhorizontal lines through 0. The horizontal lines, whose points have coordinates $(x, y, 0)$, naturally correspond to points at infinity. Moreover, there is a natural way to determine which points at infinity "belong to" a given curve.

Each curve C in $\mathbb{R}^2$, expressed by an equation

$$p(X, Y) = 0 \qquad (1)$$

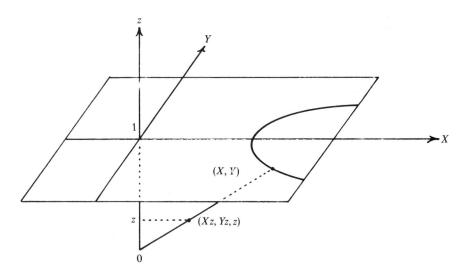

Figure 7.17

say, can be reexpressed by the equation

$$p\left(\frac{x}{z},\frac{y}{z}\right) = 0 \tag{2}$$

for $z \neq 0$. If p is a polynomial of degree n, we can extend (2) to all values of z by multiplying through by z^n, giving

$$z^n p\left(\frac{x}{z},\frac{y}{z}\right) = \bar{p}(x,y,z) = 0 \tag{3}$$

where $\bar{p}$ is a *homogeneous* polynomial of degree n in x, y, z (i.e., if (x,y,z) is a solution of (3), so is (tx, ty, tz)—as it should be, since these triples are coordinates of the same point). Equation (3) is satisfied by all points $(X, Y) = (x/z, y/z)$ of C, together with possible other points obtained for $z = 0$. The latter form horizontal lines approached by the lines corresponding to points of C as $z \to 0$, that is, as X or $Y \to \infty$, so it is natural to regard them as *the points at infinity of C.*

In geometric terms, we have enlarged the (X, Y) plane $\mathbb{R}^2$ to the *real projective plane* $P_2(\mathbb{R})$ by reinterpreting each point (X, Y) as the line from 0 to (X, Y) and completing this set of lines to the set of all lines through 0. The horizontal lines, which have no interpretation in the (X, Y) plane, are interpreted as points at infinity. In the process, each algebraic curve C in the (X, Y) plane is enlarged to its *projective completion* $\bar{C}$ (with equation $\bar{p}(x,y,z) = 0$) by including the points at infinity that are the limits of its ordinary points. We can model $P_2(\mathbb{R})$ by a surface if each line through 0 is replaced by its intersection with the unit sphere, namely, a pair of *antipodal* (diametrically opposite) points. The points at infinity then become antipodal pairs on the equator $z = 0$, which shows they are the same as all other points. A line L in $\mathbb{R}^2$, being given by a linear equation $P(X, Y) = 0$, has as completion the *projective line* $\bar{L}$ with homogeneous linear equation $\bar{p}(x,y,z) = 0$. Thus the points of $\bar{L}$ lie in a plane through 0 and hence are the antipodal pairs on a great circle. The *line at infinity*, $z = 0$, is simply the antipodal pairs on the equator, and hence the same as any other projective line.

A projective line can be visualized as a great semicircle (which contains one representative from each antipodal pair) with its ends identified. Thus it is a closed curve, and Kepler and Desargues were not far wrong in thinking of a projective line as a circle. The projective plane, however, is not a sphere but something more peculiar, as was noticed by Klein [1874]. On a sphere, any simple closed curve separates the surface into two parts. A "small" closed curve in $P_2(\mathbb{R})$, that is, one strictly contained in a hemisphere of the model, also separates $P_2(\mathbb{R})$, but a "large" one does not. The equator, for instance, does not separate the upper hemisphere from the lower, because these hemi-spheres are the *same place* under the antipodal point identification! A less paradoxical view of this is seen by going back to the model of $P_2(\mathbb{R})$ whose elements are lines through 0. The lines through the equator do not separate the lines through the upper hemisphere from the lines through the lower hemisphere, because these are the same lines.

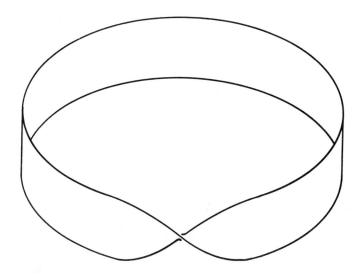

Figure 7.18

1. Show that the point at infinity of the line $aX + bY + c = 0$ has homogeneous coordinates $(tb, -ta, 0)$, $t \neq 0$. Conclude that all lines in a family of parallels have the same point at infinity.

2. Show that any two projective lines have one point in common.

3. Let X, Y denote the x, y coordinates in the plane $z = 1$ (as before), and let X', Z' denote the x, z coordinates in the plane $y = 1$. Show that the curves $Y = X^3$, $(Z')^2 = (X')^3$ have the same equation in the homogeneous coordinates x, y, z.

4. Deduce that $Y = X^3$ is mapped onto $(Z')^2 = (X')^3$ by projection from 0 of the plane $z = 1$ onto the plane $y = 1$.

5. Show that a strip of $P_2(\mathbb{R})$ surrounding a projective line is a Möbius band (Fig. 7.18).

7.6. Bézout's Theorem Revisited

As we saw in Section 6.5, a precise account of points at infinity is needed to obtain Bézout's theorem that a curve of degree m meets a curve of degree n in mn points. The projective completion does this. The preceding exercises show that lines (curves of degree 1) meet in $1 \times 1 = 1$ point. In general, if C_m is a curve with homogeneous equation of degree m,

$$p_m(x, y, z) = 0 \qquad\qquad (1)$$

and if C_n is a curve with homogeneous equation of degree n,

$$p_n(x, y, z) = 0 \qquad\qquad (2)$$

one wishes to show that the equation

$$r_{mn}(x, y) = 0 \qquad\qquad (3)$$

which results from eliminating z between (1) and (2) is homogeneous of degree mn. This is not hard to do (see exercises), but it seems that a homogeneous formulation of Bézout's theorem, with a rigorous proof that the resultant r_{mn} has degree mn, was not given until the late 1800s.

There is an obvious condition to be included in the hypothesis of Bézout's theorem: the curves C_m and C_n have no common component. The algebraic equivalent of this condition is that the polynomials p_m, p_n have no nonconstant common factor. Then the form of Bézout's theorem which can be proved with the help of homogeneous coordinates is *curves C_m, C_n with homogeneous equations $p_m(x, y, z) = 0$, $p_n(x, y, z) = 0$ of degrees m, n and no common component have intersections given by the solutions of a homogeneous equation $r_{mn}(x, y) = 0$ of degree mn.*

A useful consequence of Bézout's theorem is that curves C_m, C_n of degrees m, n with *more* than mn intersections have a common component.

EXERCISES

1. Suppose that

$$p_m(x, y, z) = a_0 z^m + a_1 z^{m-1} + \cdots + a_m$$
$$p_n(x, y, z) = b_0 z^n + b_1 z^{n-1} + \cdots + b_n$$

are homogeneous polynomials of degrees m, n. Thus $a_i(x, y)$ is homogeneous of degree i, $b_j(x, y)$ homogeneous of degree j. By multiplying p_m and p_n by suitable powers of z, show that the equations

$$p_m = 0 \qquad \text{and} \qquad p_n = 0$$

are equivalent to a system of $m + n$ homogeneous linear equations in the variables $z^{m+n-1}, \ldots, z^2, z^1, z^0$, which in turn is equivalent to

$$r_{mn}(x, y) \equiv \begin{vmatrix} a_0 & a_1 & \cdots a_m 0 \cdots & 0 \\ 0 & a_0 a_1 \cdots a_m 0 \cdots & 0 \\ \vdots & \ddots & \ddots & \vdots \\ & & & 0 \\ 0 \cdots 0 a_0 & \cdots & a_m \\ b_0 b_1 & & b_n 0 \cdots 0 \\ 0 b_0 & \cdots & b_n & \vdots \\ \vdots & \ddots & & \ddots & 0 \\ 0 \cdots 0 b_0 & \cdots & b_n \end{vmatrix} = 0.$$

2. Show that a polynomial $p(x, y)$ is homogeneous of degree $k \Leftrightarrow p(tx, ty) = t^k p(x, y)$.

3. Show $r_{mn}(tx, ty) = t^{mn} r_{mn}(x, y)$. *Hint:* Multiply the rows of $r_{mn}(tx, ty)$ by suitable powers of t to arrange that each element in any column contains the same power of t. Then remove these factors from the columns so that $r_{mn}(x, y)$ remains.

7.7. Pascal's Theorem

Pascal's *Essay on Conics* [1640] was written in late 1639, when Pascal was 16. He probably had heard about projective geometry from his father, who was a friend of Desargues. The *Essay* contains the first statement of a famous result which became known as Pascal's theorem or the *mystic hexagram*. The theorem states that the pairs of opposite sides of a hexagon inscribed in a conic section meet in three collinear points. (The vertices of the hexagon can occur in any order on the curve. In Fig. 7.19 the order was chosen to enable the three intersections to lie inside the curve.) Pascal's proof is not known, but he probably established the theorem for the circle first, then trivially extended it to arbitrary conics by projection.

Plücker [1847] threw new light on Pascal's theorem by showing it to be an easy consequence of Bézout's theorem. Plücker used an auxiliary theorem about cubics which can be bypassed, giving the following direct deduction from Bézout's theorem.

Let $L_1, L_2, \ldots, L_6$ be the successive sides of the hexagon. The unions of alternate sides, $L_1 \cup L_3 \cup L_5$ and $L_2 \cup L_4 \cup L_6$, can be regarded as cubic curves

$$l_{135}(x, y, z) = 0, \qquad l_{246}(x, y, z) = 0$$

where each l is the product of three linear factors. These two curves meet in

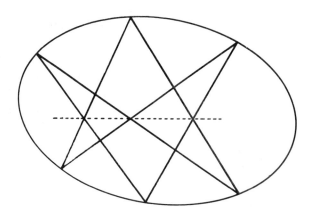

Figure 7.19

nine points: the six vertices of the hexagon and the three intersections of opposite sides. Let

$$c(x, y, z) = 0 \tag{1}$$

be the equation of the conic that contains the six vertices.

We can choose constants α, β so that the cubic curve

$$\alpha l_{135}(x, y, z) + \beta l_{246}(x, y, z) = 0 \tag{2}$$

passes through any given point P. Let P be a point on the conic, unequal to the six vertices. Then the curves (1), (2) of degrees 2, 3 have $7 > 2 \times 3$ points in common, and hence a common component by Bézout's theorem. Since c has no nonconstant factor, by hypothesis, this common component must be c itself. Hence

$$\alpha l_{135} + \beta l_{246} = cp \tag{3}$$

for some polynomial p, which must be linear since the left-hand side of (3) has degree 3 and c has degree 2. Since the curve $\alpha l_{135} + \beta l_{246} = 0$ passes through the nine points common to $l_{135} = 0$ and $l_{246} = 0$, while $c = 0$ passes through only six of them, the remaining three (the intersections of opposite sides) must be on the line $p = 0$.

EXERCISE

1. Generalize the preceding argument to show that if two degree n curves meet in n^2 points, nm of which lie on a curve of degree m, then the remaining $n(n - m)$ points lie on a curve of degree $n - m$.

7.8. Biographical Notes: Desargues and Pascal

Girard Desargues was born in Lyons in 1591 and died in 1661. He was one of nine children of Girard Desargues, a tithe collector, and Jeanne Croppet. He was evidently brought up in Lyons, but information about his early life is lacking. By 1626 he was working as an engineer in Paris and may have used his expertise in the famous siege of La Rochelle in 1628, during which a dike was built across the harbor to prevent English ships from relieving the city.

In the 1630s he joined the circle of Marin Mersenne, which met regularly in Paris to discuss scientific topics, and in 1636 contributed a chapter to a book of Mersenne on the theory of music. In the same year he published a 12-page booklet on perspective, the first hint of his ideas in projective geometry. The *Brouillon projet* [1639] was published in an edition of only 50 copies and won very little support. In fact, its reception was generally hostile, and Desargues was engaged in a pamphleteering battle for some years with his detractors (see Taton [1951], pp. 36–45). At first his only supporters were Pascal, most of whose work on projective geometry was also lost, and the engraver Abraham Bosse, who expounded Desargues' method of perspective

[1648]. Desargues had become discouraged by the attacks on his work and had left the dissemination of his ideas up to Bosse, who was not really mathematically equipped for the task. Projective geometry only secured a place in mathematics with the publication of a book by Phillipe de la Hire [1673] whose father, Laurent, had been a student of Desargues. It seems quite likely that la Hire's book influenced Newton. For this and further information on the mathematical legacy of Desargues, see Field and Gray [1987], Ch. 3.

Around 1645 Desargues turned his talents to architecture, perhaps to demonstrate to his critics the practicality of his graphical methods. He was responsible for various houses and public buildings in Paris and Lyons, excelling in complex structures such as staircases. His best known achievement in engineering, a system for raising water at the château of Beaulieu, near Paris, is also interesting from the geometrical viewpoint. It makes the first use of epicyclic curves (Section 2.5) in cogwheels, as was noted by Huygens [1671]. Huygens visited the château at a time when it was owned by Charles Perrault, the author of "Cinderella" and "Puss in Boots."

Desargues apparently returned to scientific circles in Paris toward the end of his life—Huygens heard him give a talk on the existence of geometric points on November 9, 1660—but information about this period is scanty. His will was read in Lyons on October 8, 1661, but the date and place of his death are unknown.

Blaise Pascal (Fig. 7.20) was born in Clermont-Ferrand in 1623 and died in

Figure 7.20. Blaise Pascal.

Paris in 1662. His mother, Antoinette Begon, died when he was three, and Blaise was brought up by his father, Etienne. Etienne Pascal was a lawyer with an interest in mathematics who belonged to Mersenne's circle and, as mentioned earlier, was a friend of Desargues. He has a curve named after him, the *limaçon of Pascal*. In 1631 Etienne took Blaise and his two sisters to Paris and gave up all official duties to devote himself to their education. Thus Blaise Pascal never went to school or university, but by the age of 16 he was learned in Latin, Greek, mathematics, and science. And of course he had written his *Essay on Conics* and discovered Pascal's theorem.

The *Essay on Conics* (Pascal [1640]) is a short pamphlet containing an outline of the great treatise on conics he had begun to prepare, and which is now lost. It includes a statement of Pascal's theorem for the circle. Pascal worked on his treatise until 1654, when it was nearly complete, but he never mentioned it thereafter. Leibniz saw the manuscript when he was in Paris in 1676, but no further sightings are known.

In 1640 Pascal and his sisters joined their father in Rouen, where he had become a tax official. Pascal got the idea of constructing a calculating machine to help his father in his work. He found a theoretical solution around the end of 1642, based on toothed wheels, but difficulties in the production of accurate parts delayed the appearance of the machine until 1645. This was the first working computer. The gear mechanism for addition seems rather obvious to us now, but in Pascal's day it already raised questions of the "Can a machine think?" kind. Pascal himself was sufficiently amazed by the mechanism to say that "the arithmetical machine produces effects which approach nearer to thought than all the actions of the animals. But it does nothing which would enable us to attribute will to it, as to the animals" (Pascal, *Pensées*, 340). The machine greatly impressed the French chancellor, and Pascal was granted exclusive rights to manufacture and sell it. Whether it was a commercial success is not known but for a time, at least, Pascal was diverted by the opportunity to cash in on his ideas.

The direction of Pascal's life began to shift away from such worldly concerns in 1646, when his father was treated for a leg injury by two local bonesetters. The bonesetters were Jansenists, then a fast-growing sect within the Catholic church. Their influence resulted in the conversion of the whole family to Jansenism, and Pascal began to devote more time to religious thought. For some years, though, he continued with scientific work. In 1647 he investigated the variation of barometric pressure with altitude, resulting in his *New Experiments Concerning the Vacuum*, published the same year; in 1651 he did pioneering work in hydrostatics, resulting in his *Great Experiment Concerning the Equilibrium of Fluids*, published in 1663; and in 1654 he investigated the so-called Pascal's triangle, making fundamental contributions to number theory, combinatorics, and probability theory (for more on this, see Chapter 10). In 1654 Pascal experienced a "second conversion," which led to his almost complete withdrawal from the world and science and his increasing commitment to the Jansenist cause. Only in 1658 and 1659 did he concentrate at times on mathematics (on one occasion, so the story goes, to take his mind off the pain of a toothache). His favorite topic at this stage was the cycloid, the curve generated by a point on the circumference of a circle which rolls on a straight line. Later in the seventeenth century the cycloid became important in the development of mechanics and differential geometry (see Chapters 12 and 16).

Mathematicians are of course very sorry about Pascal's withdrawal from mathematics at an early age; however, it was not just religion that gained from Pascal's conversion. The *Provincial Letters*, which he wrote to promote Jansenist ideas, and his *Pensées*, which were edited by the Jansenists after his death, became classics of French literature. Undoubtedly Pascal is the only great mathematician whose standing is equally great among writers. Moreover, his devotion to the Jansenist ideal of serving the needy had one enduring practical consequence: his idea of a public transport system. Shortly before his death in 1662, Pascal saw the inauguration of the world's first omnibus service. Coaches could be taken from the Porte Sainte-Antoine to the Luxembourg in Paris for 5 sous, with profits being directed to the relief of the poor.

CHAPTER 8

Calculus

8.1. What Is Calculus?

Calculus emerged in the seventeenth century as a system of shortcuts to results obtained by the method of exhaustion and as a method for discovering such results. The types of problem for which calculus proved suitable were finding lengths, areas, and volumes of curved figures and determining local properties such as tangents, normals, and curvature—in short, what we now recognize as problems of integration and differentiation. Equivalent problems of course arise in mechanics, where one of the dimensions is time instead of distance, hence it was calculus that made mathematical physics possible—a development we shall consider in Chapter 12. In addition, calculus was intimately connected with the theory of infinite series, initiating developments that became fundamental in number theory, combinatorics, and probability theory.

The extraordinary success of calculus was possible, in the first instance, because it replaced long and subtle exhaustion arguments by short routine calculations. As the name suggests, calculus consists of *rules for calculating* results, not their logical justification. Mathematicians of the seventeenth century were familiar with the method of exhaustion and assumed they could always fall back on it if their results were challenged, but the flood of new results became so great that there was seldom time to do so. As Huygens [1659, p. 337] wrote,

> Mathematicians will never have enough time to read all the discoveries in Geometry (a quantity which is increasing from day to day and seems likely in this scientific age to develop to enormous proportions) if they continue to be presented in a rigorous form according to the manner of the ancients.

The progress in geometry when Huygens wrote was indeed impressive, considering the very simple system of calculus then available. Virtually all that was known was the differentiation and integration of powers of x (possibly fractional) and implicit differentiation of polynomials in x, y. However, when allied to algebra and analytic geometry, this was sufficient to find tangents, maxima, and minima for all algebraic curves. And when allied to Newton's calculus of infinite series, discovered in the 1660s, the rules for powers of x formed a complete system for differentiation and integration of all functions expressible in power series.

The subsequent development of calculus is a puzzling exception to the normal process of simplification in mathematics. Nowadays we have a much less elegant system, which downplays the use of infinite series and complicates the system of rules for differentiation and integration. The rules for differentiation are still complete, given a sensible set of operations for constructing functions, but the rules for integration are pathetically incomplete. They do not suffice to integrate simple algebraic functions like $\sqrt{1 + x^3}$, or even rational functions with undetermined constants like $1/(x^5 - x - A)$.

Moreover, it is only in the last decade that we have been able to tell *which* algebraic functions are integrable by our rules. (This little-known result is expounded by Davenport [1981].)

The conclusion seems to be that, apart from streamlining the language slightly, we cannot make calculus any simpler than it was in the seventeenth century! It is certainly easier to present the history of the subject if we refrain from imposing modern ideas. This approach also has the advantage of emphasizing the highly combinatorial nature of calculus—it is about *calculation*, after all. In view of the current controversy over the relative merits of calculus and combinatorics, it may be useful to remember that most classical combinatorics was part of the algebra of series, and hence a part of calculus. We shall develop this theme at greater length in the chapter on infinite series which follows.

Much has been written on the history of calculus, and some particularly useful books are Boyer [1949], Baron [1969] and Edwards [1979]. However, historians tend to harp on the question of logical justification and to spend a disproportionate amount of time on the way it was handled in the nineteenth century. This not only obscures the boldness and vigor of early calculus, but it is overly dogmatic about the way in which calculus should be justified. Apart from the justification already available in the seventeenth century (the method of exhaustion), there is also a twentieth-century justification (the theory of infinitesimals of Robinson [1966]), and the sheer diversity of foundations for calculus suggests that we have not yet got to the bottom of it.

8.2. Early Results on Areas and Volumes

The idea of integration is often introduced by approximating the area under curves $y = x^k$ by rectangles (Fig. 8.1), say, from 0 to 1. If the base of the region is divided into n equal parts, then the heights of the rectangles are

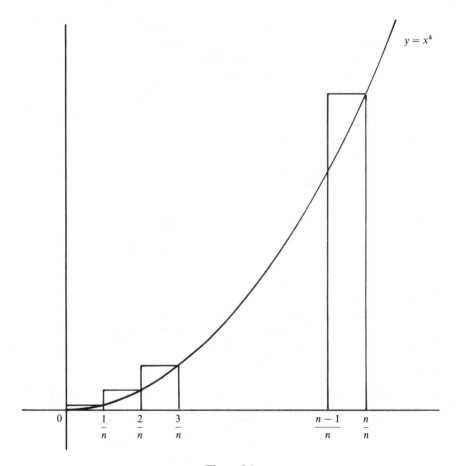

Figure 8.1

$(1/n)^k, (2/n)^k, \ldots, (n/n)^k$, and finding the area occupied by the rectangles depends on summing the series $1^k + 2^k + \cdots + n^k$. If the curve is revolved around the x axis, then the rectangles sweep out cylinders of cross-sectional area πr^2, where $r = (1/n)^k, (2/n)^k, \ldots, (n/n)^k$, which necessitates summing the series $1^{2k} + 2^{2k} + \cdots + n^{2k}$.

After the time of Archimedes, the first new results on areas and volumes were in fact based on summing these series. The Arab mathematician al-Haytham (c. 965–1039) summed the series $1^k + 2^k + \cdots + n^k$ for $k = 1, 2, 3, 4$, and used the result to find the volume of the solid obtained by rotating the parabola about its base. (See Baron [1969], p. 70, or Edwards [1979], p. 84, for al-Haytham's method of summing the series, and exercise 1 for another method.)

Cavalieri [1635] extended these results up to $k = 9$, using them to obtain the equivalent of

$$\int_0^a x^k \, dx = \frac{a^{k+1}}{k+1}$$

and conjecturing this formula for all positive integers k. This result was proved in the 1630s by Fermat, Descartes, and Roberval. Fermat even obtained the result for fractional k (see Baron [1969], p. 129, 185, and Edwards [1979], p. 116). Cavalieri is best known for his "method of indivisibles," an early method of discovery which considered areas divided into infinitely thin strips and volumes divided into infinitely thin slices. Archimedes' *Method* used similar ideas but, as mentioned in Section 4.1, this was not known until the twentieth century. Remarkably, Cavalieri's contemporary Torricelli (the inventor of the barometer) speculated that such a method may have been used by the Greeks. Torricelli himself obtained many results using indivisibles, one being almost identical with an area determination for the parabola given by Archimedes in the *Method* (Torricelli [1644]). Another of his discoveries, which caused astonishment at the time, was that the infinite solid obtained by revolving $y = 1/x$ about the x axis from 1 to ∞ had finite volume (Torricelli [1643] and exercise 3). The philosopher Thomas Hobbes [1672] wrote of Torricelli's result that "to understand this for sense, it is not required that a man should be a geometrician or logician, but that he should be mad."

EXERCISES

1. Find $1 + 2 + \cdots + n$ by summing the identity $(m + 1)^2 - m^2 = 2m + 1$ from $m = 1$ to n. Similarly find $1^2 + 2^2 + \cdots + n^2$ using the identity

$$(m + 1)^3 - m^3 = 3m^2 + 3m + 1$$

together with the previous result. Likewise, find $1^3 + 2^3 + \cdots + n^3$ using the identity

$$(m + 1)^4 - m^4 = 4m^3 + 6m^2 + 4m + 1$$

and so on.

2. Show that the approximation to the area under $y = x^2$ by rectangles in Figure 8.1 has value $(2n + 1)n(n + 1)/6n^3$, and deduce that the area under the curve is $1/3$.

3. Show that the volume of the solid obtained by rotating the portion of $y = 1/x$ from $x = 1$ to ∞ about the x axis is finite. Show, on the other hand, that its surface area is infinite.

8.3. Methods for Maxima, Minima, and Tangents

The idea of differentiation is now considered to be simpler than integration, but historically it developed later. Apart from the construction of the tangent to the spiral $r = a\theta$ by Archimedes, no examples of the characteristic limiting process

$$\lim_{\Delta x \to 0} \frac{f(x + \Delta x) - f(x)}{\Delta x}$$

appeared until it was introduced by Fermat in 1629 for polynomials f and used to find maxima, minima, and tangents. Fermat's work, like his discovery of analytic geometry, was not published until 1679, but it became known to other mathematicians through correspondence after a more complicated tangent method was published by Descartes [1637].

Fermat's calculations involve a sleight of hand also used by Newton and others: introduction of a "small" or "infinitesimal" element E at the beginning, dividing by E to simplify, then omitting E at the end as if it were zero. For example, to find the slope of the tangent to $y = x^2$ at any value x, consider the chord between the points (x, x^2) and $(x + E, (x + E)^2)$ on it.

$$\text{slope} = \frac{(x + E)^2 - x^2}{E}$$

$$= \frac{2xE + E^2}{E}$$

$$= 2x + E$$

and we now get the slope of the tangent by neglecting E. This procedure later enraged the philosophers, who thought it was being claimed that $2x + E = 2x$ and at the same time $E \neq 0$. Of course, it is only necessary to claim that $\lim_{E \to 0}(2x + E) = 2x$, but seventeenth-century mathematicians did not know how to say this. In any case, they were too carried away with the power of the method to worry about such criticisms (and it was difficult to take philosophers seriously when they were as obstinate as Hobbes; see Section 8.2). Fermat's method applies to all polynomials $p(x)$, since the highest degree term in $p(x + E)$ is always canceled by the highest degree term in $p(x)$, leaving terms divisible by E. Fermat was also able to extend it to curves given by polynomial equations $p(x, y) = 0$. He did this in 1638 when Descartes, hoping to stump him, proposed finding the tangent to the folium.

The generality of Fermat's method entitles him to be regarded as one of the founders of calculus. He could certainly find tangents to all curves given by polynomial equations $y = p(x)$ and probably to all algebraic curves $p(x, y) = 0$. A completely explicit rule for the latter problem was found by Sluse about 1655 (but not published until Sluse [1673]) and by Hudde in 1657 (published in the 1659 edition of Descartes' *La Géométrie*; Schooten [1659]). In our notation, if

$$p(x, y) = \Sigma a_{ij} x^i y^j$$

then

$$\frac{dy}{dx} = -\frac{\Sigma i a_{ij} x^{i-1} y^j}{\Sigma j a_{ij} x^i y^{j-1}}$$

Nowadays, this result is easily obtained by implicit differentiation (exercise 1), but it can also be obtained by direct manipulation of polynomials.

EXERCISE

1. Derive the formula of Hudde and Sluse.

8.4. The *Arithmetica Infinitorum* of Wallis

Wallis's efforts to arithmetize geometry were noted in Section 6.6. In his *Arithmetica Infinitorum* (Wallis [1655']) he made a similar attempt to arithmetize the theory of areas and volumes of curved figures. Some of his results were, understandably, equivalent to results already known. For example, he gave a proof that

$$\int_0^1 x^p \, dx = \frac{1}{p+1}$$

for positive integers p by showing that

$$\frac{0^p + 1^p + 2^p + \cdots + n^p}{n^p + n^p + n^p + \cdots + n^p} \to \frac{1}{p+1} \quad \text{as} \quad n \to \infty$$

However, he made a new approach to fractional powers, finding $\int_0^1 x^{m/n} \, dx$ directly rather than by consideration of the curve $y^n = x^m$, as Fermat had done. He first found $\int_0^1 x^{1/2} \, dx, \int_0^1 x^{1/3} \, dx, \ldots$, by considering the areas complementary to those under $y = x^2, y = x^3, \ldots$ (Fig. 8.2), then guessed the results for other fractional powers by analogy with those already obtained.

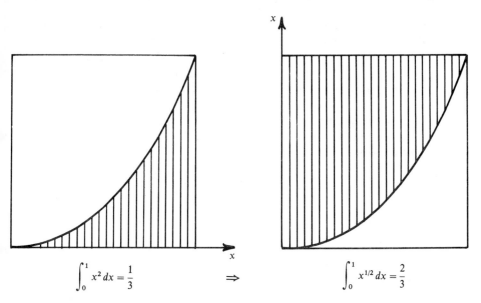

$$\int_0^1 x^2 \, dx = \frac{1}{3} \qquad \Rightarrow \qquad \int_0^1 x^{1/2} \, dx = \frac{2}{3}$$

Figure 8.2

Like other early contributors to calculus, Wallis was inarticulate about quantities that tended to zero, treating them as nonzero one minute and zero the next. For this he received a ferocious blast from his archenemy Thomas Hobbes: "Your scurvy book of *Arithmetica Infinitorum*; where your indivisibles have nothing to do, but as they are supposed to have quantity, that is to say, to be *divisibles*" ([1656], p. 301). Quite apart from this fault, which is easily remedied by limit arguments, the reasoning of Wallis is extremely incomplete by today's standards. Observing a pattern in formulas for $p = 1, 2, 3$, for example, he will immediately claim a formula for all positive integers p "by induction" and for fractional p "by interpolation." His boldness reached new heights toward the end of the *Arithmetica Infinitorum* in deriving his famous infinite product formula,

$$\frac{\pi}{4} = \frac{2}{3} \cdot \frac{4}{3} \cdot \frac{4}{5} \cdot \frac{6}{5} \cdot \frac{6}{7} \cdots$$

An exposition of his reasoning may be found in Edwards [1979], pp. 171–176, where it is described as "one of the more audacious investigations by analogy and intuition that has ever yielded a correct result."

However, we must bear in mind that Wallis was offering primarily a method of discovery, and what a discovery he made! His formula for π was not the first ever given, as Viète [1593] had discovered

$$\frac{2}{\pi} = \cos\frac{\pi}{4}\cos\frac{\pi}{8}\cos\frac{\pi}{16}\cdots$$

$$= \sqrt{\frac{1}{2}} \cdot \sqrt{\frac{1}{2}\left(1 + \sqrt{\frac{1}{2}}\right)} \cdot \sqrt{\frac{1}{2}\left[1 + \sqrt{\frac{1}{2}\left(1 + \sqrt{\frac{1}{2}}\right)}\right]} \cdots$$

However, the formula of Viète is based on a clever but simple trick (see exercises), whereas that of Wallis is of deeper significance. By relating π to the integers through a sequence of rational operations, Wallis uncovered a sequence of fractions (obtained by terminating the product at the nth factor) he called "hypergeometric." Similar sequences were later found to occur as coefficients in series expansions of many functions, which led to a broad class of functions being called "hypergeometric" by Gauss. Also, Wallis' product was closely related to two other beautiful formulas for π based on sequences of rational operations:

$$\frac{4}{\pi} = 1 + \cfrac{1^2}{2 + \cfrac{3^2}{2 + \cfrac{5^2}{2 + \cfrac{7^2}{2 + \cdots}}}}$$

and

$$\frac{\pi}{4} = 1 - \frac{1}{3} + \frac{1}{5} - \frac{1}{7} + \cdots$$

The continued fraction was obtained by Brouncker from Wallis' product and also published in Wallis [1655']. The series is a special case of the series

$$\tan^{-1}x = x - \frac{x^3}{3} + \frac{x^5}{3} - \frac{x^7}{7} + \cdots$$

discovered by Indian mathematicians in the fifteenth century (see Section 9.1) and later rediscovered by Newton, Gregory, and Leibniz. Euler [1748], p. 372, gave a direct transformation of the series for $\pi/4$ into Brouncker's continued fraction. In addition to setting off this spectacular chain reaction, Wallis' method of "interpolation" had important consequences in the work of Newton, who used it to discover the general binomial theorem (Section 9.2).

EXERCISES

1. Use the identity $\sin x = 2\sin(x/2)\cos(x/2)$ to show

$$\frac{\sin x}{2^n \sin(x/2^n)} = \cos\frac{x}{2}\cos\frac{x}{2^2}\cdots\cos\frac{x}{2^n}$$

whence

$$\frac{\sin x}{x} = \cos\frac{x}{2}\cos\frac{x}{2^2}\cos\frac{x}{2^3}\cdots$$

2. Deduce Viète's product by substituting $x = \pi/2$.

8.5. Newton's Calculus of Series

Newton made many of his most important discoveries in 1665/6, after studying the works of Descartes, Viète, and Wallis. In Schooten's edition of *La Géométrie* he encountered Hudde's rule for tangents to algebraic curves, which was virtually a complete differential calculus from Newton's viewpoint. Although Newton made contributions to differentiation that are useful to *us*, the chain rule, for example, differentiation was a minor part of *his* calculus, which depended mainly on the manipulation of infinite series. It is misleading, then, to describe Newton as a founder of calculus unless one understands calculus, as he did, as an algebra of infinite series. In this calculus, differentiation and integration are carried out term by term on powers of x and hence are comparatively trivial.

At the beginning of his main work on calculus, *A Treatise of the Methods of Series and Fluxions*, Newton clearly states his view of the role of infinite series:

> Since the operations of computing in numbers and with variables are closely similar ... I am amazed that it has occurred to no one (if you except N. Mercator with his quadrature of the hyperbola) to fit the doctrine recently established for decimal numbers in similar fashion to variables, especially since the way is then

open to more striking consequences. For since this doctrine in species has the same relationship to Algebra that the doctrine in decimal numbers has to common Arithmetic, its operations of Addition, Subtraction, Multiplication, Division and Root extraction may be easily learnt from the latter's. (Newton [1671], pp. 33–35)

The quadrature (area determination) of the hyperbola mentioned by Newton was the result that we would write as

$$\int_0^x \frac{dt}{1+t} = x - \frac{x^2}{2} + \frac{x^3}{3} - \frac{x^4}{4} + \cdots$$

first published in Mercator [1668]. Newton had discovered the same result in 1665, and it was partly his dismay in losing priority that led him to write his *Treatise* and an earlier work [1669], *On Analysis by Equations Unlimited in Their Number of Terms*. Newton also independently discovered the series for $\tan^{-1}x$, $\sin x$, and $\cos x$ in his *Analysis*, without knowing that all three series had already been discovered by Indian mathematicians (see Section 9.1).

The Mercator and Indian results were both obtained by the method of expanding a geometric series and integrating term by term. In our notation,

$$\int_0^x \frac{dt}{1+t} = \int_0^x (1 - t + t^2 - t^3 + \cdots)\,dt$$

$$= x - \frac{x^2}{2} + \frac{x^3}{3} - \frac{x^4}{4} + \cdots$$

and

$$\tan^{-1}x = \int_0^x \frac{dt}{1+t^2}$$

$$= \int_0^x (1 - t^2 + t^4 - t^6 + \cdots)\,dt$$

$$= x - \frac{x^3}{3} + \frac{x^5}{5} - \frac{x^7}{7} + \cdots$$

Newton routinely used these methods in his *Analysis* and *Treatise*, but he greatly extended their scope by algebraic manipulation. He not only obtained sums, products, quotients, and roots, as foreshadowed in his *Treatise* introduction, but his root extractions extended to the general construction of *inverse functions* by the new idea of inverting infinite series. For example, after Newton, ([1671], p. 61) found the series $x - (x^2/2) + (x^3/3) - \ldots$, for $\int_0^x dt/(1+t)$, which of course is $\log(1+x)$, he set

$$y = x - \frac{x^2}{2} + \frac{x^3}{3} - \cdots \tag{1}$$

and solved (1) for x (which we recognize to be $e^y - 1$). His method is in tabular form like the arithmetic calculations of the time but equivalent to setting

$x = a_0 + a_1 y + a_2 y^2 + \cdots$, substituting in the right-hand side of (1), and determining $a_0, a_1, a_2, \ldots$, successively by comparing with the coefficients on the left-hand side. Newton found the first few terms

$$x = y + \frac{1}{2}y^2 + \frac{1}{6}y^3 + \frac{1}{24}y^4 + \frac{1}{120}y^5 + \cdots$$

then confidently concluded that $a_n = 1/n!$ in the manner of Wallis. As he put it, "Now after the roots have been extracted to a suitable period, they may sometimes be extended at pleasure by observing the analogy of the series."

De Moivre [1698] gave a formula for inversion of series which justifies such conclusions; the impressive thing is that Newton could find such an elegant result by such a forbidding method. His discovery of the series for $\sin x$ ([1669], p. 233, 237) is even more amazing. First he used the binomial series

$$(1 + a)^p = 1 + pa + \frac{p(p-1)}{2!}a^2 + \frac{p(p-1)(p-2)}{3!}a^3 + \cdots$$

(though not with the natural choice $a = -x^2$, $p = -\frac{1}{2}$) to obtain

$$\sin^{-1}x = z = x + \frac{1}{2}\frac{x^3}{3} + \frac{1\cdot 3}{2\cdot 4}\frac{x^5}{5} + \frac{1\cdot 3\cdot 5}{2\cdot 4\cdot 6}\frac{x^7}{7} + \cdots$$

by term-by-term integration then casually stated "I extract the root, which will be

$$x = z - \frac{1}{6}z^3 + \frac{1}{120}z^5 - \frac{1}{5040}z^7 + \frac{1}{362,880}z^9 - \cdots"$$

adding, a few lines later, that the coefficient of z^{2n+1} is $1/(2n+1)!$.

EXERCISE

1. Use $\sin^{-1}x = \int_0^x dt/\sqrt{1-t^2}$ to derive the foregoing series.

8.6. The Calculus of Leibniz

Newton's epoch-making works ([1669] and [1671]) were offered to the Royal Society and Cambridge University Press but, incredible as it now seems, rejected for publication. Thus it happened that the first published paper on calculus was not by Newton but by Leibniz [1684]. This led to Leibniz's initially receiving credit for the calculus and later to a bitter dispute with Newton and his followers over the question of priority for the discovery.

There is no doubt that Leibniz discovered calculus independently, that he had a better notation, and that his followers contributed more to the spread of calculus than did Newton's. Leibniz's work lacked the depth and virtuosity of Newton's, but then Leibniz was a librarian, a philosopher, and a diplomat with only a part-time interest in mathematics. His *Nova Methodis* [1684] is a relatively slight paper, though it does lay down some important fundamentals —the sum, product, and quotient rules for differentiation—and it introduces the dy/dx notation we now use. He also introduced the integral sign, $\int$, in his *De Geometria* [1686] and proved the fundamental theorem of calculus, that integration is the inverse of differentiation. This result was known to Newton and even, in a geometric form, to Newton's teacher Barrow, but it became more transparent in Leibniz's formalism.

Leibniz's strength lay in the isolation rather than technical development of important concepts. He introduced the word "function" and was the first to begin thinking in function terms. He made the distinction between algebraic and transcendental functions and, in contrast to Newton, preferred "closed form" expressions to infinite series. Thus the evaluation of $\int f(x)\,dx$ for Leibniz was the problem of finding a known function whose derivative was $f(x)$, whereas for Newton it was the problem of expanding $f(x)$ in series, after which integration was trivial.

The search for closed forms was a wild goose chase but, like many efforts to solve intractable problems, it led to worthwhile results in other directions. Attempts to integrate rational functions raised the problem of factorization of polynomials and led ultimately to the fundamental theorem of algebra (see Chapter 13). Attempts to integrate $1/\sqrt{1 - x^4}$ led to the theory of elliptic functions (Chapter 11). As mentioned in Section 8.1, the problem of deciding which algebraic functions may be integrated in closed form has been solved only recently, though not in a form suitable for calculus textbooks, which continue to remain oblivious to most of the developments since Leibniz. (One thing has changed: it is now much easier to publish a calculus book than it was for Newton!)

8.7. Biographical Notes: Wallis, Newton, and Leibniz

John Wallis was born in 1616 in Ashford, Kent, and died in Oxford in 1703. He was one of five children of John Wallis, the rector of Ashford, and Joanna Chapman. He had two older sisters and two younger brothers. Young John Wallis was recognized as the academic talent of the family and at 14 was sent to Felsted, Essex, to attend the school of Martin Holbech, a famous teacher of the time. At school he learned Latin, Greek, and Hebrew, but he did not meet mathematics until he was home on Christmas vacation in 1631. One of his brothers was learning arithmetic to prepare for a trade, and Wallis asked him to explain it. This turned out to be the only mathematical instruction Wallis ever received, even though he later studied at Emmanuel College in Cambridge. As Wallis explained in his autobiograpy:

> Mathematicks were not, at that time, looked upon as Accademical Learning, but the business of Traders, Merchants, Sea-men, Carpenters, land-measurers, or the like; or perhaps some Almanak-makers in London. And of more than 200

Figure 8.3. John Wallis.

at that time in our College, I do not know of any two that had more of Mathematicks than myself, which was but very little; having never made it my serious studie (otherwise than as a pleasant diversion) till some little time before I was designed for a Professor in it. (Wallis [1696/1970], p.27)

At Emmanuel College, Wallis studied divinity from 1632 to 1640, when he gained a master of arts degree. College life evidently agreed with him, and he would have stayed on as a fellow, had there been a place available. He did become a fellow of Queen's College, Cambridge, for a year but, since fellows had to remain unmarried, relinquished the fellowship when he married in 1645. Thus it was that Wallis spent most of the 1640s in the ministry.

The 1640s were a decisive decade in English history, with the rise of the parliamentary opposition to Charles I and the King's execution in 1649. Partly by luck and partly by adaptation to the new political conditions, Wallis changed the direction of his life toward mathematics. Early in the conflict he found he had the very valuable ability to decipher coded messages. To quote the autobiography again:

> About the beginning of our Civil Wars, in the year 1642, a Chaplain of S^r. William Waller showed me an intercepted Letter written in Cipher.... He asked me (between jest and earnest) if I could make any thing of it.... I judged it could be no more than a new Alphabet and, before I went to bed, I found it out, which was my first attempt upon Deciphering. (Wallis [1696/1970, p. 37)

This was the first in a series of successes Wallis had in codebreaking for the Parliamentarians, which gained him not only political favor but also a reputation for mathematical skill. (For more information on Wallis' cryptography, see Kahn [1967], p. 166.) When the royalist Peter Turner was expelled from the Savilian Chair of Geometry of Oxford in 1649, Wallis was appointed in his place. At last his dormant mathematical ability had a chance to develop, and from then on he was active in mathematics almost continually until the end of his life.

Isaac Newton (Fig. 8.4) was born on Christmas Day, 1642, at Woolsthorpe, Lincolnshire. His family background and early life did not augur well for future greatness. Newton's father, also named Isaac, was fairly well off but illiterate, and he died three months before Newton was born. His mother, Hannah Ayscough, remarried when Newton was three, only to abandon him on the insistence of his stepfather. The boy was left in the care of the Ayscough family, a circumstance that helped his education (Hannah's brother William had studied at Cambridge, and eventually directed Newton there) but did not compensate emotionally for the absence of his father and mother. Newton became intensely neurotic, secretive, and suspicious in later life; he never married and tended to make enemies rather than friends.

The young Newton was more interested in building intricate machines, such as model windmills, than academic studies, though once he set his mind to it he became top of his school. In 1661 he entered Trinity College, Cambridge, as a sizar. Sizars had to earn their keep as servants to the wealthier students, and it was indicative of his mother's meanness that he had to become one, for she could afford to support him but chose not to. Newton's early studies were in Aristotle, the standard curriculum of the time. The first thinker to make an impression on him was Descartes, whose works were then creating a stir in Cambridge. By 1664, in a series of notes he called *Quaestiones quaedam Philosophicae*, Newton was absorbed with questions of mechanics, optics, and the physiology of vision. He was also struck by Descartes' geometry, preferring it

Figure 8.4. Isaac Newton.

to Euclid, which in his first encounter "he despised ... as a trifling book" (according to later reminiscences of de Moivre).

The years 1664 to 1666 were the most important in Newton's mathematical development and perhaps the most creative period in the life of any mathematician. In 1664 he devoured the mathematics of Descartes, Viète, and Wallis and began his own investigations. Late in 1664 he conceived the idea of curvature, from which much of differential geometry was to grow (see Chapter 16). The university was closed in 1665, which was the disastrous plague year in much of England. Newton returned to Woolsthorpe, where his mathematical reflections became an all-consuming passion. Fifty years later, Newton recalled the time as follows:

> In the beginning of the year 1665 I found the Method of approximating series & the Rule for reducing any dignity of any Binomial into such a series. The same year in May I found the method of Tangents of Gregory & Slusius, & in November had the direct method of fluxions & the next year in January had the Theory of Colours & in May following I had entrance into ye inverse method of fluxions. And in the same year I began to think of gravity extending to ye orb of the Moon & ... from Keplers rule of the periodical times of the Planets ... I deduced that the forces wch keep the Planets in their Orbs must [be] reciprocally as the squares of their distances from the centers. ... All this was in the two plague years of 1665–1666. For in those days I was in the prime of my age for invention & minded Mathematicks & Philosophy more then [sic] at any time since. (Whiteside [1966], p. 32)

In addition to the achievements mentioned, Newton's discoveries in this period included the series for $\log(1 + x)$ and, at least in preliminary form, the classification of cubic curves.

As we have seen, Newton's first attempts to publish his results were unsuccessful; nevertheless, there were some who read them and recognized his genius. In 1669 the Lucasian Professor of Mathematics at Trinity, Isaac Barrow, resigned to devote himself to theology, and Newton was appointed to the chair on Barrow's recommendation. Newton held the position until 1696, when he made the puzzling decision to accept the position of master of the Mint in London. The outstanding achievement of his Lucasian professorship was the classic *Principia* [1687] or, to give it its full title, *Philosophiae Naturalis Principia Mathematica* (Mathematical principles of natural philosophy).

The *Principia*, which developed the theory of gravitation based on Newton's inverse square law of 1665, owes its existence to a visit by Edmund Halley to Cambridge in 1684. The hypothesis of the inverse square law was in the air at this stage—Wren, Hooke, and Halley himself had thought of it—but a mathematical derivation of its consequences was lacking. Halley asked Newton what curve a planet would describe under this law and was delighted to learn that Newton had calculated it to be an ellipse. When asked to supply his demonstration, Newton had some trouble reconstructing it, eventually sending Halley a nine-page paper, *De motu corporum in gyrum* (On the motion of bodies in an orbit), three months later. *De motu* was the *Principia* in embryonic form.

Realizing the importance of Newton's results, Halley communicated them to the Royal Society and urged Newton to expand them for publication. His prodding came at just the right time. The excitement over Newton's early discoveries had died down, and for the preceding six or seven years he had been wasting his time on alchemical experiments. With his interest in mathematics rekindled, Newton devoted the next 18 months almost exclusively to *Principia* "so intent, so serious upon his studies, y^t he eat very sparingly, nay, ofttimes he has forget to eat at all," as a Cambridge contemporary noticed (see Westfall [1980], p. 406). When Book I was delivered to the Royal Society in April 1686, they were still reluctant to publish, and it took heroic efforts from Halley to bring them round. He not only risked his own money on the venture, but he had to coax Newton to go through with it, as Newton flew into tantrums when Hooke raised his own claims of priority. Finally in 1687 the *Principia* was published and Newton's fame was secure, at least in Britain.

In the early 1690s Newton worked on revising the *Principia* and bringing some of his earlier investigations into order. As we have seen, the final form of his classification of cubic curves dates from this period. In 1693 he had a nervous breakdown, and this may have been a factor that influenced him to leave Cambridge for the Mint in 1696. He did not completely abandon science, becoming president of the Royal Society in 1703, but his mathematical activity was mainly confined to the priority dispute with Leibniz over the invention of the calculus. Newton died in 1727 and was buried in Westminster Abbey. Westfall [1980] is an excellent recent biography.

Gottfried Wilhelm Leibniz (Fig. 8.5) was born in Leipzig in 1646 and died in Hannover in 1716. His father, Friedrich, was professor of moral philosophy at Leipzig and his mother, Katherina Schmuck, also came from an academic family. From the age of six Leibniz was given free access to his father's library, and he became a voracious reader. At 15 he entered the University of Leipzig and received a doctorate in law from Altdorf in 1666 (Leipzig refused him

Figure 8.5. Gottfried Wilhelm Leibniz.

a doctorate because of his youth). During 1663, on a summer visit to the University of Jena, he learned a little of Euclid, but otherwise his studies were in law and philosophy, the subjects that were to be the basis of his subsequent career. The lack of early practice in mathematics left its mark on Leibniz's later mathematical style, in which good ideas are sometimes insufficiently developed through lack of technical skill. Often he seemed to lack not only the technique but the patience to develop the ideas conceived by his wide-ranging imagination. It now appears that Leibniz was a pioneer in combinatorics, mathematical logic, and topology, but his ideas in these fields were too fragmentary to be of use to his contemporaries.

An interest in logic led Leibniz to his first mathematical venture, the essay *Dissertatio de arte combinatoria* [1666]. His aim was "a general method in which all truths of reason would be reduced to a kind of calculation." Leibniz foresaw that permutations and combinations would be involved, but he did not make enough progress to interest seventeenth-century mathematicians in the project. The dream of a universal logical calculus was revived in the nineteenth century but finally shattered by the results of Gödel [1931] (see Chapter 20). Nevertheless, Leibniz benefited greatly from his work on combinatorics; it led him to his ideas in calculus.

Following his doctorate in law, Leibniz had commenced a legal career in the service of the elector of Mainz. In 1672 his duties took him to Paris, where he met Huygens and for the first time gained a firm grasp of mathematics. The years 1672 to 1676 were crucial in Leibniz' mathematical life and have been covered in detail by Hofmann [1974]. Beginning with "Pascal's triangle," which he had used in his *Dissertatio* [1666], Leibniz became interested in the differences between successive terms of series. Using differences, he developed a method of interpolation for functions which, as we shall see in Section 9.2, was also the independent discovery of Newton and Gregory. Leibniz showed his discovery to Huygens, who encouraged Leibniz to use differences in the summation of infinite series by posing the problem of evaluating $\sum_{n=1}^{\infty} 1/n(n+1)$. Leibniz succeeded (after some time) and went on to use the same method successfully in other cases. This was his introduction to the infinite processes of the calculus and also, perhaps, the origin of his preference for "closed form" solutions. In 1673 he advanced to a higher level with the discovery of

$$\frac{\pi}{4} = 1 - \frac{1}{3} + \frac{1}{5} - \frac{1}{7} + \cdots$$

and

$$\frac{1}{4}\log 2 = \frac{1}{2.4} + \frac{1}{6.8} + \frac{1}{10.12} + \cdots$$

by term-by-term integration. By 1676 he had virtually completed his formulation of the calculus, including the fundamental theorem, the dx notation, and the integral sign.

The first period of Leibniz's mathematical activity came to an end in 1676. He had failed to obtain an academic position in Paris or London and, seeking a better salary, he moved to Hannover to enter the service of the duke of Brunswick-Lüneberg. His main duties were to act as adviser, librarian, and consultant on certain engineering works. When the duke died in 1679 his successor commissioned Leibniz to compile a genealogy of the House of Brunswick in order to bolster the family's dynastic claims. Leibniz threw himself into this project with a zeal that is hard to admire, given the purpose of the genealogy, though it did enable him to travel, visit libraries, and meet scholars throughout Europe. He helped found the journal *Acta Eruditorum* in 1682 and used it to publish his discoveries in calculus, as well as those of his brilliant successors, James and John Bernoulli. This led to the rapid spread of Leibniz's notation and methods throughout the Continent.

With the succession of a new Duke of Brunswick in 1698, Leibniz fell somewhat from favor, though he retained his job and, with the support of other family members, founded the Berlin Academy in 1700 and became its first president. His final years were embittered by the priority dispute over the calculus and his employer's neglect. He was still doggedly trying to complete the history of the House of Brunswick when he died in 1716. His secretary was the only person to attend his funeral, and the history was not published until 1843.

CHAPTER 9

Infinite Series

9.1. Early Results

Infinite series were present in Greek mathematics, though the Greeks tried to deal with them as finitely as possible by working with arbitrary finite sums $a_1 + a_2 + \cdots + a_n$ instead of infinite sums $a_1 + a_2 + \cdots$. However, this is just the difference between potential and actual infinity. There is no question that Zeno's paradox of the dichotomy (Section 4.1), for example, concerns the decomposition of the number 1 into the infinite series

$$\frac{1}{2} + \frac{1}{2^2} + \frac{1}{2^3} + \frac{1}{2^4} + \cdots$$

and that Archimedes found the area of the parabolic segment (Section 4.4) essentially by summing the infinite series

$$1 + \frac{1}{4} + \frac{1}{4^2} + \frac{1}{4^3} + \cdots = \frac{4}{3}$$

Both these examples are special cases of the result we express as summation of a geometric series

$$a + ar + ar^2 + ar^3 + \cdots = \frac{a}{1 - r} \qquad \text{when } |r| < 1$$

The first nongeometric examples of infinite series appeared in the Middle Ages. Richard Suiseth (or Swineshead), known as the Calculator, in *Liber calculationum*, written around 1350, used a very lengthy verbal argument (reproduced in Boyer [1949], p. 78) to show that

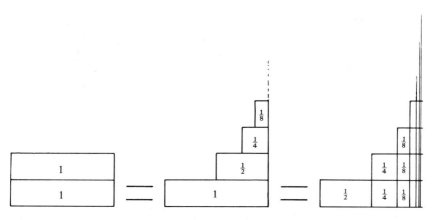

Figure 9.1

$$\frac{1}{2} + \frac{2}{2^2} + \frac{3}{2^2} + \cdots + \frac{n}{2^n} + \cdots = 2$$

At about the same time, Oresme [c. 1350], pp. 413–421, summed this and similar series by a geometric method indicated in Figure 9.1:

$$2 = \frac{1}{2} + \frac{2}{2^2} + \frac{3}{2^3} + \frac{4}{2^4} + \cdots$$

Actually Oresme gives only the last picture in the figure, but it seems likely he arrived at it by cutting up an area of two square units as shown, judging from his opening remark: "A finite surface can be made as long as we wish, or as high, by varying the extension without increasing the size." The region constructed by Oresme, incidentally, is perhaps the first example of the phenomenon encountered by Torricelli (Section 8.2) in his hyperbolic solid of revolution—infinite extent but finite content.

Another important discovery of Oresme [c. 1350'] was the divergence of the series

$$1 + \frac{1}{2} + \frac{1}{3} + \frac{1}{4} + \frac{1}{5} + \cdots$$

His proof was by an elementary argument which is now standard:

$$1 + \left(\frac{1}{2}\right) + \left(\frac{1}{3} + \frac{1}{4}\right) + \left(\frac{1}{5} + \frac{1}{6} + \frac{1}{7} + \frac{1}{8}\right) + \cdots$$

$$> 1 + \left(\frac{1}{2}\right) + \left(\frac{1}{4} + \frac{1}{4}\right) + \left(\frac{1}{8} + \frac{1}{8} + \frac{1}{8} + \frac{1}{8}\right) + \cdots$$

$$= 1 + \frac{1}{2} + \frac{1}{2} + \frac{1}{2} + \cdots$$

Thus by repeatedly doubling the number of terms collected in successive groups, we can indefinitely obtain groups of sum $> \frac{1}{2}$, enabling the sum to grow beyond all bounds.

As mentioned in Section 8.4, Indian mathematicians found the series

$$\tan^{-1}x = x - \frac{x^3}{3} + \frac{x^5}{5} - \frac{x^7}{7} + \cdots$$

with its important special case

$$\frac{\pi}{4} = 1 - \frac{1}{3} + \frac{1}{5} - \frac{1}{7} + \cdots$$

in the fifteenth century. The series for π was the first satisfactory answer to the classical problem of squaring the circle, for although the expression is infinite (as it would have to be, in view of Lindemann's theorem on the transcendence of π), the rule for generating successive terms is as finite and transparent as it could possibly be. It is sad that the Indian series became known in the West too late to have any influence or even, as yet, to obtain proper credit for its discoverer. Rajagopal and Rangachari [1977, 1986] showed that the series for $\tan^{-1}x$, $\sin x$, and $\cos x$ were known in India before 1540, and probably before 1500, but their exact dates and discoverers are uncertain.

9.2. Power Series

The Indian series for $\tan^{-1}x$ was the first example, apart from geometric series such as $1 + x + x^2 + x^3 + \cdots = 1/(1 - x)$, of a *power series*, that is, the expansion of a function $f(x)$ in powers of x. The idea of power series turned out to be fruitful not only in the representation of functions but even in the study of numerical series. Most of the interesting numerical series turned out to be instances of power series for particular values of x, for example, the series for $\pi/4$ is the $x = 1$ instance of the series for $\tan^{-1}x$.

The theory of power series began with the publication of the series

$$\log(1 + x) = x - \frac{x^2}{2} + \frac{x^3}{3} - \frac{x^4}{4} + \cdots$$

by N. Mercator [1668]. As we have seen, this was obtained by integrating the geometric series

$$\frac{1}{1 + x} = 1 - x + x^2 - x^3 + \cdots$$

term by term. Now the most important transcendental functions—logs, exponentials, and the related circular and hyperbolic functions—are obtained by integration and inversion from algebraic functions, and fairly simple algebraic functions at that. For example, e^y is the inverse function of $y = \log x$, and

$$\log(1 + x) = \int_0^x \frac{dt}{1 + t}$$

$\sin y$ is the inverse function of $y = \sin^{-1} x$ and

$$\sin^{-1} x = \int_0^x \frac{dt}{\sqrt{1 - t^2}}$$

$$\tan^{-1} x = \int_0^x \frac{dt}{1 + t^2}$$

and so on. Thus the key to finding power series is finding series expansions of simple algebraic functions. Once this is done, term-by-term integration and Newton's method of series inversion (Section 8.5) yield power series for all the common functions.

Rational functions, such as $1/(1 + t^2)$, can be expanded using geometric series; the crucial step was accomplished by Newton [1665] when he discovered the general binomial theorem,

$$(1 + x)^p = 1 + px + \frac{p(p - 1)}{2!} x^2 + \frac{p(p - 1)(p - 2)}{3!} x^3 + \cdots$$

yielding the expansion of functions such as $1/\sqrt{1 - t^2} = (1 - t^2)^{-1/2}$. This theorem was also discovered independently by James Gregory [1670]. Both Newton and Gregory were inspired by the loose heuristic method of interpolation used by Wallis [1655'], but they refined it into a result now known as the *Gregory–Newton interpolation formula*:

$$f(a + h) = f(a) + \frac{h}{b} \Delta f(a) + \frac{(h/b)(h/b - 1)}{2!} \Delta^2 f(a) + \cdots \tag{1}$$

where

$$\Delta f(a) = f(a + b) - f(a)$$
$$\Delta^2 f(a) = \Delta f(a + b) - \Delta f(a) = f(a + 2b) - 2f(a + b) + f(a)$$
$$\Delta^3 f(a) = \Delta^2 f(a + b) - \Delta^2 f(a) = f(a + 3b) - 3f(a + 2b) + 3f(a + b) - f(a)$$

$$\cdots\cdots$$

This wonderful formula determines the value of f at an arbitrary point $a + h$ from the values at an infinite arithmetic sequence of points a, $a + b$, $a + 2b$, ... The first n terms give an nth-degree polynomial in h taking the same values as f at a, $a + b$, ..., $a + nb$. Hence the formula is valid for any f that is the limit of its own approximating polynomials. This means all functions representable by power series, provided the points a, $a + b$, $a + 2b$, ..., are sensibly chosen (e.g., the points π, 2π, 3π, ..., are a bad choice for $\sin x$, since the x axis is a polynomial curve through all of them).

Newton discovered the formula (1) after his special investigations on inter-polation that led to the binomial theorem. Gregory discovered the general formula first and then used it to derive the binomial theorem (see exercises below), all independently of Newton. It even appears that Gregory used the interpolation theorem to discover Taylor's theorem 44 years before Taylor. There is strong evidence that Gregory used Taylor's series for other results ([1671]; see Turnbull [1939], pp. 356–357), and Taylor's series

$$f(a + h) = f(a) + hf'(a) + \frac{h^2}{2!}f''(a) + \cdots \tag{2}$$

is just the limiting case of (1) as $b \to 0$. Indeed, this is how it was derived by Taylor [1715]. The passage from (1) and (2) is simple if one assumes plausible limiting behavior for the infinite sum. Notice that

$$\frac{\Delta f(a)}{b} = \frac{f(a + b) - f(a)}{b} \to f'(a) \qquad \text{as } b \to 0$$

and similarly

$$\frac{\Delta^2 f(a)}{b^2} \to f''(a), \qquad \frac{\Delta^3 f(a)}{b^3} \to f'''(a)$$

and so on. We write (1) as

$$f(a + h) = f(a) + h\frac{\Delta f(a)}{b} + \frac{h(h - b)}{2!}\frac{\Delta^2 f(a)}{b^2} + \cdots$$

and observe that the nth term $\to (h^n/n!)f^{(n)}(a)$ as $n \to \infty$. Assuming that the limit of the infinite sum is the sum of these limits, we then get Taylor's series (2) as the limit of (1) as $b \to 0$.

EXERCISES

1. Show that

$$\Delta^n f(a) = \sum_{i=0}^{n} (-1)^i \binom{n}{i} f(a + ib)$$

where $\binom{n}{i}$ is the ordinary binomial coefficient.

2. If $a = 0$, $b = 1$, $f(x) = (1 + k)^x$, show $\Delta^n f(0) = k^n$ using the finite binomial series

$$(1 + h)^n = \sum_{i=0}^{n} \binom{n}{i} h^i$$

3. Deduce the general binomial series

$$(1 + k)^x = 1 + xk + \frac{x(x - 1)}{2!}k^2 + \frac{x(x - 1)(x - 2)}{3!}k^3 + \cdots$$

using the Gregory–Newton interpolation formula.

9.3. An Interpolation on Interpolation

The importance of interpolation in the development of calculus seems to have been greatly underestimated. The topic rarely appears in calculus books today, and then only as a numerical method. Yet three of the most important founders of calculus, Newton, Gregory, and Leibniz, began their work with interpolation, and we have seen how this led to two of their most important results, the binomial theorem and Taylor's theorem (For Leibniz's work, see Hofmann [1974].) With the relegation of interpolation to numerical methods, this connection has been lost. Of course, interpolation *is* a numerical method in practice, when one uses only a few terms of the Gregory–Newton series, but the full series is exact and hence of much greater interest. It was this interest in infinite expansions per se that set off Newton, Gregory, and Leibniz (as well as Wallis) from their predecessors in interpolation.

Interpolation goes back to ancient times as a method for estimating the values of functions between known values. But perhaps the first to glimpse the possibility of exact interpolation were Thomas Harriot (1560–1621) and Henry Briggs (1556–1630). The bulk of Harriot's work has not been published or even put in proper order, but a formula has been found in his papers which is equivalent to the first five terms of the Gregory–Newton series (see Lohne [1965]). Lohne dates this work of Harriot at 1611. Briggs may have learned something about interpolation from Harriot when the two were both at Oxford around 1620. Briggs' *Arithmetica Logarithmica* [1624], which is concerned with the calculation of logarithms, uses series for interpolation, and in the process gives the first instance of the binomial theorem for a fractional exponent

$$(1 + x)^{1/2} = 1 + \frac{1}{2}x - \frac{1 \cdot 1}{2 \cdot 4}x^2 + \frac{1 \cdot 1 \cdot 3}{2 \cdot 4 \cdot 6}x^3 - \frac{1 \cdot 1 \cdot 3 \cdot 5}{2 \cdot 4 \cdot 6 \cdot 8}x^4 + \cdots$$

Gregory knew of Briggs' work, and Newton certainly *could* have known of it, though no strong evidence that he did has yet been found. For more information on the history of interpolation, see Whiteside [1961] and Goldstine [1977].

9.4. Summation of Series

The results on infinite series that we have seen so far are mostly decompositions or expansions rather than summations. That is, one begins with a "known" quantity or function and decomposes it into an infinite series. Solutions of the converse problem, summation of a given series, were comparatively rare. Archimedes' summation of $1 + 1/4 + 1/4^2 + \cdots$, was one. Perhaps the next were of series such as $1/1.2 + 1/2.3 + \cdots + 1/n(n + 1) + \cdots$, given by Mengoli [1650]. The series $\sum 1/n(n + 1)$ is easily summed because of the happy accident that

$$\frac{1}{n(n+1)} = \frac{1}{n} - \frac{1}{n+1}$$

hence

$$\frac{1}{1.2} + \frac{1}{2.3} + \cdots + \frac{1}{n(n+1)} = \left(1 - \frac{1}{2}\right) + \left(\frac{1}{2} - \frac{1}{3}\right) + \cdots + \left(\frac{1}{n} - \frac{1}{n+1}\right)$$

$$= 1 - \frac{1}{n+1}$$

By letting $n \to \infty$ we then obtain the sum 1 for the infinite series.

The first really tough summation problem was $1 + 1/2^2 + 1/3^3 + 1/4^2 + \cdots$. Mengoli tackled this without success, as did the brothers James and John Bernoulli in a series of papers [1689–1704]. The Bernoulli brothers were able to sum similar series, rediscovering Mengoli's $\sum 1/n(n+1)$ and also summing $\sum 1/(n^2 - 1)$, but for $\sum 1/n^2$ itself they could obtain only trivial results such as

$$\frac{1}{2^2} + \frac{1}{4^2} + \frac{1}{6^2} + \cdots = \frac{1}{4}\left(1 + \frac{1}{2^2} + \frac{1}{3^2} + \cdots\right)$$

The solution was finally obtained by Euler [1734/5], long after James Bernoulli's death, and John Bernoulli exclaimed: "In this way my brother's most ardent wish is satisfied ... if only my brother were still alive!" (John Bernoulli, *Opera*, Vol. 4, p. 22). In fact, after hearing that the sum was $\pi^2/6$, John Bernoulli himself discovered a proof, which turned out to be the same as Euler's.

Euler (1707–1783) was probably the greatest virtuoso of series manipulation and his first summation of $1 + 1/2^2 + 1/3^2 + \cdots$, was one of his most audacious arguments. (Later he gave more rigorous proofs.) Consider the equation

$$\frac{\sin\sqrt{x}}{\sqrt{x}} = 1 - \frac{x}{3!} + \frac{x^2}{5!} - \frac{x^3}{7!} + \cdots = 0 \qquad (1)$$

easily obtained from the sine series of Section 8.5. This equation has roots $x_1 = \pi^2$, $x_2 = (2\pi)^2$, $x_3 = (3\pi)^2$, ..., but *not* 0, because $\sin\sqrt{x}/\sqrt{x} \to 1$ as $x \to 0$. Now if an *algebraic* equation

$$1 + a_1 x + a_2 x^2 + \cdots + a_n x^n = 0$$

has roots $x = x_1, x_2, \ldots, x_n$, then

$$1 + a_1 x + \cdots + a_n x^n = \left(1 - \frac{x}{x_1}\right)\left(1 - \frac{x}{x_2}\right)\cdots\left(1 - \frac{x}{x_n}\right) \qquad (2)$$

by Descartes' factor theorem (5.7). Also

$$\frac{1}{x_1} + \frac{1}{x_2} + \cdots + \frac{1}{x_n} = -\text{coefficient of } x = -a_1$$

since each x term in the expansion of the right-hand side of (2) comes from a term $-x/x_i$ in one factor multiplied by 1's in all the other factors. Assuming this is also true of the "infinite polynomial" equation (1), we get

$$\frac{1}{x_1} + \frac{1}{x_2} + \frac{1}{x_3} + \cdots = -\text{coefficient of } x = -\left(-\frac{1}{3!}\right)$$

that is,

$$\frac{1}{\pi^2} + \frac{1}{(2\pi)^2} + \frac{1}{(3\pi)^2} + \cdots = \frac{1}{6}$$

Hence

$$1 + \frac{1}{2^2} + \frac{1}{3^2} + \cdots = \frac{\pi^2}{6}$$

Q.E.D.!

9.5. Fractional Power Series

The introduction of power series helped to make mathematicians conscious of the function concept (see also Section 12.4) by drawing attention to the generality of the expression $a_0 + a_1 x + a_2 x^2 + \cdots$. However, not every function $f(x)$ is expressible as a power series $a_0 + a_1 x + a_2 x^2 + \cdots$. This is obvious in the case of functions that tend to infinity as $x \to 0$, since the power series has value a_0 when $x \to 0$. For other functions, such as $f(x) = x^{1/2}$, the behavior at 0 disallows a power series expansion for a more subtle reason. These functions have *branching behavior* at 0, they are *many-valued*, and hence they are not functions in the strict sense. The function $x^{1/2}$, for example, is two-valued because each number has two square roots, one the negative of the other.

Such behavior is not reflected in a power series $a_0 + a_1 x + a_2 x^2 + \cdots$, which can be assigned only one value for each value of x. All fractional powers of x are many-valued—$x^{1/3}$ is three-valued, $x^{1/4}$ is four-valued, and so on—and many-valued behavior is typical of algebraic functions in general. We say that y is an *algebraic function* of x if x and y satisfy a polynomial equation $p(x, y) = 0$. It follows from the impossibility of solving most polynomial equations by radicals (Section 5.7) that algebraic functions are not generally expressible by radicals, that is, by finite expressions built from $+, -, \times, \div$, and fractional powers.

Nevertheless, it was the remarkable discovery of Newton [1671] that any algebraic function y can be expressed as a *fractional power series* in x:

$$y = a_0 + a_1 x^{r_1} + a_2 x^{r_2} + a_3 x^{r_3} + \cdots$$

where $r_1, r_2, r_3, \ldots$, are rational numbers. Furthermore, the series can be rewritten in the form

$$a_0 + b_1 x^{s_1}(c_{00} + c_{01}x + c_{02}x^2 + \cdots)$$
$$+ b_2 x^{s_2}(c_{10} + c_{11}x + c_{12}x^2 + \cdots)$$
$$\vdots$$
$$+ b_n x^{s_n}(c_{n0} + c_{n1}x + c_{n2}x^2 + \cdots)$$

that is, as a finite sum of ordinary power series with fractional powers of x as multipliers. This means that in the neighborhood of $x = 0$, the behavior of y is like that of a finite sum of fractional powers.

For example, if $y^2(1 + x) = x$, we have

$$y = \frac{x^{1/2}}{1 + x}$$
$$= x^{1/2}(1 - x + x^2 - x^3 + \cdots)$$

and near the origin, y has behavior similar to $x^{1/2}$; in particular there are two values of y for each x. Newton's contribution was an ingenious algorithm for obtaining the successive powers of x. The fractional powers themselves were not properly understood until the variables x and y were taken to be complex. This was done in the nineteenth century, and on this basis a more rigorous derivation of Newton's series was given by Puiseux [1850]. For this reason, the fractional power series expansions of algebraic functions are now called *Puiseux expansions*.

EXERCISE

1. Any ordinary power series expansion of $x^{1/2}$ would have to be of the form

$$x^{1/2} = a_1 x + a_2 x^2 + a_3 x^3 + \cdots$$

because $x^{1/2} = 0$ when $x = 0$. Now square both sides and deduce a contradiction.

9.6. Generating Functions

Fibonacci [1202] introduced a famous sequence now known as the *Fibonacci sequence*

$$1, 2, 3, 5, 8, 13, 21, 34, 55, \ldots$$

in which each term (after the first two) is the sum of two preceding terms. Despite this simple law of formation, there is no obvious formula for the nth term of the sequence. Such a formula was only discovered more than 500 years later, by de Moivre [1730], and in doing so de Moivre introduced a powerful new application of infinite series, the method of generating functions. This method, which is of great importance in combinatorics, probability, and number theory, will be illustrated using the Fibonacci sequence itself.

It is technically convenient to begin the sequence with $F_0 = 0$ and $F_1 = 1$ then take subsequent terms as above by defining

$$F_{n+2} = F_{n+1} + F_n \quad \text{for} \quad n \geqslant 0$$

This is an example of a *linear recurrence relation*, and it was to solve such relations in probability theory that de Moivre introduced generating functions. The generating function for the Fibonacci sequence is

$$f(x) = F_0 + F_1 x + F_2 x^2 + F_3 x^3 + \cdots$$

We notice that

$$xf(x) = F_0 x + F_1 x^2 + F_2 x^3 + \cdots$$

$$x^2 f(x) = \qquad F_0 x^2 + F_1 x^3 + \cdots$$

Hence

$$f(x) - xf(x) - x^2 f(x) = F_0 + F_1 x - F_0 x$$
$$+ (F_2 - F_1 - F_0)x^2$$
$$+ (F_3 - F_2 - F_1)x^3$$
$$+ \cdots$$

that is, $f(x)(1 - x - x^2) = F_0 + F_1 x - F_0 x = x$ because all coefficients $F_{n+2} - F_{n+1} - F_n = 0$ by definition of the Fibonacci sequence. Thus

$$f(x) = \frac{x}{1 - x - x^2}$$

and using the roots $(-1 \pm \sqrt{5})/2 = 2/(1 \mp \sqrt{5})$ of $1 - x - x^2 = 0$ to factorize the denominator we get

$$f(x) = \frac{x}{[1 - ((1 + \sqrt{5})/2)x][1 - ((1 - \sqrt{5})/2)x]}$$

Then splitting into partial fractions

$$f(x) = \frac{1}{\sqrt{5}}\left[\frac{1}{1 - ((1 + \sqrt{5})/2)x} - \frac{1}{1 - ((1 - \sqrt{5})/2)x}\right]$$

and using the geometric series expansions

$$\frac{1}{1 - ((1 + \sqrt{5})/2)x} = 1 + \frac{1 + \sqrt{5}}{2}x + \left(\frac{1 + \sqrt{5}}{2}\right)^2 x^2 + \cdots$$

$$\frac{1}{1 - ((1 - \sqrt{5})/2)x} = 1 + \frac{1 - \sqrt{5}}{2}x + \left(\frac{1 - \sqrt{5}}{2}\right)^2 x^2 + \cdots$$

we finally get

$$f(x) = \frac{1}{\sqrt{5}}\left[\frac{1 + \sqrt{5}}{2} - \frac{1 - \sqrt{5}}{2}\right]x + \cdots$$
$$+ \frac{1}{\sqrt{5}}\left[\left(\frac{1 + \sqrt{5}}{2}\right)^n - \left(\frac{1 - \sqrt{5}}{2}\right)^n\right]x^n + \cdots$$

Equating this with the definition of $f(x)$ gives

$$F_n = \frac{1}{\sqrt{5}}\left[\left(\frac{1+\sqrt{5}}{2}\right)^n - \left(\frac{1-\sqrt{5}}{2}\right)^n\right]$$ (1)

No wonder a formula for F_n was hard to find! One would not have expected the irrational $\sqrt{5}$ to be involved in the integer-valued function F_n. The explanation is that the Fibonacci sequence actually defines $\sqrt{5}$, because $F_{n+1}/F_n \to (1+\sqrt{5})/2$ (the golden ratio) as $n \to \infty$, so (1) in effect defines the individual terms of the Fibonacci sequence in terms of the sequence as a whole (or, if one prefers, in terms of the behavior of the sequence at infinity). The remarkable fact that the definition of F_n becomes explicit, rather than recursive, when expressed in terms of $(1+\sqrt{5})/2$ is due to the simplicity of the generating function $f(x)$, which encodes the whole sequence.

The recursive property of Fibonacci numbers used in de Moivre's proof is that they satisfy a linear recurrence relation; that is, F_n is expressed as a fixed linear combination of earlier terms in the sequence. The proof is easily generalized to show that the generating function $\sum a_n x^n$ of any sequence $\{a_n\}$ defined by a linear recurrence relation is rational. Also, the proof can be reversed to show that the power series of any rational function has coefficients that satisfy a linear recurrence relation. Thus rational functions can be characterized in terms of their power series, a fact that was noticed by Kronecker [1881], Section IX.

EXERCISES

1. Show $F_{n+1}/F_n \to (1+\sqrt{5})/2$ as $n \to \infty$.

2. Show that F_n = nearest integer to $(1/\sqrt{5})[(1+\sqrt{5})/2]^n$.

3. Using $1/(1 + F_n/F_{n+1}) = F_{n+1}/F_{n+2}$, or otherwise, show that

$$1 + \cfrac{1}{1 + \cfrac{1}{1 + \cdots}} = \frac{1+\sqrt{5}}{2}$$

9.7. The Zeta Function

The purpose of a generating function is to encode a complicated sequence by a function (of a real or complex variable) that is in some ways simpler. The method of encoding need not be as direct as taking the nth term of the sequence to be the coefficient of x^n. For example, a famous formula of Euler [1748], p. 288, encodes the sequence 2, 3, 5, 7, 11, ..., of prime numbers in the following sum of powers of 1, 2, 3, 4, ...:

$$\zeta(s) = 1 + \frac{1}{2^s} + \frac{1}{3^s} + \frac{1}{4^s} + \cdots$$

(the *zeta function*). Euler's formula is

$$\frac{1}{(1 - 1/2^s)}\frac{1}{(1 - 1/3^s)}\frac{1}{(1 - 1/5^s)}\frac{1}{(1 - 1/7^s)}\frac{1}{(1 - 1/11^s)}\cdots$$

$$= 1 + \frac{1}{2^s} + \frac{1}{3^s} + \frac{1}{4^s} + \cdots$$

The factors on the left-hand side are $(1 - 1/p_n^s)$, where p_n is the nth prime. We expand each such factor as a geometric series

$$1 + \frac{1}{p_n^s} + \frac{1}{p_n^{2s}} + \frac{1}{p_n^{3s}} + \cdots$$

Multiplying all these series together, we get the reciprocal of each possible product of primes, to the sth power, exactly once. That is, the left-hand side is the sum

$$1 + \sum \frac{1}{p_1^{m_1 s} p_2^{m_2 s} \cdots p_r^{m_r s}} = 1 + \sum \frac{1}{(p_1^{m_1} p_2^{m_2} \cdots p_r^{m_r})^s}$$

in which each product $p_1^{m_1} p_2^{m_2} \dots p_r^{m_r}$ of primes occurs exactly once. But each natural number $\geqslant 2$ is expressible in just one way as a product of primes (Section 3.3), hence the latter sum equals the right-hand side

$$1 + \frac{1}{2^s} + \frac{1}{3^s} + \frac{1}{4^s} + \cdots$$

Initially the exponent $s > 1$ was there only to ensure convergence. We saw in Section 9.1 that $\zeta(s)$ diverges when $s = 1$; it converges when $s > 1$. Riemann [1859] discovered that $\zeta(s)$ becomes much more powerful when s is taken to be a complex variable. In recognition of this, $\zeta(s)$ is often called the *Riemann zeta function*. Euler's result of Section 9.4 can be rephrased as $\zeta(2) = \pi^2/6$. The values of $\zeta(4)$, $\zeta(6)$, $\zeta(8)$, ..., were also found by Euler and turn out to be rational multiples of π^4, π^6, π^8, ..., respectively. The values of $\zeta(3)$, $\zeta(5)$, ..., are unknown, though Apéry [1981] showed that $\zeta(3)$ is irrational. The most famous conjecture about $\zeta(s)$, and one of the most sought-after results in mathematics today, is the so-called *Riemann hypothesis*: $\zeta(s) = 0$ only when $\mathrm{Re}(s) = \frac{1}{2}$.

EXERCISE

1. (Euler) Show that if there are only finitely many primes $p_1, \dots, p_n$, then

$$\frac{1}{1 - 1/p_1} \cdot \frac{1}{1 - 1/p_2} \cdot \dots \cdot \frac{1}{1 - 1/p_n} = 1 + \frac{1}{2} + \frac{1}{3} + \frac{1}{4} + \cdots$$

Deduce that there are infinitely many primes.

9.8. Biographical Notes: Gregory and Euler

James Gregory was born in 1638 in Drumoak, near Aberdeen, the youngest of three sons of John Gregory, the town's minister. He received his early education from his mother, Janet Anderson, whose uncle Alexander had been secretary to Viète and editor of Viète's posthumously published works. The middle brother, David, also had mathematical ability and after their father's death in 1651, he encouraged James in his subsequent studies at grammar school and Marischal College in Aberdeen. Marischal College now possesses the only known portrait of James Gregory (Fig. 9.2).

Figure 9.2. James Gregory. (Marischal College)

Gregory's first major achievement was the invention of the reflecting tele-
scope, which he described in his book *Optica Promota* of 1663. Unfortunately,
he failed to get a satisfactory instrument constructed, and his design was
overtaken by the simpler type invented by Newton. In the meantime, Gregory
had decided to improve his scientific knowledge on the Continent, and he
spent most of 1664 to 1668 studying mathematics in Italy. His teacher was
Stefano degli Angeli (1623–1697) of Padua, from whom Gregory learned
the methods of Cavalieri. The influence of the Italian school was evident in
Gregory's geometric approach to integration problems in his first mathematical works, *Vera circuli et hyperbolae quadratura* [1667] and *Geometriae
pars universalis* [1668], but so too was Gregory's originality. The books
received glowing reviews in London and, when Gregory went there on his
return from Italy, he was elected to the Royal Society.

The *Geometriae pars universalis* was mainly a systematization of the results
in differentiation and integration then known, but it included the first published proof of the fundamental theorem of calculus. Important as this was,
the theorem was not Gregory's alone, as Newton and Leibniz discovered
it independently. What really set Gregory apart from other seventeenth-
century mathematicians was the *Vera Quadratura* (True quadrature), an extraordinarily bold and imaginative attempt to prove that the numbers π and e
are transcendental.

As mentioned in Section 2.3, transcendence of e and π was not proved until
the nineteenth-century, and certainly not by seventeenth-century methods,
so it is understandable that Gregory's attempt fell short. Nevertheless, it is
full of brilliant ideas: the unification of circular and hyperbolic functions
(without use of complex numbers), the concept of convergence, and the distinction between algebraic and transcendental functions. Gregory showed that
areas cut off from both the circle and the hyperbola (giving π and various
logarithms as special cases) could be obtained as the limit of alternate geometric
and harmonic means:

$$i_{n+1} = \sqrt{i_n I_n}$$

$$\frac{1}{I_{n+1}} = \frac{1}{2}\left(\frac{1}{i_{n+1}} + \frac{1}{I_n}\right)$$

$$\lim_{n \to \infty} i_n = \lim_{n \to \infty} I_n = I$$

If $i_0 = 2$ and $I_0 = 4$, then I (the *geometric-harmonic mean* of 2 and 4) is π. If,
on the other hand, $i_0 = 99/20$ and $I_0 = 18/11$, then I is log 10. These examples
of Gregory illustrate the way his geometric-harmonic mean embraces both
circular and hyperbolic functions. The alternating procedure used to define
the mean had an interesting echo in the work of Gauss, who investigated the
analogously defined *arithmetic-geometric mean* in the 1790s, with far-reaching
results (Section 11.6).

In 1669 Gregory returned to Scotland to take up the chair of mathematics
at St. Andrew's. He married a young widow, Mary Burnet, the daughter of
artist George Jameson, who was also descended from the Anderson family.

James and Mary had two daughters and a son, who became professor of medicine in Aberdeen. The rather impressive Gregory family tree may be found in Turnbull's short biography of Gregory [1939].

Gregory stayed at St. Andrew's for five years, during which he obtained his important results on series. However, his contact with other scientists was restricted to letters from London, and on hearing of Newton's related results he assumed he had been anticipated and did not publish. The lack of contact and hostility to mathematics at St. Andrew's led him to accept the offer of a chair at Edinburgh in 1674. Alas, he had been in Edinburgh barely a year when he collapsed, apparently from a stroke, while showing the moons of Jupiter to a group of students. He died a few days later, in October 1675, too soon for the world to have understood the importance of his work.

Leonhard Euler was born in Basel in 1707 and died in St. Petersburg in 1783. His father, Paul, studied theology at the University of Basel, where he also attended the mathematics lectures of James Bernoulli. After graduation he became a Protestant minister and married a minister's daughter, Margarete Bruckner. Leonhard was the first of their six children. The family was quite poor and, soon after Euler's birth, moved to a village outside Basel where they lived in a two-room house. Euler received his first mathematical instruction at home from his father. He later moved back to Basel to attend secondary school, but mathematics was not taught there, so he took some private lessons from a university student.

At 13, Euler entered the University of Basel, which had become the mathematical center of Europe under John Bernoulli, the younger brother and successor of James. Bernoulli advised Euler to study mathematics on his own and made himself available on Saturday afternoons to help with any difficulties. Euler's official studies were in philosophy and law. After receiving his master's degree in philosophy in 1723, he followed his father's wish by entering the department of theology. However, he was falling increasingly under the spell of mathematics and realized he would have to drop the idea of becoming a minister.

There were few opportunities for mathematicians in Switzerland, and in 1727 Euler left Basel for St. Petersburg. John Bernoulli's sons, Nicholas and Daniel, had been appointed to the new Academy of Sciences there, and they persuaded the authorities to find a place for Euler. Euler had already shown promise with a couple of papers in *Acta Eruditorum* and an honorable mention in the Paris Academy competition of 1727, but in St. Petersburg he surpassed all expectations, producing top-quality work at a rate that has astounded mathematicians ever since. The early years in St. Petersburg with the Bernoullis must have been a young mathematician's dream. Yet it is equally true that Euler's productivity was unaffected by later setbacks, including the loss of his sight. He filled half the pages published by the St. Petersburg Academy from 1729 until over 50 years after his death (!), and he also accounted for half the production of the Berlin Academy between 1746 and 1771.

The first major changes in Euler's life in St. Petersburg occurred in 1733, when Daniel Bernoulli returned to Basel. Euler then became professor of mathematics but also had to take over the Department of Geography. In the same year, he married a compatriot, Katharina Gsell, the daughter of an artist who taught in St. Petersburg. They were eventually to have 13 children, 5 of whom reached maturity. Euler's duties in geography included the preparation of a map of Russia, a task that strained his eyes and perhaps led to the fever that destroyed the sight of his right eye in 1738. Figure 9.3 is a portrait from his good side.

Figure 9.3. Leonhard Euler.

By 1740 the political situation in St. Petersburg had become unsettled and Euler moved to Berlin, where Frederick the Great had just reorganized the Berlin Academy. Euler became director of the mathematical section, and stayed in Berlin for 25 years. Some of his most famous works date from this period, in particular the *Introductio in analysin infinitorum* [1748] and the *Letters à une princesse d'Allemagne sur divers sujets de physique et de philosophie*, one of the classics of popular science. However, Euler was not comfortable in Berlin. There were quarrels over the leadership of the Academy, and the cynical Frederick tended to sneer at the pious and unassuming Euler. In 1762 Catherine the Great came to the throne in Russia, and the St. Petersburg Academy, with which Euler had maintained contact throughout, began to look attractive again.

In 1766 he moved back to St. Petersburg with his family (as a bonus, his eldest son gained the chair of physics there). Soon after his arrival Euler suffered an illness that destroyed most of his remaining sight, and in 1771 he became completely blind. If anything, blindness concentrated Euler's mind more wonderfully. He had always had an extraordinary memory—knowing Virgil's *Aeneid* by heart, for example—and with assistance from two of his sons and other collaborators his flow of publications continued at a greater rate than ever. His *Algebra* [1770] was dictated to his valet, yet it became the most successful mathematics textbook since Euclid's *Elements*.

One of Euler's most admirable qualities was a willingness to explain how his discoveries were made. Mathematicians of the eighteenth-century were less secretive than their sixteenth- and seventeenth-century predecessors, but Euler was unique in revealing his preliminary guesses, experiments, and partial proofs. Some of the most interesting of these exposés are presented in Polya's book [1954] on plausible reasoning. Chapter 6 of the book, for example, includes a translation of the memoir in which Euler announced the pentagonal number theorem. It is impossible to summarize all of Euler's contributions to mathematics here, though several of the highlights will be presented in the chapters that follow. The best symmary available is in Youschkevitch's article on Euler in the *Dictionary of Scientific Biography*.

CHAPTER 10

The Revival of Number Theory

10.1. Number Theory between Diophantus and Fermat

Some important results in number theory were discovered in the Middle Ages, though they failed to take root until they were rediscovered in the seventeenth-century or later. Among these were the discovery of Pascal's triangle and the "Chinese remainder theorem" by Chinese mathematicians and formulas for permutations and combinations by Levi ben Gershon [1321]. The Chinese remainder theorem will not be discussed here, as it did not reemerge until after the period we are about to cover. A full account of its history may be found in Libbrecht [1973], Ch. 5. Pascal's triangle, on the other hand, began to flourish in the seventeenth-century after a long dormancy, so it is of interest to see what was known of it in medieval times and what Pascal did to revive it.

The Chinese used Pascal's triangle as a means of generating the binomial coefficients, that is, the coefficients occurring in the formulas

$$(a + b)^2 = \qquad\qquad a^2 + 2ab + b^2$$

$$(a + b)^3 = \qquad a^3 + 3a^2b + 3ab^2 + b^3$$

$$(a + b)^4 = a^4 + 4a^3b + 6a^2b^2 + 4ab^3 + b^4$$

and so on. When tabulated as follows (with two trivial rows added at the top, corresponding to the powers 0 and 1 of $a + b$),

$$1$$
$$1 \quad 1$$
$$1 \quad 2 \quad 1$$
$$1 \quad 3 \quad 3 \quad 1$$
$$1 \quad 4 \quad 6 \quad 4 \quad 1$$
$$1 \quad 5 \quad 10 \quad 10 \quad 5 \quad 1$$
$$1 \quad 6 \quad 15 \quad 20 \quad 15 \quad 6 \quad 1$$
$$1 \quad 7 \quad 21 \quad 35 \quad 35 \quad 21 \quad 7 \quad 1$$

and so on, the kth element $\binom{n}{k}$ of the nth row is the sum $\binom{n-1}{k-1} + \binom{n-1}{k}$ of the two elements above it in the $(n-1)$th row, as follows from the formula (exercise 1)

$$(a + b)^n = (a + b)^{n-1}a + (a + b)^{n-1}b$$

The triangle appears to a depth of six in Yáng Huí [1261] and to a depth of eight in Zhū Shìjié [1303] (Fig. 10.1). Yáng Huí attributes the triangle to Jiǎ Xiàn, who lived in the eleventh century.

The number $\binom{n}{k}$ appears in medieval Hebrew writings as the number of combinations of n things taken k at a time. Levi ben Gershon [1321] gives the formula

$$\binom{n}{k} = \frac{n!}{(n-k)!k!}$$

together with the fact that there are $n!$ permutations of n elements. In his treatment of permutations and combinations Levi ben Gershon comes very close to using mathematical induction, if not actually inventing it. As we now formulate this method of proof, a property $P(n)$ of natural numbers n is proved to hold for all n if one can prove $P(1)$ (the base step) and, for arbitrary n, $P(n) \Rightarrow P(n + 1)$ (the induction step). Rabinovitch [1969] offered an exposition of some of Levi Ben Gershon's proofs that certainly seems to show a division into base step and induction step, but the induction step needs some notational help to become a proof for truly arbitrary n. Levi ben Gershon does not say "Consider n elements $a, b, c, d, \ldots, e$," as we might, but only "Let the elements be a, b, c, d, e," since he does not have the device of ellipses.

In view of these excellent results, why do we call the table of binomial coefficients "Pascal's triangle"? It is of course not the only instance of a mathematical concept being named after a rediscoverer rather than a discoverer, but in any case Pascal deserves credit for more than just rediscovery. In his *Traité du triangle arithmétique* [1654], Pascal united the algebraic and combinatorial theories by showing that the elements of the arithmetic triangle could be interpreted in two ways: as the coefficients of $a^{n-k}b^k$ in $(a + b)^n$ and as the numbers of combinations of n things k at a time. In effect, he showed that

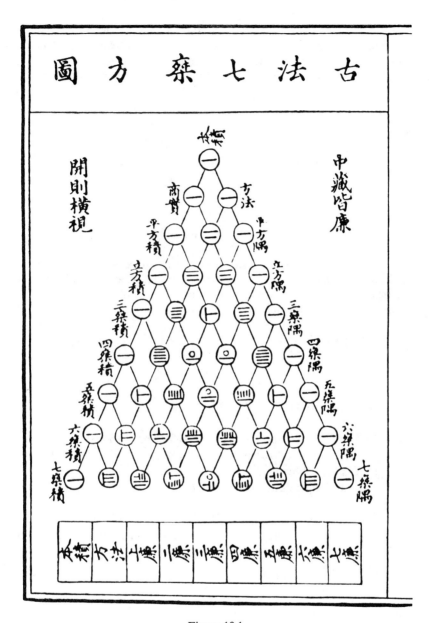

Figure 10.1

$(a + b)^n$ is a *generating function* for the numbers of combinations. As an application, he founded the mathematical theory of probability by solving the problem of divison of stakes (exercise 2), and as a method of proof he used mathematical induction for the first time in a really conscious and unequivocal way. Altogether, quite some progress!

In going to Pascal's work in 1654 we have overshot the end of the pre-Fermat period in number theory, since Fermat was already active in this field in the 1630s. However, it is convenient to have some background of binomial coefficients established, since Fermat's early work appears in this setting.

EXERCISES

1. Use the identity

$$(a + b)^n = (a + b)^{n-1}a + (a + b)^{n-1}b$$

to prove

$$\binom{n}{k} = \binom{n-1}{k-1} + \binom{n-1}{k}$$

2. Suppose that a game between players I and II has to be called off with n plays remaining, k of which I has to win in order to carry off the stakes. Assuming I has an even chance of winning each play, show that the ratio of his chance of winning the stakes to that of II's winning is

$$\binom{n}{n} + \binom{n}{n-1} + \cdots + \binom{n}{k} : \binom{n}{k-1} + \binom{n}{k-2} + \cdots + \binom{n}{0}$$

10.2. Fermat's Little Theorem

The best known theorem actually proved by Fermat [1640], known as his "little" or "lesser" theorem to distinguish it from his "last" or "great" theorem (next section), is the following:

If p is prime and n is relatively prime to p, then

$$n^{p-1} \equiv 1 \pmod{p}$$

Equivalent statements of the conclusion, which avoid using the "congruent mod p" language unknown in Fermat's time, are

$$n^{p-1} - 1 \text{ is divisible by } p$$

or

$$n^p - n \text{ is divisible by } p$$

(the latter because $n^p - n = n(n^{p-1} - 1)$ is divisible by p only if $n^{p-1} - 1$ is, since p is prime and does not divide n).

Fermat's little theorem has recently become indispensable in certain areas of applied mathematics, such as cryptography, so it is thought-provoking to learn that it originated in one of the least applied problems in mathematics, the construction of perfect numbers. As we saw in Section 3.2, this depends on the construction of prime numbers of the form $2^m - 1$, and it was initially for this reason that Fermat became interested in conditions for $2^m - 1$ to have divisors. At the same time (mid-1630s) he was investigating properties of the binomial coefficients, and the combination of these two interests very likely led to the discovery of his little theorem, at least for $n = 2$. His actual proof is unknown, but various authors (e.g., Weil [1984], p. 56) have pointed out that the theorem follows immediately from the fact that $\binom{p}{1}, \binom{p}{2}, \ldots, \binom{p}{p-1}$, for p prime, are divisible by p:

$$2^p = (1 + 1)^p = 1 + \binom{p}{1} + \binom{p}{2} + \cdots + \binom{p}{p-1} + 1$$

hence

$$2^p - 2 = \binom{p}{1} + \binom{p}{2} + \cdots + \binom{p}{p-1}$$

is divisible by p, and therefore so is $2^{p-1} - 1$.

But how does one prove that $\binom{p}{1}, \binom{p}{2}, \ldots, \binom{p}{p-1}$ are divisible by p? This follows easily from the Levi ben Gershon formula

$$\binom{p}{k} = \frac{p!}{(p-k)!k!}$$

which shows that the prime p is a factor of the numerator but not the denominator. The denominator nevertheless divides the numerator, since $\binom{p}{k}$ is an integer, so the factor p must remain intact after the division has taken place. Fermat may not have had precisely this formula, since he did not yet have Pascal's combinatorial interpretation of the binomial coefficients, but he did have the formula

$$n\binom{n+m-1}{m-1} = m\binom{n+m-1}{m}$$

which implies it and from which the divisibility property can be extracted (see Weil [1984], p. 47).

Thus far we have a proof of Fermat's theorem for $n = 2$. Weil [1984] suggests two possible routes to the general theorem from this point. One is by iteration of the binomial theorem, a method that was used in the first published proof of Fermat's theorem by Euler [1736]. The other is by direct application of the *multinomial theorem*, the method of the earliest known proof, which is in an unpublished paper of Leibniz from the late 1670s (see Weil [1984], p. 56).

Just as the coefficient of $a^{p-k}b^k$ in $(a + b)^p$ is $p!/(p - k)!k!$, the coefficient of $a_1^{q_1}a_2^{q_2}\ldots a_n^{q_n}$ in $(a_1 + a_2 + \cdots + a_n)^p$ is $p!/q_1!q_2!\ldots q_n!$, where $q_1 + q_2 + \cdots + q_n = p$ (exercise 1). This *multinomial coefficient* is divisible by p, by the same argument as before, provided no $q_i = p$. Thus the coefficients of all but

$a_1^p, a_2^p, \ldots, a_n^p$ in $(a_1 + a_2 + \cdots + a_n)^p$ are divisible by the prime p. It follows that

$$(\underbrace{1 + 1 + \cdots + 1)^p}_{n \text{ times}} = \underbrace{1^p + 1^p + \cdots + 1^p}_{n \text{ times}} + \text{terms divisible by } p$$

that is, that $n^p - n$ is divisible by p. Then if n itself is relatively prime to p (hence not divisible by p), we have $n^{p-1} - 1$ divisible by p, or the general Fermat theorem.

EXERCISE

1. Prove the formula for the multinomial coefficient given above by showing that the formula gives the number of ways of combining p things into disjoint subsets of sizes $q_1, q_2, \ldots, q_n$.

10.3. Fermat's Last Theorem

> On the other hand, it is impossible for a cube to be written as a sum of two cubes or a fourth power to be written as a sum of two fourth powers or, in general, for any number which is a power greater than the second to be written as a sum of two like powers. I have a truly marvellous demonstration of this proposition which this margin is too narrow to contain. (Fermat [1670], p. 241)

This remark, written in the margin of his copy of Bachet's *Diophantus* when he was studying that work in the late 1630s, is the second item in Fermat's *Observations on Diophantus*, published posthumously in 1670. Fermat was responding to Diophantus' treatment of the problem of expressing a square as the sum of two squares. As we saw in Chapter 1, this is the problem of finding Pythagorean triples (a, b, c) or, equivalently, of finding the rational points $(a/c, b/c)$ on the circle $x^2 + y^2 = 1$.

"Fermat's last theorem," the claim that there are no triples (a, b, c) of positive integers such that

$$a^n + b^n = c^n$$

when n is an integer > 2, has become the most famous unsolved problem in mathematics. Many mathematicians have contributed solutions for particular values of n: Euler for $n = 3$, Fermat himself for $n = 4$ (see Section 10.4), Legendre and Dirichlet for $n = 5$, Lamé for $n = 7$, Kummer for all prime $n < 100$ except 37, 59, 67. A thorough account of these early results may be found in Edwards [1977]. Of course it is sufficient to prove the theorem for prime exponents, since a counterexample

$$a^n + b^n = c^n$$

for a nonprime exponent $n = mp$, where p is prime, would also be a counterexample

$$(a^m)^p + (b^m)^p = (c^m)^p$$

for the prime exponent p. It is still not known whether Fermat's last theorem holds for infinitely many prime exponents.

In recent times progress has been made in a different direction by Faltings [1983], in proving that for each exponent n there are at most *finitely many* *counterexamples* to Fermat's last theorem. This is a consequence of Faltings' much more general theorem, settling a conjecture of Mordell [1922] that each curve of genus > 1 has at most finitely many rational points. The concept of genus will be explained in Chapter 14. For the moment we mention only that the "Fermat curve,"

$$x^n + y^n = 1$$

has genus 0 when $n = 2$, genus 1 when $n = 3$, and genus > 1 otherwise. Curves of genus 0 can have infinitely many rational points, as we saw in Section 1.3 for the circle $x^2 + y^2 = 1$. So can curves of genus 1, though this is not the case for the Fermat curves $x^3 + y^3 = 1$ and $x^4 + y^4 = 1$; the proofs of Euler and Fermat show that these curves have only the trivial rational points at which one of x, y is 0. Faltings' theorem shows that each remaining Fermat curve $x^n + y^n = 1$ has only finitely many rational points (and hence that $a^n + b^n = c^n$ has only finitely many integer solutions). Unfortunately, his proof gives no way of bounding the search for these rational points, so at present we have no way of deciding whether they include any beyond the trivial ones.

10.4. Rational Right-angled Triangles

> The area of a right-angled triangle the sides of which are rational numbers cannot be a square number. This proposition, which is my own discovery, I have at length succeeded in proving, though not without much labour and hard thinking. I give the proof here, as this method will enable extraordinary developments to be made in the theory of numbers. (Fermat [1670], p. 271)

This is number 45 of Fermat's *Observations on Diophantus*, responding to the problem posed by Bachet: to find a right-angled triangle whose area equals a given number. It is important not only for the theorem and method announced but also because it is followed by the only reasonably complete proof left by Fermat in number theory. As a bonus, the proof implicitly settles Fermat's last theorem for $n = 4$ (see exercise 3) and is an excellent illustration of his "method" of infinite descent, which did indeed lead to extraordinary developments in the theory of numbers. In what follows, the statements that make up Fermat's proof, appearing in small type, are expanded and expressed in modern notation following the reconstruction of Zeuthen [1903], p. 163. We use the translation of Fermat given by Heath [1910], p. 293, in his version of the reconstruction.

If the area of a right-angled triangle were a square, there would exist two biquadrates the difference of which would be a square number. Consequently there would exist two square numbers the sum and difference of which would be squares.

By choosing a suitable unit of length, we can express the sides of a rational right triangle as a Pythagorean triple of relatively prime integers $p^2 - q^2$, $2pq$, $p^2 + q^2$, as noted in Section 1.2. Since these are relatively prime, so are p, q and since $2pq$ is even, $p^2 - q^2$ and its factors $p + q$, $p - q$ must be odd. Also, $p + q$, $p - q$ must be relatively prime to p, q. Then if the area $pq(p + q)(p - q)$ is a square, its factors, being relatively prime, must all be squares:

$$p = r^2, \qquad q = s^2, \qquad p + q = r^2 + s^2 = t^2, \qquad p - q = r^2 - s^2 = u^2 \quad (1)$$

Thus the sum and difference of the squares r^2, s^2 are also squares, and

$$(r^2 + s^2)(r^2 - s^2) = t^2 u^2 = v^2$$

so $r^4 - s^4 = v^2$.

Therefore we should have a square number which would be equal to the sum of a square and the double of another square, while the squares of which this sum is made up would themselves have a square number for their sum.

From (1) we have

$$2s^2 = t^2 - u^2$$

that is,

$$t^2 = u^2 + 2s^2 \qquad\qquad\qquad (2)$$

Also from (1),

$$u^2 + s^2 = r^2$$

But if a square is made up of a square and the double of another square, its side, as I can very easily prove, is also made up of a square and the double of another square.

Since $t^2 - u^2 = 2s^2$ from (2), $(t + u)(t - u) = 2s^2$ is even. Then one of $t + u$, $t - u$ is even, and consequently so is the other. Put

$$t + u = 2w$$
$$t - u = 2x \qquad\qquad\qquad (3)$$

Then,

$$s^2 = 2wx$$

Tracing back through (3), (2), (1) we see that any common factor of w, x would also be common to t, u, to t^2, u^2, to r^2, s^2, and hence to p, q. Thus w, x are relatively prime and since wx is twice a square, we have either

$$w = y^2, \qquad x = 2z^2$$

or

$$w = 2z^2, \qquad x = y^2$$

In either case,

$$t = w + x = y^2 + 2z^2 \tag{4}$$

From this we conclude that the said side is the sum of the sides about the right angle in a right-angled triangle, and that the simple square contained in the sum is the base, and the double of the other square the perpendicular.

If we let y^2, $2z^2$ be the sides of a right triangle, then the hypotenuse h satisfies

$$h^2 = (y^2)^2 + (2z^2)^2$$
$$= \tfrac{1}{2}[(y^2 + 2z^2)^2 + (y^2 - 2z^2)^2]$$
$$= \tfrac{1}{2}(t^2 + u^2) \qquad \text{[by (3) and (4)]}$$
$$= r^2 \text{[by (1)]}$$

Hence $h = r$ and the triangle is rational.

This right-angled triangle will thus be formed from two squares, the sum and difference of which will be squares. But both these squares can be shown to be smaller than the squares originally assumed to be such that both their sum and their difference are squares.

The original squares with sum and difference equal to squares were $p = r^2$, $q = s^2$, which came from the perpendicular sides $p^2 - q^2$ and $2pq$ of the rational right triangle whose area was assumed to be a square. We now have a rational (indeed integral) right triangle with perpendicular sides y^2, $2z^2$, whose area y^2z^2 is also a square. This triangle is smaller, since its hypotenuse r is less than side $2pq$ of the original triangle, and hence it gives a smaller pair of (integer) squares p', q', whose sum and difference are squares.

Thus, if there exist two squares such that the sum and difference are both squares, there will also exist two other integer squares which have the same property but a smaller sum. By the same reasoning we find a sum still smaller than the last found, and we can go on *ad infinitum* finding integer square numbers smaller and smaller with the same property. This is, however, impossible because there cannot be an infinite series of numbers smaller than any given integer we please.

This contradiction means that the initial assumption of a rational right triangle with square area was false. The versions of Zeuthen and Heath proceed more directly to a contradiction than Fermat by observing that the descent from the hypothetical initial triangle to the one with area y^2z^2 can be iterated to give an infinite descending sequence of integer areas. Weil [1984], p. 77, shortens the proof further.

The logical principle involved in Fermat's method of descent is of course the same as that on which mathematical induction is based: any set of natural numbers has a least member. However, the circumstances in which the two methods can be applied are quite different. With induction, one needs a suitable hypothesis on which to make the induction step; with descent, one needs a suitable quantity on which to descend. In practice, descent is a much more special method, being associated with geometric properties of certain curves, the genus 1 curves we shall meet in Section 10.6 and later chapters (see also Weil [1984], p. 140). The general problem raised by Bachet—deciding which areas n are the areas of rational right triangles—is in fact intimately connected with the theory of genus 1 curves, and its recent resurgence is beautifully covered by Koblitz [1985].

EXERCISES

1. Show that the existence of squares whose sum and difference are both squares implies the existence of a rational right triangle with square area.

2. Show that a nontrivial integer solution of $r^4 - s^4 = v^2$ implies the existence of a rational right triangle with square area. *Hint*: Consider $r^2 s^2 (r^4 - s^4) = r^2 s^2 v^2$.

3. Deduce Fermat's last theorem for the exponent $n = 4$.

4. Show that if $r^4 - s^4 = v^2$ in integers, then there are integers a, b such that

$$r^2 = a^2 + b^2$$

$$s^2 = 2ab = 4cd(c^2 - d^2) \qquad a = c^2 - d^2, \quad b = 2cd$$

$$v = a^2 - b^2$$

   Also show that if r, s, v have no common factor, then neither have c, d, $c^2 - d^2$.

5. Deduce that c, d and $c^2 - d^2$ are squares and hence

$$e^4 - f^4 = g^2$$

   for an integer pair (e, f) smaller than (r, s).

10.5. Rational Points on Cubics of Genus 0

One reason for doubting that Fermat had a correct proof of his "last" theorem is the fact that most of his work deals with curves of genus ≤ 1 (curves of degree ≤ 4 to be specific), which indeed were the only ones accessible to seventeenth- and eighteenth-century techniques. Admittedly, we do not know for certain what Fermat's methods were, and he did not talk in terms of finding rational points on curves. Nevertheless, this is the most natural way to interpret his solutions of Diophantine equations and to link them with earlier and later results in the same vein by Diophantus and Euler, respectively. We have already described methods for finding rational points on curves of degree 2 (in Section 1.3) and 3 (in Section 3.4). Now we shall reexamine them from the point of view of genus, which becomes increasingly important as curves of higher degree are considered. In this section we confine our attention to genus 0.

One of the properties of a curve C of degree 2 that we observed in Section 1.3 is that a rational line L through a rational point P on C meets C in a second rational point Q, provided the equation of C has rational coefficients. Also, one obtains all rational points Q on C in this way to rotating L about C. There is another important consequence of this construction, not depending on the rationality of C or L. It is that by expressing the x and y coordinates of Q in terms of the slope t of L we obtain *parametrization of C by rational functions* (recall that a rational function need not have rational coefficients). For example, this construction on the circle $x^2 + y^2 = 1$ in Section 1.3 gave the parametrization

$$x = \frac{1 - t^2}{1 + t^2}$$

$$y = \frac{2t}{1 + t^2}$$

(Fig. 10.2). Genus 0 curves can be defined as those that admit parametrization by rational functions. I shall now show that genus 0 includes some cubic curves by applying a similar construction to the folium of Descartes.

The folium was defined in Section 6.3 as the curve with equation

$$x^3 + y^3 = 3axy \tag{1}$$

The origin 0 is an obvious rational point on the folium; moreover, 0 is a *double point* of the curve, as Figure 10.3 makes clear. The line $y = tx$ through 0 therefore meets the folium at one other point P, and varying t gives all other points P on the curve. By finding the coordinates of P as functions of t, we

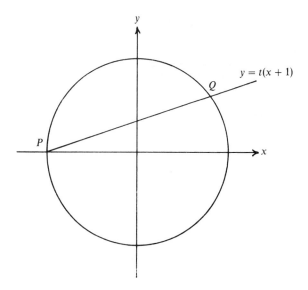

Figure 10.2

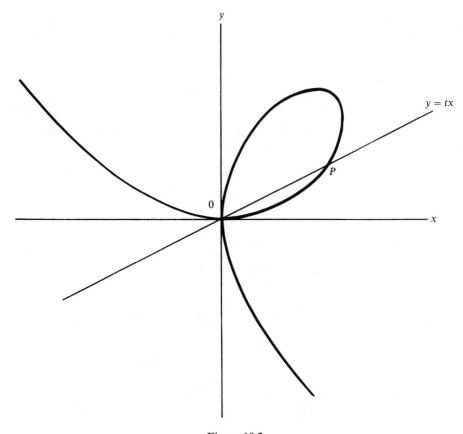

Figure 10.3

therefore obtain a parametrization. To find P we substitute $y = tx$ in (1), obtaining

$$x^3 + t^3x^3 = 3axtx$$

hence

$$x(1 + t^3) = 3at$$

and

$$x = \frac{3at}{1 + t^3} \qquad (2)$$

whence

$$y = \frac{3at^2}{1 + t^3} \qquad (3)$$

(These parametric equations were pulled out of the air in exercise 6.3.1.) A similar argument applies to any other cubic with a double point, or more generally to any curve of degree $n + 1$ with an n-tuple point; hence all such curves are of genus 0.

EXERCISE

1. Show that if x and y are rational, then so is t in (2) and (3). Conclude that when a is rational, the rational points on the folium (1) are precisely those with rational t-values.

10.6. Rational Points on Cubics of Genus 1

We cannot yet give a precise definition of genus 1, but it so happens that this is the genus of all cubic curves that are not of genus 0. We know from Section 10.5 that cubics of genus 1 cannot have double points, and in fact they also cannot have cusps, because both these cases lead to rational parametrizations. What we have yet to exhibit are the functions that do parametrize cubics of genus 1. Such functions, the *elliptic functions*, were not defined until the nineteenth century, and they were first used by Clebsch [1864] to parametrize cubics.

Many clues to the existence of elliptic functions were known before this, but at first they seemed to point in other directions. Initially, the mystery was how Diophantus and Fermat generated solutions of Diophantine equations. Newton's [late 1670s] interpretation of their results by the chord-tangent construction (Section 3.5) cleared up this first mystery—or would have if anyone had noticed it at the time. But before mathematicians really became conscious of the chord-tangent construction, they had to explain some puzzling relations between integrals of functions such as $1/\sqrt{ax^3 + bx^2 + cx + d}$, found by Fagnano [1718] and Euler [1768]. Eventually Jacobi [1834] noticed that the chord-tangent construction explained this mystery too. Jacobi's explanation was rather cryptic and, even though elliptic functions were then known in connection with integrals they did not become fully integrated into number theory and the theory of curves until the appearance of Poincaré [1901].

The analytic origins of elliptic functions will be explained in the next chapter. In this section we shall prepare to link up with this theory by deriving the algebraic relation between collinear points on a cubic curve. (A much deeper treatment of the whole story appears in Weil [1984].)

We start with Newton's form of the equation for a cubic curve (Section 6.4):

$$y^2 = ax^3 + bx^2 + cx + d \tag{1}$$

Figure 10.4 shows this curve when $y = 0$ for three distinct real values of x. In Section 3.5 we obtained that if a, b, c, d are rational, and P_1, P_2 are rational

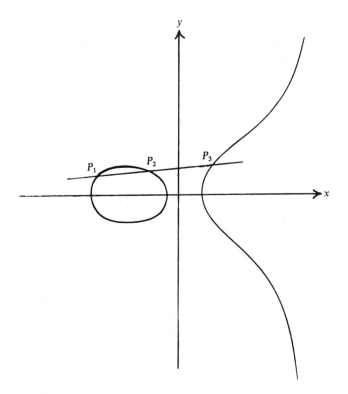

Figure 10.4

points on the curve, then the straight line through P_1, P_2 meets the curve at a third rational point P_3. If

$$y = tx + k \tag{2}$$

is the equation of this straight line, then the result of substituting (2) in (1) is an equation

$$ax^3 + bx^2 + cx + d - (tx + k)^2 = 0 \tag{3}$$

for the x coordinates, x_1, x_2, x_3, of the three points P_1, P_2, P_3. But if the roots of (3) are x_1, x_2, x_3, its left-hand side must have the form

$$a(x - x_1)(x - x_2)(x - x_3)$$

In particular, the coefficient of x^2 must be

$$-a(x_1 + x_2 + x_3)$$

Comparing this with the coefficient of x^2 in (3), we find

$$b - t^2 = -a(x_1 + x_2 + x_3)$$

hence

$$x_3 = -(x_1 + x_2) - \frac{b - t^2}{a} \qquad (4)$$

If $P_1 = (x_1, y_1)$, $P_2 = (x_2, y_2)$, then $t = (y_2 - y_1)/(x_2 - x_1)$, hence substituting this in (4) we finally obtain

$$x_3 = -(x_1 + x_2) - \frac{b - [(y_2 - y_1)/(x_2 - x_1)]^2}{a} \qquad (5)$$

giving x_3 as an explicit rational combination of the coordinates of P_1, P_2. If P_1, P_2 are rational points, then this shows that x_3 (and hence $y_3 = tx_3 + k$) is also rational, as we already knew.

What is unexpected is that (5) is also an *addition theorem* for elliptic functions. This has the consequence that the curve can be parametrized by elliptic functions $x = f(u)$, $y = g(u)$ such that (5) is precisely the equation expressing $x_3 = f(u_1 + u_2)$ in terms of $f(u_1) = x_1$, $f(u_2) = x_2$, $g(u_1) = y_1$, $g(u_2) = y_2$. Thus the straight-line construction of x_3 from x_1 and x_2 can also be interpreted as addition of the parameter values, u_1 and u_2, of x_1 and x_2. The first addition theorems were found by Fagnano [1718] and Euler [1768] by means of transformations of integrals. Euler realized there was a connection between such transformations and number theory, but he could never quite put his finger on it. Even earlier, Leibniz had suspected such a connection when he wrote:

> I ... remember having suggested (what could seem strange to some) that the progress of our integral calculus depended in good part upon the development of that type of arithmetic which, so far as we know, Diophantus has been the first to treat systematically. (Leibniz [1702])

Jacobi [1834] apparently saw the connection for the first time after receiving a volume of Euler's works on the transformation of integrals, but considerable clarification of elliptic functions was needed before Jacobi's insight became generally available. We shall look at some of the main steps in this process of clarification in Chapters 11 and 15.

10.7. Biographical Notes: Fermat

Pierre Fermat was born in Beaumont, near Toulouse, in 1601 and died in Castres, also near Toulouse, in 1665. His life—like his mathematics—is not known in detail, but it seems to have been relatively uneventful. Fermat's father, Dominique, was a wealthy merchant and lawyer, his mother, Claire de Long, came from a prominent family, and they had two sons and two daughters. Pierre went to school in Beaumont, commenced university studies at Toulouse, and completed them with a law degree from Orléans in 1631.

Thus Fermat's academic progress was far from meteoric, and not necessarily because he was distracted by mathematics, either. As far as we know, his earliest work was the analytic geometry of 1629 and, in the opinion of Weil [1984], his number theory did not mature until Fermat was in his late thirties.

On the evidence available, Fermat seems to defy the usual clichés about mathematical genius: he didn't start young, didn't work with passionate intensity, and was generally unwilling to publish his results (though he did sometimes boast about them). It is true that few mathematicians of Fermat's era actually did mathematics for a living, but Fermat was the purest of amateurs. It seems that mathematics never caused any interruption to his professional life.

Figure 10.5. Pierre de Fermat.

In fact, after getting his law degree in 1631 he married a distant cousin on his mother's side, Louise de Long, collected a generous dowry, and settled into a comfortable legal career. His position entitled him to be addressed as Monsieur de Fermat, hence the name Pierre de Fermat by which he subsequently became known. He and Louise had five children, the oldest of whom, Clement-Samuel, edited his father's mathematical works (Fermat [1670]). Probably the most dramatic, and terrifying, experience of Fermat's life was his contracting the plague during an outbreak of the disease in Toulouse in 1652 or 1653. He was at first reported to be dead but was one of the lucky few who recovered.

During the 1660s Fermat was in ill health. A meeting with Pascal in 1660 had to be called off because neither was well enough to travel. As a result, Fermat missed the only chance he ever had to meet a major mathematician. He never traveled far from Toulouse and all his work was done by correspondence, mostly with members of Mersenne's circle in Paris. After 1662 his letters cease to refer to scientific work, but he was signing legal documents until three days before his death. He died in Castres while on the court circuit, and he was buried there. However, in 1675 his remains were transferred to the Fermat family vault in the Church of the Augustines in Toulouse.

Fermat's apparent refusal to put mathematics ahead of his work makes the depth and range of his mathematical achievement all the more perplexing. Perhaps we will never know enough about Fermat to understand his mathematical thought, but the attempts which have been made so far raise hopes that more can be done. Mahoney [1973] gives a survey of all of Fermat's mathematics but fails to do justice to the number theory. The review by Weil [1973] makes this, and other defects of the book, painfully clear. Weil [1984] partly rectifies the situation with a brilliant analysis of Fermat' number theory, but the other facets of Fermat's mathematics have yet to be analyzed with comparable insight.

CHAPTER 11

Elliptic Functions

11.1. Elliptic and Circular Functions

The story of elliptic functions is one of the most curious in the history of mathematics, beginning with a complicated analytic idea—integrals $\int R[t, \sqrt{p(t)}]\, dt$, where R is a rational function and p is a polynomial of degree 3 or 4—and reaching a climax with a simple geometric idea—the torus surface. Perhaps the best way to understand it is to compare it with a fictitious history of circular functions which begins with the integral $\int dt/\sqrt{1 - t^2}$ and ends with the discovery of the circle. Unlikely as this fiction is, it was paralleled by the actual development of elliptic functions between the 1650s and the 1850s.

The late recognition of the geometric nature of elliptic functions was due to late recognition of the existence and geometric nature of complex numbers. In fact, the later history of elliptic functions unfolds alongside the development of complex numbers, which will be the subject of Chapters 13 to 15. In the present chapter we are concerned mainly with the history up to 1800, before complex numbers entered in a really essential way. However, there are some subplots of the main story that do not require complex numbers for their understanding and nicely show the parallel with the fictitious history of circular functions. It is convenient to relate one of these now, because it illustrates the parallel in a simplified way and also ties up a loose end from Chapter 10—the parametrization of cubic curves.

11.2. Parametrization of Cubic Curves

To see how to construct parametrizing functions for a cubic curve, we first reconstruct the parametrizing functions

$$x = \sin u$$

$$y = \cos u$$

for the circle $x^2 + y^2 = 1$, pretending that we do not know this curve geometrically but only as an algebraic relation between x and y.

The sine function can be defined as the inverse f of $f^{-1}(x) = \sin^{-1} x$, which in turn is definable as the integral

$$f^{-1}(x) = \int_0^x \frac{dt}{\sqrt{1 - t^2}}$$

Finally, the integral can be viewed as an outgrowth of the equation $y^2 = 1 - x^2$, because the integrand $1/\sqrt{1 - x^2}$ is simply $1/y$. Why do we use this integrand rather than any other to define $u = f^{-1}(x)$ and hence obtain x as a function $f(u)$? The answer is that we then obtain y as $f'(u)$, hence x, y are both functions of the parameter u. This is confirmed by the following calculation:

$$f'(u) = \frac{dx}{du} = 1 \bigg/ \frac{du}{dx}$$

and

$$\frac{du}{dx} = \frac{d}{dx} \int_0^x \frac{dt}{\sqrt{1 - t^2}} = \frac{1}{\sqrt{1 - x^2}} = \frac{1}{y}$$

hence $y = f'(u)$ (which of course is $\cos u$).

Exactly the same construction can be used to parametrize any relation of the form $y^2 = p(x)$. We put

$$u = g^{-1}(x) = \int_0^x \frac{dt}{\sqrt{p(t)}}$$

to get $x = g(u)$, and then find that $y = g'(u)$ by differentiation of u. Thus in a sense it is trivial to parametrize curves of the form $y^2 = p(x)$ (which we know from Section 7.4 to include all cubic curves, up to a projective transformation of x and y). As we shall see in the next section, the integrals $\int dt/\sqrt{p(t)}$ had been studied since the 1600s for p a polynomial of degree 3 or 4; however, no one thought to invert them until about 1800. Jacobi had a deep knowledge of both the integrals *and* inversion when he wrote his cryptic paper [1834] pointing out the relation between integrals and rational points on curves (cf. Sections 10.6 and 11.5). Thus it seems likely he understood the preceding parametrization, though such a parametrization was first given explicitly by Clebsch [1864].

11.3. Elliptic Integrals

Integrals of the form $\int R[t, \sqrt{p(t)}]\, dt$, where R is a rational function and p is a polynomial of degree 3 or 4 without multiple factors, are called *elliptic integrals*, because the first example occurs in the formula for the arc length of the ellipse. (The functions obtained by inverting elliptic integrals are called *elliptic functions*, and the curves that require elliptic functions for their parametrization are called *elliptic curves*. This drift in the meaning of "elliptic" is rather unfortunate because the ellipse, being parametrizable by rational functions, is not an elliptic curve!)

Elliptic integrals arise in many important problems of geometry and mechanics, for example, as arc lengths of the ellipse and hyperbola, period of the simple pendulum, and deflection of a thin elastic bar (see Chapter 12 and, e.g., Melzak [1976], pp. 253–269). When these problems first arose in the late seventeenth century they posed the first obstacle to Leibniz's program of integration in "closed form" or "by elementary functions." As mentioned in Section 8.6, Leibniz considered the proper solution of an integration problem $\int f(x)\,dx$ to be a known function $g(x)$ such that $g'(x) = f(x)$. The functions then "known," and now called "elementary," were those composed from algebraic, circular, and exponential functions and their inverses. All efforts to express elliptic integrals in these terms failed, and as early as 1694 James Bernoulli conjectured that the task was impossible. The conjecture was eventually confirmed by Liouville [1833], in the course of showing that a large class of integrals are nonelementary. In the meantime, mathematicians had discovered so many properties of elliptic integrals, and the elliptic functions obtained from them by inversion, that they could be considered known even if not elementary.

The key that unlocked many of the secrets of elliptic integrals was the curve known as the *lemniscate of Bernoulli* (Fig. 11.1). This curve was mentioned briefly in Section 2.5 as one of the spiric sections of Perseus. It has cartesian equation

$$(x^2 + y^2)^2 = x^2 - y^2$$

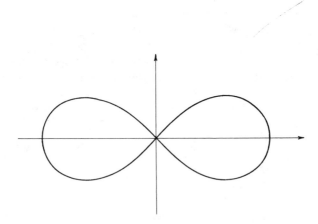

Figure 11.1

and polar equation

$$r^2 = \cos 2\theta$$

The first to consider it in its own right was James Bernoulli [1694]. He showed that its arc length was expressed by the elliptic integral $\int_0^x dt/\sqrt{1 - t^4}$, subsequently known as the *lemniscatic integral*, and thus he gave this formal expression a concrete geometric interpretation. Many later developments in the theory of elliptic integrals and functions grew out of interplay between the lemniscate and the lemniscatic integral. Being the simplest elliptic integral, or at any rate the most analogous to the arcsine integral $\int_0^x dt/\sqrt{1 - t^2}$, the lemniscatic integral $\int_0^x dt/\sqrt{1 - t^4}$ was the most amenable to manipulation. It was often possible, after some property had been extracted from the lemniscatic integral, to extend the argument to more general elliptic integrals.

The most notable example of this methodology was in the discovery of the addition theorems, which we discuss in the next section.

EXERCISES

1. Deduce the cartesian equation of the lemniscate from its polar equation

$$r^2 = \cos 2\theta$$

2. Use the polar equation of the lemniscate and the formula for the element of arc in polar coordinates,

$$ds = \sqrt{(rd\theta)^2 + dr^2}$$

to deduce that arc length of the lemniscate is given by

$$s = \int \frac{d\theta}{r}$$

3. Conclude, by changing the variable of integration to r that the total length of the lemniscate is $4\int_0^1 dr/\sqrt{1 - r^4}$.

11.4. Doubling the Arc of the Lemniscate

An addition theorem is a formula expressing $f(u_1 + u_2)$ in terms of $f(u_1)$ and $f(u_2)$ [and perhaps also $f'(u_1)$ and $f'(u_2)$]. For example, the addition theorem for the sine function is

$$\sin(u_1 + u_2) = \sin u_1 \cos u_2 + \sin u_2 \cos u_1$$

Since the derivative, $\cos u$, of $\sin u$ equals $\sqrt{1 - \sin^2 u}$, we can also write the addition theorem as

$$\sin(u_1 + u_2) = \sin u_1 \sqrt{1 - \sin^2 u_2} + \sin u_2 \sqrt{1 - \sin^2 u_1}$$

showing that $\sin(u_1 + u_2)$ is an algebraic function of $\sin u_1$ and $\sin u_2$.

To simplify the comparison with elliptic functions we consider the following special case of the sine addition theorem:

$$\sin 2u = 2 \sin u \sqrt{1 - \sin^2 u} \tag{1}$$

If we let

$$u = \sin^{-1}x = \int_0^x \frac{dt}{\sqrt{1 - t^2}}$$

then

$$2u = 2 \int_0^x \frac{dt}{\sqrt{1 - t^2}}$$

But from (1) we also have

$$2u = \sin^{-1}(2x\sqrt{1 - x^2})$$

so

$$2 \int_0^x \frac{dt}{\sqrt{1 - t^2}} = \int_0^{2x\sqrt{1-x^2}} \frac{dt}{\sqrt{1 - t^2}} \tag{2}$$

Bearing in mind that $\sin^{-1}x = \int_0^x dt/\sqrt{1 - t^2}$ represents the angle u seen in Figure 11.2, equation (2) tells us that the angle (or arc length) u is doubled by

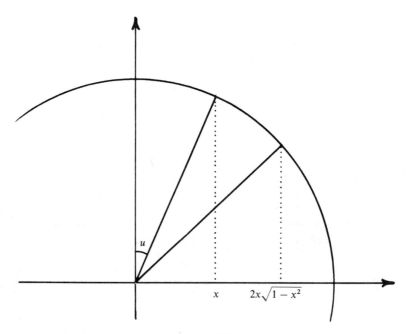

Figure 11.2

going from x to $2x\sqrt{1-x^2}$. The latter number, since it is obtained from x by rational operations and square roots, is constructible from x by ruler and compasses (confirming the geometrically obvious fact that an angle can be duplicated by ruler and compasses).

All this has a remarkable parallel in the properties of the lemniscate and its arc length integral $\int_0^x dt/\sqrt{1-t^4}$. The discovery of a formula for doubling the arc of the lemniscate by Fagnano [1718] showed that geometric information could be extracted from the previously intractable elliptic integrals, and we can also view it as the first step toward the theory of elliptic functions. In our notation, Fagnano's formula was

$$2\int_0^x \frac{dt}{\sqrt{1-t^4}} = \int_0^{2x\sqrt{1-x^4}/(1+x^4)} \frac{dt}{\sqrt{1-t^4}} \tag{3}$$

Since $2x\sqrt{1-x^4}/(1+x^4)$ is obtained from x by rational operations and square roots, (3) shows, like (2), that the arc can be doubled by ruler and compass construction.

Fagnano derived his formula by two substitutions which, as Siegel [1969], p. 3, points out, are analogous to a natural substitution for the arcsine integral (see exercises below).

EXERCISES

1. Show that the substitution $t = 2v/(1+v^2)$ gives $\sqrt{1-t^2} = (1-v^2)/(1+v^2)$ and hence that $dt/\sqrt{1-t^2} = 2dv/(1+v^2)$.

2. Show that the substitution $t^2 = 2v^2/(1+v^4)$ gives $\sqrt{1-t^4} = (1-v^4)/(1+v^4)$ and hence that

$$\frac{dt}{\sqrt{1-t^4}} = \sqrt{2}\frac{dv}{\sqrt{1+v^4}}$$

Similarly show that the substitution $v^2 = 2w^2/(1-w^4)$ gives

$$\frac{dv}{\sqrt{1+v^4}} = \sqrt{2}\frac{dw}{\sqrt{1-w^4}}$$

3. Calculate the result $t = 2w\sqrt{1-w^4}/(1+w^4)$ of the two substitutions above and deduce the Fagnano duplication formula.

11.5. General Addition Theorems

Fagnano's duplication formula remained a little-known curiosity until Euler received a copy of Fagnano's works on December 23, 1751, a date later described by Jacobi as "the birth day of the theory of elliptic functions." Euler was the first to see that Fagnano's substitution trick was not just a curious fluke but a revelation into the behavior of elliptic integrals. With his superb manipulative skill Euler was quickly able to extend it to very general additional theorems. First to the addition theorem for the lemniscatic integral,

$$\int_0^x \frac{dt}{\sqrt{1-t^4}} + \int_0^y \frac{dt}{\sqrt{1-t^4}} = \int_0^{(x\sqrt{1-y^4}+y\sqrt{1-x^4})/(1+x^2y^2)} \frac{dt}{\sqrt{1-t^4}}$$

then to $\int dt/\sqrt{p(t)}$, where $p(t)$ is an arbitrary polynomial of degree 4. An ingenious reconstruction of Euler's train of thought, by analogy with the arcsine addition theorem

$$\int_0^x \frac{dt}{\sqrt{1-t^2}} + \int_0^y \frac{dt}{\sqrt{1-t^2}} = \int_0^{x\sqrt{1-y^2}+y\sqrt{1-x^2}} \frac{dt}{\sqrt{1-t^2}}$$

has been given by Siegel [1969], pp. 1–10. Brilliant as his results were, Euler was dealing only with elliptic integrals, *not* with elliptic functions, their inverses, so one could still quibble with Jacobi's assessment. But one has to remember that Jacobi could see an elliptic function a mile off, probably more easily than we can see that the arcsine addition theorem is really a theorem about sines!

It should be mentioned that Euler's addition theorems do not cover all elliptic integrals. The general form $\int R[t, \sqrt{p(t)}] \, dt$ does, however, reduce to just three kinds, of which Euler's are the first and most important. The classical theory of elliptic integrals of the different kinds, with their various addition and transformation theorems, was systematized by Legendre [1825]. Ironically, this was just before the appearance of elliptic functions, which made much of Legendre's work obsolete.

These early investigations exploited some of the formal similarities between $\int dt/\sqrt{p(t)}$, where p is a polynomial of degree 4, and $\int dt/\sqrt{q(t)}$, where q is a quadratic. There is no real difference if p is of degree 3, as an easy transformation shows (exercise 1). This is why $\int dt/\sqrt{p(t)}$ is also called an elliptic integral when p is of degree 3. In fact, it eventually turned out that the most convenient integral to use as a basis for the theory of elliptic functions is $\int dt/\sqrt{4t^3 - g_2 t - g_3}$, whose inverse is known as the Weierstrass $\wp$-function.

The addition theorem for this integral is

$$\int_0^{x_1} \frac{dt}{\sqrt{4t^3 - g_2 t - g_3}} + \int_0^{x_1} \frac{dt}{\sqrt{4t^3 - g_2 t - g_3}} = \int_0^{x_3} \frac{dt}{\sqrt{4t^3 - g_2 t - g_3}}$$

where x_3 is none other than the x- coordinate of the third point on

$$y^2 = 4x^3 - g_2 x - g_3$$

of the straight line through (x_1, y_1) and (x_2, y_2) (cf. Section 10.6). Now that we know, from Section 11.2, that his curve is parametrized by $x = \wp(u), y = \wp'(u)$, defined by inverting the integral, some connection between the geometry of the curve and the addition theorem is understandable. But the stunning simplicity of the relationship seems to demand a deeper explanation. This lies in the realm of complex numbers, which we shall enter briefly in the next section and more thoroughly in Sections 15.4 and 15.5.

EXERCISE

1. Show that the substitution $t = 1/u$ transforms

$$\frac{dt}{\sqrt{(t-a)(t-b)(t-c)}}$$

into

$$\frac{-du}{\sqrt{u(1-ua)(1-ub)(1-uc)}}$$

11.6. Elliptic Functions

The idea of inverting elliptic integrals to obtain elliptic functions is due to Gauss, Abel, and Jacobi. Gauss had the idea in the late 1790s but did not publish it; Abel had the idea in 1823 and published it in 1827, independently of Gauss. Jacobi's independence is not quite so clear. He seems to have been approaching the idea of inversion in 1827, but he was only stung into action by the appearance of Abel's paper. At any rate, his ideas subsequently developed at an explosive rate, and he published the first book on elliptic functions, the *Fundamenta Nova Theoriae Functionum Ellipticarum*, two years later (Jacobi [1829]).

Gauss first considered inverting an elliptic integral in 1796, in the case of $\int dt/\sqrt{1-t^3}$. The following year he inverted the lemniscatic integral and made better progress. Defining the "lemniscatic sine function" $x = sl(u)$ by

$$u = \int_0^x \frac{dt}{\sqrt{1-t^4}}$$

he found that this function was periodic, like the sine, with period

$$2\tilde{\omega} = 4 \int_0^1 \frac{dt}{\sqrt{1-t^4}}$$

Gauss also noticed that $sl(u)$ invites complex arguments, since it follows from $i^2 = -1$ that

$$\frac{d(it)}{\sqrt{1-(it)^4}} = i\frac{dt}{\sqrt{1-t^4}}$$

hence $sl(iu) = isl(u)$ and the lemniscatic sine has a second period $2i\tilde{\omega}$. Thus Gauss discovered *double periodicity*, one of the key properties of elliptic functions, though at first he did not realize its universality. The scope and importance of elliptic functions hit him on May 30, 1799, when he discovered an extraordinary numerical coincidence. His diary entry of that day reads:

> We have established that the arithmetic-geometric mean between 1 and $\sqrt{2}$ is $\pi/\tilde{\omega}$ to 11 places; the demonstration of this fact will surely open up an entirely new field of analysis.

Gauss had been fascinated by the arithmetic-geometric mean since 1791, when he was 14. The arithmetic-geometric mean of two positive numbers a and b is the common limit, agM(a, b) of the two sequences $\{a_n\}$ and $\{b_n\}$ defined by

$$a_0 = a, \qquad b_0 = b$$

$$a_{n+1} = \frac{a_n + b_n}{2}, \qquad b_{n+1} = \sqrt{a_n b_n}$$

(For more information on the theory and history of the agM function, see Cox [1984].)

It is indeed true that agM$(1, \sqrt{2}) = \pi/\tilde{\omega}$, as Gauss soon proved, and the "entirely new field of analysis" he created from the merger of these ideas was extraordinarily rich. It encompassed elliptic functions in general, the theta functions later rediscovered by Jacobi, and the modular functions later rediscovered by Klein. The theory was not clearly improved until the 1850s, when Riemann showed that double periodicity becomes obvious when elliptic integrals are placed in a suitable geometric setting.

Unfortunately, Gauss released virtually none of his results on elliptic functions. Apart from publishing an expression for agM(a, b) as an elliptic integral (Gauss [1818]), he did nothing until Abel's results appeared in 1827—then promptly claimed them as his own. He wrote to Bessel [1828]:

> I shall most likely not soon prepare my investigations on the transcendental functions which I have had for many years—since 1798.... Herr Abel has now, as I see, anticipated me and relieved me of the burden in regard to one third of these matters.

It was gratuitous of Gauss to claim he had more results than Abel, because Abel also had results unknown to Gauss. True, Gauss had priority on the key ideas of inversion and double periodicity, but priority isn't everything, as Gauss himself perhaps knew. His own cherished discovery of the relation between agM and elliptic integrals had not only been found earlier, but even published by Lagrange [1785].

EXERCISES

1. Show that $sl'(u) = \sqrt{1 - sl^4(u)}$.

2. Deduce from the Euler addition theorem (Section 11.4) that

$$sl(u + v) = \frac{sl(u)sl'(v) + sl(v)sl'(u)}{1 + sl^2(u)sl^2(v)}$$

11.7. A Postscript on the Lemniscate

The duplication of the arc of the lemniscate had some interesting consequences for the lemniscate itself. Fagnano showed, by similar arguments, that a quadrant of the lemniscate could be divided into two, three, or five equal arcs by ruler and compass (see Ayoub [1984]). This raised a question: for which n can the lemniscate be divided into n equal parts by ruler and compasses? Recall from Section 2.3 that the corresponding question for the circle had been answered by Gauss [1801], Art. 366. The answer is $n = 2^m p_1 p_2 \ldots p_k$, where each p_i is a prime of the form $2^{2^h} + 1$. In the introduction to his theory, Gauss claims

> The principles of the theory which we are going to explain actually extend much further than we will indicate. For they can be applied not only to circular functions but just as well to other transcendental functions, e.g. to those which depend on the integral $\int (1/\sqrt{1 - x^4})\,dx$. (Art. 335)

However, his surviving papers do not include any result on the lemniscate as incisive as his result on the circle. There is only a diary entry of March 21, 1797, stating divisibility of the lemniscate into five equal parts.

The answer to the problem of dividing the lemniscate into n equal parts was found by Abel [1827], transforming Gauss' obscurity into crystal clarity: division by ruler and compasses is possible for *precisely the same n* as for the circle. This wonderful result serves, perhaps better than any other, to underline the unifying role of elliptic functions in geometry, algebra, and number theory. A modern proof of it may be found in Rosen [1981].

11.8. Biographical Notes: Abel and Jacobi

Niels Henrik Abel was born in the small town of Finnöy, on the southwestern coast of Norway, in 1802 and died in Oslo in 1829. In his short life he managed to win the esteem of the best mathematicians in Europe, but he fell victim to official indifference, terrible family burdens, and tuberculosis. His heartbreaking story is not unlike that of his great contemporary in another field, the poet John Keats (1797–1823).

Like several mathematicians before him (Wallis, Gregory, Euler), Abel was the son of a Protestant minister. His father, Sören, distinguished himself in theology and philology at the University of Copenhagen and was a supporter of the new literary and social movements of his time. Sören's liberality, particularly toward the consumption of alcohol, was unfortunately not matched by good judgment, and his marriage to Anne Marie Simonsen in 1799 eventually led to disaster. The beautiful Anne Marie was a talented pianist and singer but completely irresponsible and later openly unfaithful to her husband. The family held together during Abel's early years, when he was educated by his father, but both parents were becoming frequently drunk and unstable by 1815, when Niels and his older brother, Hans Mathias, were sent to the Cathedral School in Oslo.

At first school was not much better than home. Some of its best teachers had gone to the recently opened Oslo University, and discipline had deteriorated to the point where fights between staff and students were common. The mathematics teacher, Bader, was particularly brutal, beating even good students like Abel, and injuring one boy severely enough to cause his death. This led to Bader's dismissal (without his being brought to court, however) and to the appointment of a new mathematics teacher, Bernt Michael Holmboe, in 1818. Although not a creative mathematician, Holmboe knew his subject and was an inspiring teacher. He introduced Abel to Euler's calculus texts, and Abel soon abandoned all other reading for the works of Newton, Lagrange, and Gauss, among others. By 1819 Holmboe was writing in his report book: "With the most excellent genius he combines an insatiable interest and desire for mathematics, so that if he lives he probably will become a great mathematician" (see Ore [1957], p. 33). Ore informs us that the last three words are a revision, probably of the phrase "the world's foremost mathematician," which Holmboe may have been asked to tone down by the school principal. Why Holmboe chose to balance the phrase with the ominous "if he lives" is a mystery, though uncomfortably close to correct prophecy.

During his last two years at the Cathedral School, around 1820, Abel believed he had discovered the solution of the quintic equation. The mathematicians in Oslo were skeptical but unable to fault Abel's argument, so it was sent to the Danish mathematician Ferdinand Degen. Degen, too, was unable to find an error, but he prudently asked Abel for more details and a numerical illustration. When Abel attempted to compute one he discovered his error. However, Degen also had another suggestion: Abel would do better to apply his energy to "the elliptic transcendentals."

Meanwhile, Abel's family was disintegrating. Hans Mathias, after a promising start at the Cathedral School, slipped to the bottom of the class and was sent home, eventually to become feeble-minded. His father drank himself to death in 1820, leaving the family penniless. Niels Henrik, now the oldest responsible member of the family, took steps which were to save his sister Elisabeth and younger brother Peder. He found another home for Elisabeth and took Peder with him when he entered the University of Oslo in 1821.

Before long, Abel had read most of the advanced mathematical works in the university library, and his own research began in earnest. By 1823 he had discovered the inversion that was the key to elliptic functions, proved the unsolvability of the quintic, and discovered a wonderful general theorem on integration, now known as Abel's theorem, which implicitly introduces the concept of genus. On a trip to Copenhagen in 1823 to tell Degen of these results, he met and fell in love with Christine ("Crelly") Kemp. Like Abel, she came from an educated but impoverished family; she was making a living for herself by tutoring. The remaining six years of Abel's life were consumed by the struggle for recognition of his mathematics and attempts to gain a position that paid enough to allow him to marry Crelly.

In 1824 he won a government grant to travel and meet other scientists, and he became engaged to Crelly at Christmas. She was now working in Oslo as a governess, a job that Abel had arranged for her. The grant was mainly intended to take him to Paris, but when he finally set off, late in 1825, he impulsively detoured to Berlin to visit friends. There he also met August Crelle, an engineer and amateur mathematician, who was about to found the first German mathematical journal. The meeting was fortuitous, as Crelle was able to give an international circulation to Abel's first important results, while Abel could supply papers of a quality that ensured the success of the new journal. In meeting influential mathematicians Abel was less lucky. He made no effort to visit Gauss while in Germany, being convinced that Gauss was "absolutely unapproachable," and failed to make an impression on Cauchy in Paris, though he presented him with a copy of the memoir on Abel's theorem. During his stay in Paris, Abel discovered his theorem on the lemniscate and sat for his only known portrait (Fig. 11.3).

By the end of 1826 Abel was running out of money and eating only one meal a day. He feared he was losing touch with Crelly, as she had returned to Copenhagen and her letters were infrequent. He left Paris for Berlin on December 29, while he still had money to pay for the journey, and found a letter from Crelly waiting. Some good news at last! Crelly stood by him as

Figure 11.3. Niels Henrik Abel.

ever, and their plans for the future were revived. Abel returned to Oslo in May 1827, via Copenhagen, and arranged another job in Norway for Crelly. Unfortunately, the university was still unwilling to give him more than a temporary appointment, which paid barely enough to meet his family's debts. In September 1827 Abel's first memoir on elliptic functions was published in Crelle's journal. In the same month, Jacobi appeared on the scene with the first announcement of his results. These were results which Abel knew how to prove, and when Jacobi's proofs appeared, some months later, Abel was shocked to see Jacobi using the method of inversion without acknowledging its previous appearance in Abel's paper. Abel was initially bitter over this blow and strove to "knock out" Jacobi with a second memoir. However, he ceased to bear a grudge after he learned how much Jacobi really admired his work. Jacobi in fact admitted that his first announcement had been based on guesswork and that he had realized inversion was the key to the proof only after reading Abel.

In May 1828 Abel finally received a decent job offer from Berlin, only to have it withdrawn two months later. Crelle had been working in support of Abel, but another candidate had pushed in ahead of him. Then a group of French mathematicians petitioned the king of Norway–Sweden to use his influence on Abel's behalf, but still the University of Oslo remained unmoved. By now, time was running out. Abel's health worsened and in January 1829 he began spitting blood. Crelle renewed his efforts in Berlin, but it was too late. Abel died on April 6, 1829, just two days before the arrival of a letter from Crelle informing him of his appointment as professor in Berlin.

Carl Gustav Jacob Jacobi (Fig. 11.4) was born in Potsdam in 1804 and died in Berlin in 1851. He was the second of three sons of Simon Jacobi, a banker. The oldest son, Moritz, became a physicist and inventor of a popular pseudoscience called "galvanoplastics," which made him more famous in his time than Carl. The youngest, Eduard, carried on the family business, and there was also a sister, Therese. Jacobi's mother's name has not come down to us, though her side of the family was also important, one of her brothers taking charge of Jacobi's education until he entered secondary school in 1816. He was promoted to the top class after only a few months, but he had to remain there for four years, until he became old enough to enter university. During his school days Jacobi excelled in classics and history as well as mathematics. He studied Euler's *Introductio in analysin infinitorum* [1748] and attempted, like Abel, to solve the quintic equation.

Entering the University of Berlin in 1821, Jacobi continued his broad classical education for two years, before private study of the works of Euler, Lagrange, Laplace, and Gauss convinced him that he had time only for mathematics. He gained his first degree in 1824 and began lecturing (in differential geometry) at the University of Berlin in 1825. Despite a reputation for bluntness and sarcasm, Jacobi made rapid progress in his career. He moved to Königsberg in 1826, becoming associate professor there in 1827 and full professor in 1832. Overriding Jacobi's sometimes abrasive manner

Figure 11.4. Carl Gustav Jacob Jacobi.

was his exceptional energy and enthusiasm for both research and teaching. He managed to combine the two by lecturing up to 10 hours a week on elliptic functions, incorporating his latest discoveries. Such high-intensity instruction was unheard of then, as it is now, yet Jacobi built up a school of talented pupils.

In 1831 he married Marie Schwink, the daughter of a formerly wealthy man who had lost his fortune through speculation. Nine years later, with a growing family (eventually five sons and three daughters), Jacobi found himself in a similar predicament. His father's fortune had vanished and he had to support his widowed mother. In 1843 he suffered a breakdown from overwork, and diabetes was diagnosed. His friend Dirichlet managed to secure a grant for Jacobi to travel to Italy for the sake of his health. After eight months there, Jacobi was well enough to return. He was given permission to move to Berlin, because of its milder climate, and an increase in salary to meet the higher living costs in the capital. However, in 1849 the salary bonus was retracted. Jacobi had to move out of his house to an inn, and he sent the rest of his family to the small town of Gotha, where accommodations were cheaper. Early in 1851 he came down with influenza after visiting them. Before he had quite recovered, he was stricken with smallpox and died within a week.

Jacobi is remembered for his contributions to many fields of mathematics, including differential geometry, mechanics, and number theory as well as elliptic functions. He was a great admirer of Euler and planned the edition of Euler's works which eventually began to appear, on a reduced scale, in 1911. In fact, in many ways Jacobi was a second, if lesser Euler. He saw elliptic functions not so much as things in themselves, as Abel did, but as a source of dazzling formulas with implications in number theory. An astounding collection of formulas may be found in his major work on elliptic functions, the *Fundamenta nova* [1829]. At the same time, he was deeply impressed by Abel's ideas and selflessly campaigned to make them better known. He introduced the terms "Abelian integral" and "Abelian function" for the generalizations of elliptic integrals and functions considered by Abel as well as "Abelian theorem" for Abel's theorem, which he described as "the greatest mathematical discovery of our time."

CHAPTER 12

Mechanics

12.1. Mechanics before Calculus

The ambiguous title reflects the dual purpose of this section: to give a brief survey of the mechanics that came before calculus and to introduce the thesis that mechanics was psychologically, if not logically, a prerequisite for calculus itself. The remainder of the chapter will expand on this thesis, demonstrating how several important fields in calculus (and beyond) originated in the study of mechanical problems. Lack of space, not to mention lack of expertise, prevents my venturing far into the history of mechanical concepts, so I shall assume some understanding of time, velocity, acceleration, force, and the like, and concentrate on the mathematics which emerged from reflection on these notions. These mathematical developments will be pursued as far as the nineteenth century. More details may be found in Dugas [1957, 1958] and Truesdell [1954, 1960]. In our own century, mathematics seems to have been the motivation for mechanics rather than the other way round. The outstanding mechanical concepts of the twentieth century—relativity and quantum mechanics—would not have been conceivable without nineteenth-century advances in pure mathematics, some of which we shall discuss later.

It was mentioned in Section 4.5 that Archimedes made the only substantial contribution to mechanics in antiquity by introducing the basics of statics (balance of a lever requires equality of moments on the two sides) and hydro-statics (a body immersed in a fluid experiences an upward force equal to the weight of fluid displaced). Archimedes' famous results on areas and volumes were in fact discovered, as he revealed in his *Method*, by hypothetical balancing of thin slices of different figures. Thus the earliest nontrivial results in calculus, if by calculus one means a method for discovering results about limits, relied on concepts from mechanics.

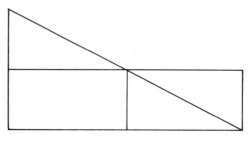

Figure 12.1

The medieval mathematician Oresme also was mentioned (Section 6.1) for his use of coordinates to give a geometric representation of functions. The relationship which Oresme represented was in fact velocity v as a function of time t. He understood that displacement is then represented by the area under the curve, and hence in the case of constant acceleration (or "uniformly deformed velocity," as he called it) the displacement equals total time × velocity at the middle instant (Fig. 12.1). This result is known as the "Merton acceleration theorem" (see, e.g., Clagett [1959], p. 255) because it originated in the work of a group of mathematicians at Merton College, Oxford, in the 1330s. The first proofs were arithmetical and far less transparent than Oresme's figure.

Although constant acceleration was understood theoretically in the 1330s, it was not clear that constant acceleration was actually a natural occurrence—namely, with falling bodies—until the time of Galileo (1564–1642). Galileo announced the equivalent result, that displacement of a body falling from rest at time $t = 0$ is proportional to t^2, in a letter [1604]. At first he was uncertain whether this derived from a velocity proportional to time $v = kt$ (i.e., constant acceleration) or proportional to distance $v = ks$, but he resolved the question correctly in favor of $v = kt$ later [1638]. By composing the uniformly increasing vertical velocity with constant horizontal velocity, Galileo derived for the first time the correct trajectory of a projectile: the parabola.

The motion or projectiles was a matter of weighty importance in the Renaissance, and presumably observed often, yet the trajectories suggested before Galileo were quite preposterous (see, e.g., Fig. 5.4). The belief, deriving from Aristotle, that motion could be sustained only by continued application of a force led mathematicians to ignore the evidence and to draw trajectories in which the horizontal velocity dwindled to zero. Galileo overthrew this mistaken belief by affirming the *principle of inertia*: a body not subject to external forces travels with constant velocity.

The principle of inertia was Newton's starting point in mechanics; indeed, it is often called Newton's first law. It is a special case of his second law, that force is proportional to mass × acceleration (Newton [1687/1934], p. 13). Under this law, the motion of a body is determined by composition of the forces acting on it. The correct law for the composition of forces, that forces add vectorially, had been discovered in the case of perpendicular forces by Stevin [1586] and in the general case by Roberval (published in Mersenne [1636]). The motion is thus determined by vector addition of the corresponding accelerations, the method Galileo used in the case of the projectile.

The determination of velocity and displacement from acceleration are of course problems of integration, so mechanics contributed a natural class of problems to calculus just at the time the subject was emerging. But more than this was true. The early practitioners of calculus believed that continuity was an essential attribute of functions, and the only way they were able to define continuity was ultimately by falling back on the dependence of a velocity or displacement on time. From this viewpoint, *all* problems of integration and differentiation were problems of mechanics, and Newton described them as such when explaining how his calculus of infinite series could be applied:

> It now remains, in illustration of this analytical art, to deliver some typical problems and such especially as the nature of curves will present. But first of all I would observe that difficulties of this sort may all be reduced to these two problems alone, which I may be permitted to propose with regard to the space traversed by any local motion however accelerated or retarded:
>
> 1. Given the length of space continuously (that is, at every time), to find the speed of motion at any time proposed.
>
> 2. Given the speed of motion continuously, to find the length of the space described at any time proposed. (Newton [1671], p. 71)

Of course we now know that the first problem requires differentiability rather than continuity for its solution, but the pioneers of calculus thought that differentiability was implied by continuity, and hence did not recognize it as a distinct notion. In fact it was a mechanical question—the problem of the vibrating string—whose investigation finally brought the distinction to light (see Section 12.4).

12.2. Celestial Mechanics

Astronomy has been a powerful stimulus to mathematics since ancient times. The epicyclic theory of Apollonius and Ptolemy introduced an interesting family of algebraic and transcendental curves, as we saw in Section 2.5, and the theory itself ruled Western astronomy until the seventeenth century. Even Copernicus (1472–1543), when he overthrew Ptolemy's earth-centered system with a sun-centered system in his *De revolutionibus orbium coelestium* [1543], was unwilling to give up epicycles. Taking the sun as the center of the system simplifies the orbits of the planets but does not make them circular, so Copernicus, accepting the Ptolemaic philosophy that orbits must be generated by circular motions, modeled them by epicycles. In fact he used more epicycles than Ptolemy.

A more important advance, from the mathematical point of view, was Kepler's introduction of elliptical orbits in his *Astronomia Nova* [1609]. When Newton explained these orbits as a consequence of the inverse square law of gravitation in the *Principia* (Newton [1687/1934], p. 56) he showed that there was a deeper level of explanation—the infinitesimal level—where simplicity could be attained even when it was not possible at the global level. The force on a given body B_1 is simply the vector sum of the forces due to the other bodies $B_2, \ldots, B_n$ in the system, determined by their masses and distances from B_1 by the inverse square law and, by Newton's second law, this determines the acceleration of B_1. The accelerations of $B_2, \ldots, B_n$ are similarly determined, hence the behavior of the system is completely determined by the inverse square law, once initial positions and velocities are given. The inverse square law is an infinitesimal law in the sense that it describes the limiting behavior of a body—its acceleration—and not its global behavior such as the shape or period of its orbit.

As we now know, it is rarely possible to describe the global behavior of a dynamical system explicitly, so Newton found the only viable basis for dynamics in directing attention to infinitesimal behavior. Unfortunately, he communicated this insight poorly by expressing it in geometrical terms, in the belief that calculus, which he had used to discover his results, was inappropriate in a serious publication. By the eighteenth century this belief had been dispelled by Leibniz and his followers, and definitive formulations of dynamics in terms of calculus were given by Euler and Lagrange. They recognized that the infinitesimal behavior of a dynamical system was typically described by a system of *differential equations* and that the global behavior was derivable from these equations, in principle, by integration.

The question remained, however, whether the inverse square law did indeed account for the observed global behavior of the solar system. In a system with only two bodies, Newton showed ([1687/1934], p. 166) that each describes a conic section relative to the other, in normal cases an ellipse as stated by Kepler. With a three-body system, such as the earth–moon–sun, no simple global description was possible, and Newton could obtain only qualitative results through approximations. With the many bodies in the solar system, extremely complex behavior was possible, and for 100 years mathematicians were unable to account for some of the phenomena actually observed.

A famous example was the so-called secular variation of Jupiter and Saturn, which was detected by Halley in 1695 from the observations then available. For several centuries Jupiter had been speeding up (spiraling toward the sun) and Saturn had been slowing down (spiraling outward). The problem was to explain this behavior and to determine whether it would continue, with the eventual destruction of Jupiter and disappearance of Saturn. Euler and Lagrange worked on the problem without success; then, in the centenary year of *Principia*, Laplace [1787] succeeded in explaining the phenomenon. He showed that the secular variation was actually periodic, with Jupiter and Saturn returning to their initial positions every 929 years. Laplace viewed this as confirmation not only of the Newtonian theory but also of the stability of the solar system, though it seems that the latter is still an open question.

Laplace introduced the term "celestial mechanics" and left no doubt that the theory had arrived with his monumental *Mécanique céleste*, a work of five volumes which appeared between 1799 and 1825. In astronomy, the theory had its finest hour in 1846, with the discovery of Neptune, whose position had been computed by Adams and Leverrier from observed perturbations in the orbit of Uranus. The difficult question of stability was taken up again in the three volume *Les Méthodes Nouvelles de la Mécanique Céleste* of Poincaré [1892, 1893, 1899]. In this work Poincaré directed attention toward asymptotic behavior, in a sense complementing Newton's infinitesimal view with a view toward infinity, and his methods have become highly influential in twentieth-century dynamics.

12.3. Mechanical Curves

When Descartes gave his reasons for restricting *La Géométrie* to algebraic curves (which he called "geometric"; see Section 6.3), he explicitly excluded certain classical curves on the rather vague grounds that they

> belong only to mechanics, and are not among those curves that I think should be included here, since they must be conceived of as described by two separate movements whose relation does not admit of exact determination. (Descartes [1637/1954], p. 44)

The curves that Descartes relegated "to mechanics" were those the Greeks had defined by certain hypothetical mechanisms, for example, the epicycles (described by rolling one circle on another) and the spiral of Archimedes (described by a point moving at constant speed along a uniformly rotating line). He was probably aware that the spiral is transcendental by virtue of the fact that it meets a straight line in infinitely many points. This is contrary to the behavior of an algebraic curve $p(x, y) = 0$, which meets a straight line $y = mx + c$ in only finitely many points, corresponding to the finitely many solutions of $p(x, mx + c) = 0$. This proof that spirals are transcendental was given explicitly by Newton [1687/1934], p. 110.

We do not know whether Descartes distinguished, say, the algebraic epicycles from the transcendental ones; nevertheless, it is broadly true that his "mechanical" curves were transcendental. This remained true with the great expansion of mechanics and calculus in the seventeenth century, and indeed most of the new transcendental curves originated in mechanics. In this section we shall look at three of the most important of them: the catenary, the cycloid, and the elastica.

The *catenary* is the shape of a hanging cord, assumed to be perfectly flexible and with mass uniformly distributed along its length. In practice, the flexibility and uniformity of mass are realized better by a hanging chain, hence the name "catenary," which comes from the Latin *catena* for chain. Hooke [1675] observed that the same curve occurs as the shape of an arch of infinitesimal stones. The catenary looks very much like a parabola and was at first conjectured to be one by Galileo. This was disproved by the 17-year-old Huygens [1646], though at the time he was unable to determine the correct curve. He did show, however, that the parabola was the shape assumed by a flexible cord loaded by weights which are uniformly distributed in the horizontal direction (as is approximately the case for the cable of a suspension bridge).

The problem of the catenary was finally solved independently by John Bernoulli [1691], Huygens [1691], and Leibniz [1691], in response to a challenge from James Bernoulli in 1690. John Bernoulli showed that the curve satisfies the differential equation

$$\frac{dy}{dx} = \frac{s}{a}$$

where a is constant and $s = $ arc length OP (Fig. 12.2). He derived this equation

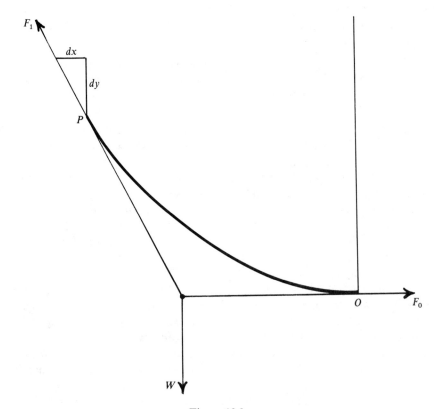

Figure 12.2

by replacing the portion OP of the chain, which is held in equilibrium by the tangential force F_1 at P and the horizontal force F_0, which is independent of P, by a point mass W equal to the weight of OP (hence proportional to s) held in equilibrium by the same forces. Comparing the directions and magnitudes of the forces gives

$$\frac{dy}{dx} = \frac{W}{F_0} = \frac{s}{a}$$

By ingenious transformations Bernoulli reduced the equation to

$$dx = \frac{a\, dy}{\sqrt{y^2 - a^2}}$$

in other words, to an integral. This solution was as simple as could be stated at the time, since x is a transcendental function of y and hence can be expressed, at best, as an integral. Today, of course, we recognize the function as one of the "standard" ones and abbreviate the solution as

$$y = a \cosh\frac{x}{a} - a$$

The *cycloid* is the curve generated by a point on the circumference of a circle rolling on a straight line. Despite being a natural limiting case in the epicyclic family, the cycloid does not seem to have been investigated until the seventeenth century, when it became a favorite curve with mathematicians. It has many beautiful geometrical properties, and even more remarkable mechanical properties. The first of these, discovered by Huygens [1659"], is that the cycloid is the *tautochrone* (equal-time curve). A particle constrained to slide along an inverted cycloid takes the same time to descend to the lowest point, regardless of its starting point.

Huygens [1673] made a classic application of this property to pendulum clocks, using a geometric property of the cycloid [1659']. If the pendulum, taken to be a weightless cord with a point mass at the end, is constrained to swing between two cycloidal "cheeks," as Huygens called them (Fig. 12.3), then the point mass will travel along a cycloid. Consequently, the period of the cycloidal pendulum is independent of amplitude. This makes it theoretically superior to the ordinary pendulum whose period, though approximately constant for small amplitudes, actually involves an elliptic function. In practice, problems such as friction make the cycloidal pendulum no more accurate than the ordinary pendulum, but its theoretical superiority shut the ordinary pendulum out of mechanics for some time. Newton's *Principia*, for example, often mentions the cycloidal pendulum but never the simple pendulum.

The second remarkable property of the cycloid is that it is the *brachistochrone*, the curve of shortest time. John Bernoulli [1696] posed the problem of finding the curve, between given points A and B, along which a point mass descends in the shortest time. He already knew that the solution was a cycloid, and solutions were found independently by James Bernoulli [1697], l'Hôpital

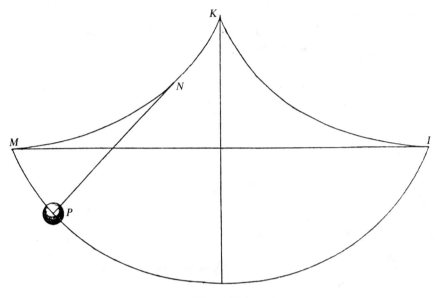

Figure 12.3

[1697], Leibniz [1697], and Newton [1697]. The problem is deeper than that of the tautochrone, because the cycloid has to be singled out from *all possible* curves between A and B. James Bernoulli's solution was the most profound because it recognized the "variable curve" aspect of the problem, and it is now considered to be the first major step in the development of the calculus of variations.

The *elastica* was another of James Bernoulli's discoveries, and likewise important in the development of another field—the theory of elliptic functions. The elastica is the curve assumed by a thin elastic rod compressed at the ends. James Bernoulli [1694] showed that the curve satisfied a differential equation which he reduced to the form

$$ds = \frac{dx}{\sqrt{1 - x^4}}$$

To interpret this integral geometrically, he introduced the lemniscate and showed that its arc length was expressed by precisely the same integral. This was the beginning of the investigations of the lemniscate integral, which included the important discoveries of Fagnano and Gauss mentioned in the last chapter. Euler's investigations of elliptic integrals were also stimulated by the elastica. Euler [1743] gave pictures of elastica which show they have periodic forms (Fig. 12.4). These drawings were the first to show the real period of elliptic functions, though of course periodicity was implicit in the first elliptic integral, the arc length of the ellipse (the real period being the circumference of the ellipse).

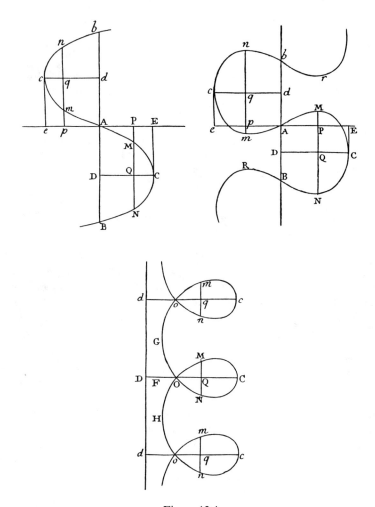

Figure 12.4

EXERCISES

1. Use a formula for arc length s to transform the differential equation

$$\frac{dy}{dx} = \frac{s}{a}$$

to

$$\frac{dx}{dz} = \frac{a}{\sqrt{1 + z^2}} \qquad (1)$$

where $z = dy/dx$.

2. Solve (1) for x and hence show that the original equation has a solution

$$y = a \cosh \frac{x}{a} + \text{const}$$

3. Show that the functions sin and cos, and hence the functions sinh and cosh, are transcendental.

12.4. The Vibrating String

The problem of the vibrating string is one of the most fertile in mathematics, being the source of such diverse fields as partial differential equations, Fourier series, and set theory. It is also remarkable in being perhaps the only setting in which the sense of hearing led to important mathematical discoveries. As we saw in Section 1.5, the Pythagoreans discovered the relationship between pitch and length by hearing the harmonious tones produced by two strings whose lengths were in a simple whole number ratio. Thus in a sense it was possible to "hear the length of a string," and some later discoveries of mathematically significant properties of the string—overtones, for example— were initially prompted by hearing (see Dostrovsky [1975]).

Various authors in ancient times suggested that the physical basis of pitch was frequency of vibration, but it was not until the seventeenth century that the precise relationship between frequency and length was discovered, by Descartes' mentor Isaac Beeckman. Beeckman [1615] gave a simple geometric argument to show that frequency is inversely proportional to length; hence the Pythagorean ratios of lengths can also be interpreted as (reciprocal) ratios of frequencies. The latter interpretation is more fundamental because frequency alone determines pitch, whereas length determines pitch only when the material, cross section, and tension of the string are fixed. The relation between frequency v, tension T, cross-sectional area A, and length l was discovered experimentally by Mersenne [1625] to be

$$v \propto \frac{1}{l} \sqrt{\frac{T}{A}}$$

The first derivation of Mersenne's law from mathematical assumptions was given by Taylor [1713], in a paper that marks the beginning of the modern theory of the vibrating string. In it he discovered the simplest possibility for the instantaneous shape of the string, the half sine wave

$$y = k \sin \frac{\pi x}{l}$$

and established generally that the force on an element was proportional to d^2y/dx^2.

The latter result was the starting point for a dramatic advance in the theory by d'Alembert [1747]. Taking into account the dependence of y on time t as well as x, d'Alembert realized that acceleration should be expressed by $\partial^2 y/\partial t^2$ and the force found by Taylor by $\partial^2 y/\partial x^2$, hence *partial* derivatives are involved. Newton's second law then gives what is now called the *wave equation*,

$$\frac{\partial^2 y}{\partial x^2} = \frac{1}{c^2}\frac{\partial^2 y}{\partial t^2}$$

writing the constant of proportionality as $1/c^2$. Undeterred by the novelty of this partial differential equation, d'Alembert forged ahead to a general solution as follows. The equation may be simplified by a change of time scale $s = ct$ to

$$\frac{\partial^2 y}{\partial x^2} = \frac{\partial^2 y}{\partial s^2} \tag{1}$$

The chain rule gives

$$d\left(\frac{\partial y}{\partial x} \pm \frac{\partial y}{\partial s}\right) = \frac{\partial^2 y}{\partial x^2}dx + \frac{\partial^2 y}{\partial x \partial s}(ds \pm dx) \pm \frac{\partial^2 y}{\partial s^2}ds$$

$$= \left(\frac{\partial^2 y}{\partial s^2} \pm \frac{\partial^2 y}{\partial x \partial s}\right)(ds \pm dx)$$

from which d'Alembert concluded that

$$\frac{\partial^2 y}{\partial s^2} + \frac{\partial^2 y}{\partial x \partial s}$$

is a function of $s + x$ and

$$\frac{\partial^2 y}{\partial s^2} - \frac{\partial^2 y}{\partial x \partial s}$$

is a function of $s - x$, whence, say,

$$\frac{\partial y}{\partial x} + \frac{\partial y}{\partial s} = \int\left(\frac{\partial^2 y}{\partial s^2} + \frac{\partial^2 y}{\partial x \partial s}\right)d(s + x) = f(s + x)$$

and similarly

$$\frac{\partial y}{\partial x} - \frac{\partial y}{\partial s} = g(s - x)$$

This gives

$$\frac{\partial y}{\partial x} = \frac{1}{2}[f(s + x) + g(s - x)]$$

$$\frac{\partial y}{\partial s} = \frac{1}{2}[f(s + x) - g(s - x)]$$

and finally,

$$y = \int \left(\frac{\partial y}{\partial x} dx + \frac{\partial y}{\partial s} ds \right)$$

$$= \int \frac{1}{2}[f(s + x)(ds + dx) - g(s - x)(ds - dx)]$$

$$= \Phi(s + x) + \Psi(s - x)$$

Reversing the argument, we see that the functions Φ and Ψ can be arbitrary, at least as long as they admit the various differentiations involved.

But how arbitrary *is* an arbitrary function? Is it as arbitrary as an arbitrarily shaped string? The vibrating string problem caught eighteenth-century mathematicians unprepared to answer these questions. They had understood a function to be something expressed by a formula, possibly an infinite series, and this had been thought to guarantee differentiability. Yet the most natural shape of the vibrating string was one with a nondifferentiable point—the triangle of the plucked string as it is released—so nature seemed to demand an extension of the concept of function beyond the world of formulas.

The confusion was heightened when Daniel Bernoulli [1753] claimed, on physical grounds, that a general solution of the wave equation *could* be expressed by a formula, the infinite trigonometric series

$$y = a_1 \sin\frac{\pi x}{l}\cos\frac{\pi ct}{l} + a_2 \sin\frac{2\pi x}{l}\cos\frac{2\pi ct}{l} + \cdots$$

This amounts to claiming that any mode of vibration results from the superposition of simple modes, a fact which he considered to be intuitively evident. The nth term in the series represents the nth mode, generalizing Taylor's formula for the fundamental mode and building in the time dependence; however, Daniel Bernoulli gave no method for calculating the coefficient a_n.

We now know that his intuition was correct and that the triangular wave form, among others, is representable by a trigonometric series. However, it was well into the nineteenth century before anything like a clear understanding of trigonometric series was obtained. The fact that the triangular wave could be represented by a series made it a bona fide function by classical standards, hence mathematicians were brought to the realization that a series representation does not guarantee differentiability. Later, continuity was also called into question, and infinitely subtle problems concerning the convergence of trigonometric series led Cantor to develop the theory of sets (see Chapter 20).

These remarkably remote consequences of what seemed at first to be a purely physical question were of course not the only fruits of the vibrating string investigations. Trigonometric series proved to be valuable all over mathematics, from the theory of heat, where Fourier applied them with such success that they became known as *Fourier series*, to the theory of numbers. Their most famous application to number theory is probably the proof of Dirichlet [1837] that any arithmetic progression $a, a + b, a + 2b, \ldots$, where $\mathrm{hcf}(a, b) = 1$, contains infinitely many primes. Pythagoras would surely have approved!

12.5. Hydrodynamics

The properties of fluid flow have been investigated since ancient times, initially in connection with practical questions such as water supply and water-powered machinery. However, nothing like a mathematical theory was obtained before the Renaissance, and until the advent of calculus it was only possible to deal with fairly coarse macroscopic quantities such as the average speed of emission from an opening in a container. Newton [1687], Book II, introduced infinitesimal methods into the study of fluids, but much of his reasoning is incomplete, based on inappropriate mathematical models, or simply wrong. As late as 1738, when the field of hydrodynamics finally got its name in the classic *Hydrodynamica* of Daniel Bernoulli, the basic infinitesimal laws of fluid motion had still not been discovered.

The first important law was discovered by Clairaut [1740], in a context that in fact was essentially static. Clairaut was interested in one of the burning questions of the time, the shape (or "figure") of the earth. Newton had argued that the earth must bulge somewhat at the equator as a result of its spin. Natural as this seems now (and indeed then, since the phenomenon was clearly observable in Jupiter and Saturn), it was opposed by the anti-Newtonian Cassini, who argued for a spindle-shaped earth, elongated toward the poles. Clairaut actually took part in an expedition to Lapland that confirmed Newton's conjecture by measurement, but he also attacked the problem theoretically by studying the conditions for equilibrium of a fluid mass.

He considered the vector field of force acting on the fluid and observed that it must be what we now call a *conservative*, or *potential* field. That is, the integral of the force around any closed path must be zero; otherwise the fluid would circulate. The condition he actually formulated was the equivalent one that the integral between any two points be independent of the path. In the special two-dimensional case where there are components P, Q of force in the x and y directions, the quantity to be integrated is

$$P\,dx + Q\,dy$$

Clairaut argued that for the integral to be path-independent, this quantity must be a complete differential

$$df = \frac{\partial f}{\partial x}dx + \frac{\partial f}{\partial y}dy$$

Consequently, $P = \partial f/\partial x$, $Q = \partial f/\partial y$ and P, Q satisfy the condition

$$\frac{\partial P}{\partial y} = \frac{\partial Q}{\partial x} \tag{1}$$

This condition is indeed necessary, but the existence of the potential f involved more mathematical subtleties than could have been foreseen at the time. Clairaut derived the corresponding equations for the components P, Q, R in the physically more natural three-dimensional case and went as far as studying the equipotential surfaces $f = $ constant. He also found a satisfying solution to the problem of the figure of the earth. When the force at a point is the resultant of gravity and the rotational force, then an ellipsoid of revolution is an equilibrium figure, with the axis of rotation being the shorter axis of the ellipse (Clairaut [1743], p. 194).

The two-dimensional equation (1), despite being physically special if not unnatural, turned out to have a deep mathematical significance. This was discovered in the dynamic situation, with P, Q taken to be components of velocity rather than force. In this case (1) still holds when the flow is time-independent and irrotational, as d'Alembert [1752] showed by an argument similar to Clairaut's. The crucial additional fact that now emerges is that P, Q satisfy a second relation

$$\frac{\partial P}{\partial x} + \frac{\partial Q}{\partial y} = 0 \tag{2}$$

derived by d'Alembert as a consequence of the incompressibility of the fluid. He considered an infinitesimal rectangle of fluid with corners (x, y), $(x + dx, y)$, $(x, y + dy)$, $(x + dx, y + dy)$, and the parallelogram into which it is carried in an infinitesimal time interval by the known velocities (P, Q), $(P + (\partial P/\partial x)\,dx$, $Q + (\partial Q/\partial x)\,dx)$, ... Equating the areas of these two parallelograms leads to (2). In the three-dimensional case one similarly gets

$$\frac{\partial P}{\partial x} + \frac{\partial Q}{\partial y} + \frac{\partial R}{\partial z} = 0$$

but the significance of (1) and (2), as d'Alembert discovered, is that they can be combined into a single fact about the complex function $P + iQ$. This flash of inspiration became the basis for the theory of complex functions developed in the nineteenth century by Cauchy and Riemann (see Section 15.1).

12.6. Biographical Notes: The Bernoullis

Undoubtedly the most outstanding family in the history of mathematics was the Bernoulli family of Basel, which included at least eight excellent mathematicians between 1650 and 1800. Three of these, the brothers James (1654–1705) and John (1667–1748) and John's son Daniel (1700–1782) were among the great mathematicians of all time, as one may guess from their contributions already mentioned in this chapter. In fact, all the mathematicians Bernoulli were important in the history of mechanics. One can trace their influence in this field in Szabó [1977], which also contains portraits of most of them, and in Truesdell [1954, 1960]. However, James, John, and Daniel are of interest from a wider point of view, in mathematics, as well as their personal lives. The Bernoulli family, with all its mathematical talent, also had more than its share of arrogance and jealousy, which turned brother against brother and father against son. In three successive generations, fathers tried to steer their sons into nonmathematical careers, only to see them gravitate back to mathematics. The fiercest conflicts occurred between James, John, and Daniel.

James, the first mathematician in the family, was the oldest son of Nicholas Bernoulli, a successful pharmacist and civic leader in Basel, and Margaretha Schönauer, the daughter of another wealthy pharmacist. There were three other sons, Nicholas, who became an artist and in 1686 painted the portrait of James seen here (Fig. 12.5), John, and Hieronymus, who took over the family business. Their father's wish was that James should study theology, which he initially did, obtaining his licentiate in 1676. However, James also began to teach himself mathematics and astronomy, and he traveled to France in 1677 to study with the followers of Descartes. In 1681 his astronomy brought him into conflict with the theologians. Inspired by the appearance of a great comet in 1680, he published a pamphlet which proposed laws governing the behavior of comets and claiming that their appearances could be predicted. His theory was not actually correct (this was six years before *Principia*) but it certainly clashed with the theology of the time, which exploited the unexpectedness of comets in claiming they were signs of divine displeasure. James decided that his future was in mathematics rather than theology, and he adopted the motto *Invito Patre, Sidera verso* (Against my father's will, I turn to the stars). He made a second study tour, to the Netherlands and England, where he met Hooke and Boyle, and began to lecture on mechanics in Basel in 1683.

He married Judith Stepanus in 1684, and they eventually had a son and daughter, neither of whom became a mathematician. In a sense, the mathematical heir of James was his nephew Nicholas (son of the painter), who

Figure 12.5. Portrait of James Bernoulli by Nicholas Bernoulli.

carried on one of James' most original lines of research, probability theory. He arranged for the posthumous publication of James' book on the subject, the *Ars Conjectandi* [1713], which contains the first proof of a law of large numbers. James Bernoulli's law describes the behavior of long sequences of trials of an experiment for which a positive outcome has a fixed probability p (such trials are now called Bernoulli trails). In a precise sense, the proportion of successful trials will be "close" to p for "almost all" sequences.

In 1687 James became professor of mathematics in Basel and, together with John (whom he had been secretly teaching mathematics), set about mastering the new methods of calculus that were then appearing in the papers of Leibniz. This proved to be difficult, perhaps more for James than John, but by the 1690s the brothers equaled Leibniz himself in the brilliance of their discoveries. James, the self-taught mathematician, was the slower but more penetrating of the two. He sought to get to the bottom of every problem, whereas John was content with any solution, and the quicker the better.

John was the tenth child of the family, and his father intended him to have a business career. When his lack of aptitude for business became clear, he was allowed to enter the University of Basel in 1683 and became a master of arts in 1685. During this time he also attended his brother's lectures and, as mentioned earlier, learned mathematics from him privately. Their rivalry did not come to the surface until the catenary contest of 1690, but James may have felt uneasy about his younger brother's talent as early as 1685. In that year he persuaded John to take up the study of medicine, making the highly optimistic forecast that it offered great opportunities for the application of mathematics. John went into medicine quite seriously, obtaining a licentiate in 1690 and a doctorate in 1695, but by that time he was more famous as a mathematician. With the help of Huygens he gained the chair of mathematics in Groningen, and thus became free to concentrate on his true calling.

The great applications of mathematics to medicine did not eventuate, though John Bernoulli did make an amusing application of geometric series which still circulates today as a piece of physiological trivia. In his *De Nutritione* [1699] he used the assumption that a fixed proportion of bodily substance, homogeneously distributed, is lost each day and replaced by nutrition, to calculate that almost all the material in the body would be renewed in three years. This result provoked a serious theological dispute at the time, since it implied the impossibility of resurrecting the body from all its past substance.

John Bernoulli made several important contributions to calculus in the 1690s, outside mechanics. One was the first textbook in the subject, the *Analyse des infiniment petits*. This was published under the name of his student, the Marquis l'Hôpital [1696], apparently in return for generous financial compensation. Another contribution, made jointly with Leibniz, was the technique of partial differentiation. The two kept this discovery secret for 20 years in order to use it as a "secret weapon" in problems about families of curves (see Engelsmann [1984]). Other discoveries still remain outside the territory usually explored in calculus, for example,

$$\int_0^1 x^x \, dx = 1 - \frac{1}{2^2} + \frac{1}{3^3} - \frac{1}{4^4} + \cdots$$

This startling result of John Bernoulli [1697'] can be proved using a suitable series expansion of x^x and integration by parts (see exercises).

The rivalry between James and John turned to open hostility in 1697 over the *isoperimetric problem*, the problem of finding the curve of given length which encloses the greatest area. James correctly recognized that this was a calculus of variations problem but withheld his solution, whereas John persisted in publicizing an incorrect solution and claiming that James had no solution at all. James presented his solution to the Paris Academy in 1701, but it somehow remained in a sealed envelope until after his death. Even when the solution was made public in 1706, John refused to admit his own error or the superiority of James' analysis.

John was married to Dorothea Falkner, the daughter of a parliamentary deputy in Basel, and through his father-in-law's influence was awarded the chair of Greek in Basel in 1705. This enabled him to return to Basel from Groningen, but his real goal was the chair of mathematics, not Greek. James was then in ill health, and his last days were embittered by the belief that John was plotting to take his place, using the Greek offer as a steppingstone. This is precisely what happened, for when James died in 1705 John became professor of mathematics.

With the death of James and the virtual retirement of Leibniz and Newton, John enjoyed about 20 years as the leading mathematician in the world. He was particularly proud of his successful defense of Leibniz against the supporters of Newton:

> When in England war was declared against M. Leibniz for the honour of the first invention of the new calculus of the infinitely small, I was despite my wishes involved in it; I was pressed to take part. After the death of M. Leibniz the contest fell to me alone. A crowd of English antagonists fell upon my body. It was my lot to meet the attacks of Messrs Keil, Taylor, Pemberton, Robins and others. In short I alone like the famous Horatio Cocles kept at bay at the bridge the entire English army. (Translation by Pearson [1978], p. 235)

His portrait from this era shows the Bernoulli arrogance at its peak (Fig. 12.6).

John Bernoulli finally met his match at the hands of his own pupil Euler in 1727. There was no open warfare, just a polite exchange of correspondence on the logarithms of negative numbers, but it revealed that John Bernoulli understood some of his own results less well than Euler did. John Bernoulli persisted in his stubborn misunderstanding for another 20 years, while Euler went on to develop his brilliant theory of complex logarithms and exponentials (see Section 15.1). John Bernoulli seems not to have minded his pupil's success at all; instead, he became consumed with jealousy over the success of his son Daniel.

Daniel Bernoulli (Fig. 12.7) was the middle of John's three sons, all of whom became mathematicians. The oldest, Nicholas (called Nicholas II by historians to distinguish him from the first mathematician Nicholas), died of a fever in St. Petersburg in 1725 at the age of 30. The youngest, John II, was the least distinguished of the three, but he fathered the next generation of Bernoulli mathematicians, James II and John III. Daniel's path to mathematics was very similar to his father's. During his teens he was tutored by his older brother; his father wanted him to go into business, but when that career failed Daniel was permitted to study medicine.

He gained his doctorate in 1721 and made several attempts to win the chair of anatomy and botany in Basel, finally succeeding in 1733. By that time, however, he had drifted into mathematics, with such success that he had been called to the St. Petersburg Academy. During his years there (1725–1733) he

Figure 12.6. John Bernoulli.

conceived his ideas on modes of vibration and produced the first draft of his *Hydrodynamica*. Although he missed finding the basic partial differential equations of hydrodynamics, the *Hydrodynamica* made other important advances. One was the systematic use of a principle of conservation of energy; another was the kinetic theory of gases, including the derivation of Boyle's law that is now standard.

Unfortunately, publication of the *Hydrodynamica* was delayed until 1738.

Figure 12.7. Daniel Bernoulli.

This left Daniel's priority open to attack, and the one to take advantage of him was his own father. The self-styled Horatius of the priority dispute between Leibniz and Newton attempted the most blatant priority theft in the history of mathematics by publishing a book on hydrodynamics in 1743 and dating it 1732. Daniel was devastated, and wrote to Euler:

> Of my entire *Hydrodynamics*, not one iota of which do in fact I owe to my father, I am all at once robbed completely and lose thus in one moment the fruits of the work of ten years. All propositions are taken from my *Hydrodynamics*, and then my father calls his writings *Hydraulics, now for the first time disclosed*, 1732, since my *Hydrodynamics* was printed only in 1738. (Daniel Bernoulli [1743], from the translation in Truesdell [1960])

The situation was not quite as clear-cut as Daniel claimed (a detailed assessment is in Truesdell [1960]), but at any rate John Bernoulli's move backfired. His reputation was so tarnished by the episode that he did not even receive credit for parts of his work that *were* original. Daniel went on to enjoy fame and a long career, becoming professor of physics in 1750 and lecturing to enthusiastic audiences until 1776.

EXERCISES

1. Use integration by parts to show that

$$\int_0^1 x^n (\log x)^n \, dx = \frac{(-1)^n n!}{(n+1)^{n+1}}$$

2. Deduce that

$$\int_0^1 x^x \, dx = 1 - \frac{1}{2^2} + \frac{1}{3^3} - \frac{1}{4^4} + \cdots$$

using a series expansion of $x^x = e^{x \log x}$.

CHAPTER 13

Complex Numbers in Algebra

13.1. Impossible Numbers

Over the last few chapters it has often been claimed that certain mysteries—
de Moivre's formula for $\sin n\theta$ (Section 5.6), the factorization of polynomials (Section 5.7), the classification of cubic curves (Section 7.4), branch
points (Section 9.5), genus (Section 10.3), and the behavior of elliptic functions (Sections 10.6 and 11.6)—are clarified by the introduction of complex numbers. That complex numbers do all this and more is one of the
miracles of mathematics. At the beginning of their history, complex numbers
$a + b\sqrt{-1}$ were considered to be "impossible numbers," tolerated only
in a limited algebraic domain because they seemed useful in the solution
of cubic equations. But their significance turned out to be geometric and
ultimately led to the unification of algebraic functions with conformal
mapping, potential theory, and another "impossible" field, noneuclidean
geometry. This resolution of the paradox of $\sqrt{-1}$ was so powerful, unexpected, and beautiful that only the word "miracle" seems adequate to
describe it.

In the present chapter we shall see how complex numbers emerged from the
theory of equations and enabled its fundamental theorem to be proved—at
which point it became clear that complex numbers had meaning far beyond
algebra. Their impact on curves and function theory, which is where conformal
mapping and potential theory come in, will be described in Chapters 14 and
15. Noneuclidean geometry had entirely different origins but arrived at the
same place as function theory in the 1880s, thanks to complex numbers. This
unexpected meeting will be described in Chapter 17, after some geometric
preparations in Chapter 16.

13.2. Quadratic Equations

The usual way to introduce complex numbers in a mathematics course is to point out that they are needed to solve certain quadratic equations, such as the equation $x^2 + 1 = 0$. However, this did not happen when quadratic equations first appeared, since at that time there was no *need* for all quadratic equations to have solutions. Many quadratic equations are implicit in Greek geometry, as one would expect when circles, parabolas, and the like, are being investigated, but one does not demand that every geometric problem have a solution. If one asks whether a particular circle and line intersect, say, then the answer can be yes or no. If yes, the quadratic equation for the intersection has a solution; if no, it has no solution. An "imaginary solution" is uncalled for in this context.

Even when quadratic equations appeared in algebraic form, with Diophantus and the Arab mathematicians, there was initially no reason to admit complex solutions. One still wanted to know only whether there were real solutions, and if not the answer was simply—no solution. This is plainly the appropriate answer when quadratics are solved by geometrically completing the square (Section 5.2), as was still done up to the time of Cardano. A square of negative area did not exist in geometry. The story might have been different had mathematicians used symbols more and dared to consider the symbol $\sqrt{-1}$ as an object in its own right, but this did not happen until quadratics had been overtaken by cubics, at which stage complex numbers became unavoidable, as we shall now see.

13.3. Cubic Equations

The del Ferro–Tartaglia–Cardano solution of the cubic equation

$$y^3 = py + q$$

is

$$y = \sqrt[3]{\frac{q}{2} + \sqrt{\left(\frac{q}{2}\right)^2 - \left(\frac{p}{3}\right)^3}} + \sqrt[3]{\frac{q}{2} - \sqrt{\left(\frac{q}{2}\right)^2 - \left(\frac{p}{3}\right)^3}}$$

as we saw in Section 5.5. The formula involves complex numbers when $(q/2)^2 - (p/3)^3 < 0$. However, it is not possible to dismiss this as a case with no solution, *because a cubic always has at least one real root* (since $y^3 - py - q$ is positive for sufficiently large positive y and negative for sufficiently large negative y). Thus the Cardano formula raises the problem of reconciling a real value, found by inspection, say, with an expression of the form

$$\sqrt[3]{a + b\sqrt{-1}} + \sqrt[3]{a - b\sqrt{-1}}$$

Cardano did not face up to this problem in his *Ars Magna* [1545]. He did, it is true, once mention complex numbers, but in connection with a quadratic equation, and accompanied by the comment that these numbers were "as subtle as they are useless" (Cardano [1545], Ch. 37, Rule II).

The first work to take complex numbers seriously and achieve the necessary reconciliation, was Bombelli [1572]. Bombelli worked out the formal algebra of complex numbers, with the particular aim of reducing expressions $\sqrt[3]{a + b\sqrt{-1}}$ to the form $c + d\sqrt{-1}$. His method enabled him to show the reality of some expressions resulting from Cardano's formula. For example, the solution of

$$x^3 = 15x + 4$$

is

$$x = \sqrt[3]{2 + 11\sqrt{-1}} + \sqrt[3]{2 - 11\sqrt{-1}}$$

according to the formula. On the other hand, inspection gives the solution

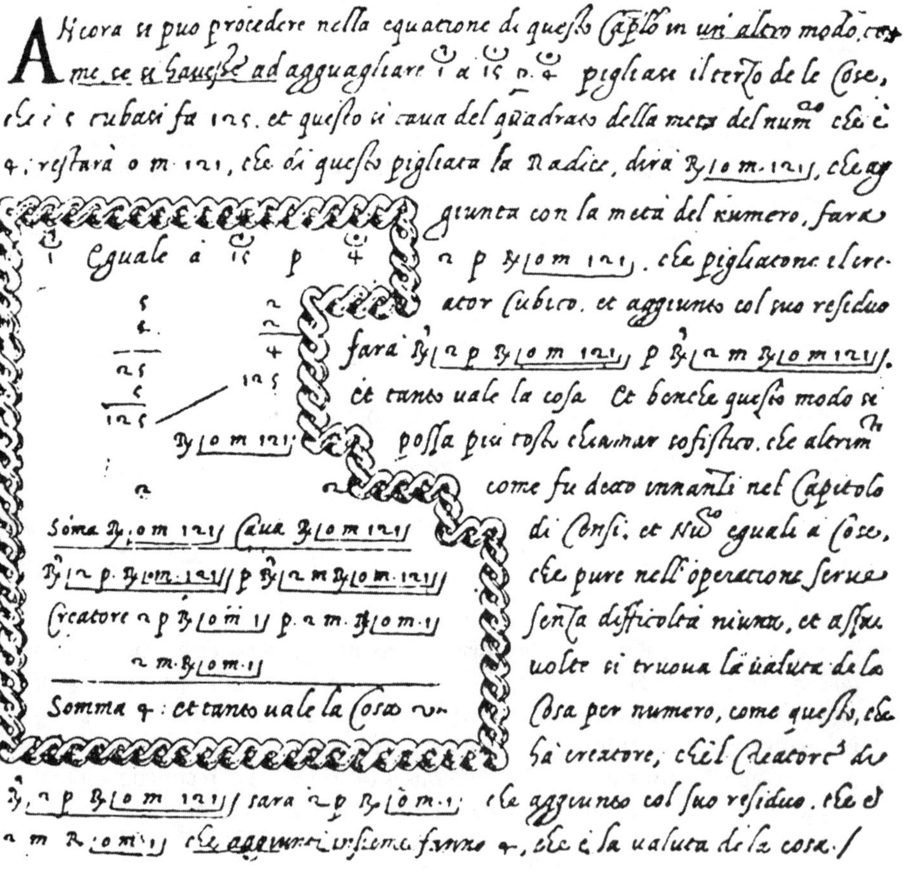
Ancora si può procedere nella equatione di questo Capitolo in un'altro modo, come se si haueße ad agguagliare 1 cubo a 15 tanti p. 4 pigliasi il terzo de le Cose, che è 5 cubasi fa 125. et questo si caua del quadrato della metà del numero che è 4; restarà 0 m. 121, che di questo pigliata la Radice, dirà R[0 m. 121], che aggiunta con la metà del numero, farà 2 p R[0 m 121]. che pigliatone il creator Cubico. et aggiunto col suo residuo farà R³[2 p R[0 m 121]] p R³[2 m R[0 m 121]]. et tanto uale la cosa. Et benche questo modo se possa piu tosto chiamar sofistico. che altrimenti come fu detto innanti nel Capitolo di Cubi. et Numero eguali a Cose, che pure nell'operatione serue senza difficoltà niuna, et assai uolte si truoua la ualuta de la Cosa per numero, come questo, che ha creatore, che il Creatore di R³[2 p R[0 m 121]] sarà 2 p R[0 m 1]; che aggiunto col suo residuo. che è 2 m R[0 m 1] che aggiunti insieme fanno 4, che è la ualuta de la cosa.

Figure 13.1

$x = 4$. Bombelli had the hunch that the two parts of x in the Cardano formula were of the form $2 + n\sqrt{-1}$, $2 - n\sqrt{-1}$, and he found by cubing these expressions formally (using $(\sqrt{-1})^2 = -1$) that indeed

$$\sqrt[3]{2 + 11\sqrt{-1}} = 2 + \sqrt{-1}$$
$$\sqrt[3]{2 - 11\sqrt{-1}} = 2 - \sqrt{-1}$$

hence the Cardano formula also gives $x = 4$.

Figure 13.1 is a facsimile of the manuscript page on which Bombelli stated his result. It is not hard to pick out the preceding expressions when one allows for the notation and the fact that $11\sqrt{-1}$ is written as $\sqrt{0 - 121}$.

Much later, Hölder [1896] showed that any algebraic formula for the solution of the cubic must involve square roots of quantities which become negative for particular values of the coefficients. A proof of Hölder's result may be found in van der Waerden [1949], p. 180.

13.4. Wallis' Attempt at Geometric Representation

Despite Bombelli's successful use of complex numbers, most mathematicians regarded them as impossible, and of course even today we call them "imaginary" and use the symbol i for the imaginary unit $\sqrt{-1}$. The first attempt to give complex numbers a concrete interpretation was made by Wallis [1673]. This attempt was unsatisfactory, as we shall see, but nevertheless an interesting "near miss." Wallis wanted to give a geometric interpretation to the roots of the quadratic equation, which we shall write as

$$x^2 + 2bx + c^2 = 0, \qquad b, c \geqslant 0$$

The roots are

$$x = -b \pm \sqrt{b^2 - c^2}$$

and hence real when $b \geqslant c$. In this case the roots can be represented by points P_1, P_2 on the real number line which are determined by the geometric construction in Figure 13.2. When $b < c$, lines of length b attached to Q are too short to reach the number line, so the points P_1, P_2 "cannot be had in the line," and Wallis seeks them "out of that line ... (in the same Plain)." He is on the right track, but he arrives at unsuitable positions for P_1, P_2 by sticking too closely to his first construction. Figure 13.3 compares his representation of P_1, $P_2 = -b \pm i\sqrt{c^2 - b^2}$ when $b < c$ with the modern representation. Apparently Wallis thought $+$ and $-$ should continue to correspond to "right" and "left," though this has the unacceptable consequence that $i = -i$ (let $b \to 0$ in his representation). This was an understandable oversight, since in Wallis' time even negative numbers were still under suspicion, and there was confusion about the meaning of $(-1) \times (-1)$, for example. Confusion was compounded by the introduction of square roots, and as late as 1770 Euler gave a "proof" in his *Algebra* that $\sqrt{-2} \times \sqrt{-3} = \sqrt{6}$ (Euler [1770], p. 43).

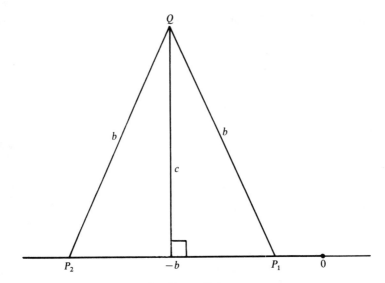

Figure 13.2

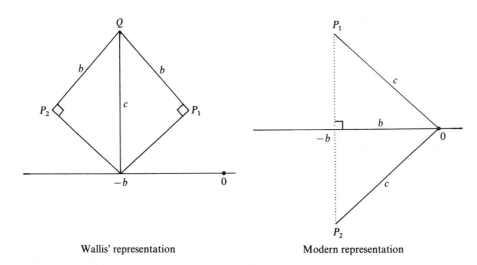

Wallis' representation   Modern representation

Figure 13.3

13.5. Angle Division

In Section 5.6 we saw how Viète related angle trisection to the solution of cubic equations, and how Leibniz [1975] and de Moivre [1707] solved the angle n-section equation by the Cardano-type formula

$$x = \frac{1}{2}\sqrt[n]{y + \sqrt{y^2 - 1}} + \frac{1}{2}\sqrt[n]{y - \sqrt{y^2 - 1}} \tag{1}$$

We also saw how this and Viète's formulas for $\cos n\theta$ and $\sin n\theta$ could easily be explained by the formula

$$(\cos\theta + i\sin\theta)^n = \cos n\theta + i\sin n\theta \qquad (2)$$

usually associated with de Moivre. Actually, de Moivre never stated (2) explicitly. The closest he came was to give a formula for $(\cos\theta + i\sin\theta)^{1/n}$ ([1730]; see Smith [1929] for a series of extracts from de Moivre's works on angle division). It seems that the clues in the algebra of circular functions were not strong enough to reveal (2) until a deeper reason for it had been brought to light by calculus.

Complex numbers made their entry into the theory of circular functions in a paper on integration by John Bernoulli [1702]. Observing that $\sqrt{-1} = i$ makes possible the partial fraction decomposition

$$\frac{1}{1 + z^2} = \frac{1/2}{1 + zi} + \frac{1/2}{1 - zi}$$

he saw that integration would give an expression for $\tan^{-1}z$ as an imaginary logarithm, though he did not write down the expression in question and was evidently puzzled as to what it could mean. In Section 15.1 we shall see how Euler clarified John Bernoulli's discovery and developed it into the beautiful theory of complex logarithms and exponentials. What is relevant here is that John Bernoulli took up the idea again [1712], and this time he carried out the integration to obtain an algebraic relation between $\tan n\theta$ and $\tan\theta$. His argument is as follows. Given

$$y = \tan n\theta, \qquad x = \tan\theta$$

we have

$$n\theta = \tan^{-1}y = n\tan^{-1}x$$

hence, taking differentials,

$$\frac{dy}{1 + y^2} = \frac{n\,dx}{1 + x^2}$$

or

$$dy\left(\frac{1}{y + i} - \frac{1}{y - i}\right) = n\,dx\left(\frac{1}{x + i} - \frac{1}{x - i}\right)$$

Integration gives

$$\log(y + i) - \log(y - i) = n\log(x + i) - n\log(x - i)$$

that is,

$$\log\frac{y + i}{y - i} = \log\left(\frac{x + i}{x - i}\right)^n$$

whence

$$(x - i)^n(y + i) = (x + i)^n(y - i) \qquad (3)$$

This formula was the first of the de Moivre type actually to use i explicitly and the first example of a phenomenon later articulated by Hadamard: the shortest route between two truths in the real domain sometimes passes through the complex domain. Solving (3) for y as a function of x expresses $\tan n\theta$ as a rational function of $\tan \theta$, which is difficult to obtain using real formulas alone. In fact, it is easy to show from (3) that y is the quotient of the polynomials consisting of alternate terms in $(x + 1)^n$, provided with alternate $+$ and $-$ signs (see exercise 1).

During the eighteenth century, mathematicians were ambivalent about $\sqrt{-1}$. They were willing to use it en route to results about real numbers but doubted whether it had a concrete meaning of its own. Cotes [1714] even used $a + \sqrt{-1}b$ to represent the point (a, b) in the plane (as Euler did later), apparently without noticing that (a, b) was a valid *interpretation* of $a + \sqrt{-1}b$. Since results about $\sqrt{-1}$ were suspect, they were often left unstated when it was possible to state an equivalent result about reals. This may explain why de Moivre stated (1) but not (2). Another example of the avoidance of results about $\sqrt{-1}$ is the remarkable theorem on the regular n-gon discovered by Cotes in 1716 and published posthumously in Cotes [1722].

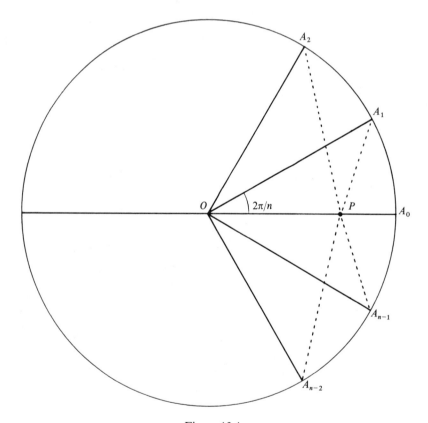

Figure 13.4

If $A_0, \ldots, A_{n-1}$ are equally spaced points on the unit circle with center O, and if P is a point on OA_0 such that $OP = x$, then (Fig. 13.4)

$$PA_0 \cdot PA_1 \cdot \cdots \cdot PA_{n-1} = 1 - x^n$$

This theorem not only relates the regular n-gon to the polynomial $x^n - 1$ but in fact geometrically realizes the *factorization of $x^n - 1$ into real linear and quadratic factors.* By symmetry one has $PA_1 = PA_{n-1}, \ldots$, hence

$$PA_0 \cdot PA_1 \cdots PA_{n-1} = \begin{cases} PA_0 \cdot PA_1^2 \cdot PA_2^2 \cdots PA_{(n-1)/2}^2 & n \text{ odd} \\ PA_0 \cdot PA_1^2 \cdot PA_2^2 \cdots PA_{n/2-1}^2 PA_{n/2} & n \text{ even} \end{cases}$$

$PA_0 = 1 - x$ is a real linear factor, as is $PA_{n/2}$ when n is even, and it follows from the cosine rule in triangle OPA_k that

$$PA_k^2 = 1 - 2x \cos \frac{2k\pi}{n} + x^2$$

The easiest route from here to the theorem is by splitting PA_k^2 into complex linear factors and using de Moivre's theorem, though we can only speculate that this was Cotes' method, since he stated the theorem without proof. There is a second half to Cotes' theorem, which similarly decomposes $1 + x^n$ into real linear and quadratic factors. These factorizations were needed to integrate $1/(1 \pm x^n)$ by resolution into partial fractions, which was in fact Cotes' main objective. Such problems were then high on the mathematical agenda, and they motivated subsequent research into the factorization of polynomials, in particular the first attempts to prove the fundamental theorem of algebra.

EXERCISE

1. Show that if $y = \tan 4\theta$, $x = \tan \theta$, then

$$y = \frac{x^4 - 6x^2 + 1}{4x^3 - 4x}$$

(John Bernoulli [1712]).

13.6. The Fundamental Theorem of Algebra

The fundamental theorem of algebra is the statement that every polynomial equation $p(z) = 0$ has a solution in the complex numbers. As Descartes observed (Section 5.7), a solution $z = a$ implies that $p(z)$ has a factor $z - a$. The quotient $q(z) = p(z)/(z - a)$ is then a polynomial of lower degree; hence if every polynomial equation has a solution, we can also extract a factor from $q(z)$, and if $p(z)$ has degree n, we can go on to factorize $p(z)$ into n linear factors. The existence of such a factorization is of course another way to state the fundamental theorem.

It was observed by d'Alembert [1746] that if $z = u + iv$ is a solution of $p(z) = 0$, then so is its conjugate $\bar{z} = u - iv$. Thus the imaginary linear factors of $p(z)$ can always be combined in pairs to form real quadratic factors:

$$(z - u - iv)(z - u + iv) = z^2 - 2uz + (u^2 + v^2)$$

This gave another equivalent of the fundamental theorem, at least for polynomials with real coefficients: each polynomial $p(z)$ can be expressed as a product of real linear and quadratic factors. The theorem was usually stated in this way during the eighteenth century, when its main purpose was to make possible the integration of rational functions (see Section 13.5). This also avoided mention of $\sqrt{-1}$.

It has often been said that attempts to prove the fundamental theorem began with d'Alembert [1746], and that the first satisfactory proof was given by Gauss [1799]. This opinion should not be accepted without question, as the source of it is Gauss himself. Gauss [1799] gave a critique of proofs from d'Alembert on, showing that they all had serious weaknesses, then offered a proof of his own. His intention was to convince readers that the new proof was the first valid one, even though it used one unproved assumption (which will be discussed further in the next section). The opinion as to which of two incomplete proofs is more convincing can of course change with time, and I believe that Gauss [1799] might be judged differently today. We can now fill the gaps in d'Alembert [1746] by appeal to standard methods and theorems, whereas there is still no easy way to fill the gap in Gauss [1799].

Both proofs depend on the geometric properties of the complex numbers and the concept of continuity for their completion. The basic geometrical insight—that the complex number $x + iy$ can be identified with the point (x, y) in the plane—mysteriously eluded all mathematicians until the end of the eighteenth century. This was one of the reasons that d'Alembert's proof was unclear, and the use of this insight by Argand [1806] was an important step in d'Alembert's reinstatement. Gauss seems to have had the same insight but concealed its role in his proof, perhaps believing that his contemporaries were not ready to view the complex numbers as a plane.

As for the concept of continuity, neither Gauss nor d'Alembert understood it very well. Gauss [1799] seriously understated the difficulties involved in the unproved step, claiming that "no one, to my knowledge, has ever doubted it. But if anybody desires it, then on another occasion I intend to give a demonstration which will leave no doubt" (translation from Struik [1969], p. 121). Perhaps to preempt criticism, he gave a second proof [1816], in which the role of continuity was minimized. The second proof is purely algebraic except for use of a special case of the intermediate value theorem. Gauss assumed that a polynomial function $p(x)$ of a real variable x takes all values between $p(a)$ and $p(b)$ as x runs from a to b. The first to appreciate the importance of continuity for the fundamental theorem of algebra was Bolzano [1817], who proved the continuity of polynomial functions and attempted a proof of the intermediate value theorem. The latter proof was unsatisfactory because

Bolzano had no clear concept of real number on which to base it, but it did point in the right direction. When a definition of real number emerged in the 1870s (e.g., with Dedekind cuts; Section 4.2), Weierstrass [1874] rigorously established the basic properties of continuous functions, such as the intermediate value theorem and extreme value theorem. This completed not only the second proof of Gauss but also the proof of d'Alembert, as we shall see in the next section.

EXERCISE

1. Show that $p(\bar{z}) = \overline{p(z)}$ for any polynomial $p(z)$ with real coefficients. Deduce that the complex roots of $p(z) = 0$ occur in conjugate pairs.

13.7. The Proofs of d'Alembert and Gauss

The key to d'Alembert's proof is a proposition now known as *d'Alembert's lemma*: if $p(z)$ is a polynomial function and $p(z_0) \neq 0$, then any neighborhood of z_0 contains a point z_1 such that $|p(z_1)| < |p(z_0)|$.

The proof of this lemma offered by d'Alembert depended on solving the equation $w = p(z)$ for z as a fractional power series in w. As mentioned in Section 9.5, such a solution was claimed by Newton [1671], but it was only made clear and rigorous by Puiseux [1850]. Thus d'Alembert's argument did not stand on solid ground, and in any case it was unnecessarily complicated.

A simple elementary proof of d'Alembert's lemma was given by Argand [1806]. Argand was one of the co-discoverers of the geometric representation of complex numbers (probably the first was Wessel [1797], but his work remained almost unknown for 100 years), and he offered the following proof as an illustration of the effectiveness of the representation.

The value of $p(z_0) = x_0 + iy_0$ is interpreted as the point (x_0, y_0) in the plane, so that $|p(z_0)|$ is the distance of (x_0, y_0) from the origin. We wish to find a Δz such that $p(z_0 + \Delta z)$ is nearer to the origin than $p(z_0)$. If

$$p(z) = a_0 z^n + a_1 z^{n-1} + \cdots + a_n$$

then

$$p(z_0 + \Delta z) = (a_0 z_0^n + a_1 z_0^{n-1} + \cdots + a_n)$$
$$+ \Delta z(na_0 z_0^{n-1} + (n-1)a_1 z_0^{n-2} + \cdots + a_{n-1})$$
$$+ \text{ terms in } (\Delta z)^2, (\Delta z)^3, \ldots$$
$$= p(z_0) + A\Delta z + \varepsilon$$

where A is constant and $|\varepsilon|$ is small compared to $|A\Delta z|$ when $|\Delta z|$ is small. It is then clear (Fig. 13.5) that by choosing the direction of Δz so that $A\Delta z$ is opposite in direction to $p(z_0)$, we get $|p(z_0 + \Delta z)| < |p(z_0)|$. This completes the proof of d'Alembert's lemma.

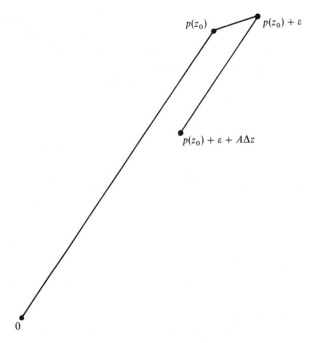

Figure 13.5

To complete the proof of the fundamental theorem of algebra, take an arbitrary polynomial p and consider the continuous function $|p(z)|$. Since $p(z) \simeq a_0 z^n$ for $|z|$ large, $|p(z)|$ increases with $|z|$ outside a sufficiently large circle $|z| = R$. We now get a z for which $|p(z)| = 0$ from the extreme value theorem of Weierstrass [1874]: a continuous function on a closed bounded set assumes maximum and minimum values. By this theorem, $|p(z)|$ takes a minimum value for $|z| \leqslant R$. The minimum is $\geqslant 0$ by definition, and if it is > 0 we get a contradiction by d'Alembert's lemma: either a point z with $|z| \leqslant R$ where $|p(z)|$ takes a value less than its minimum or a point z with $|z| > R$ where $|p(z)|$ is less than its values on $|z| = R$. Thus there is a point z where $|p(z)| = 0$ and hence $p(z) = 0$.

The proof of Gauss also used the fact that $p(z)$ behaves like its highest degree term $a_0 z^n$ for $|z|$ large and likewise relied on a continuity argument to show that $p(z) = 0$ inside some circle $|z| = R$. Gauss considered the real and imaginary parts of $p(z)$, $\text{Re}[p(z)]$, and $\text{Im}[p(z)]$ and investigated the curves

$$\text{Re}[p(z)] = 0 \quad \text{and} \quad \text{Im}[p(z)] = 0$$

(These are easily seen to be algebraic curves $p_1(x, y) = 0$ and $p_2(x, y) = 0$ by expanding the powers $z^k = (x + iy)^k$ and collecting real and imaginary terms.) His aim was to find a point where these curves meet, because at such

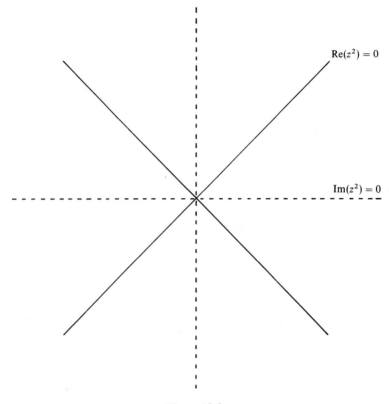

$$\text{Re}(z^2) = 0$$

$$\text{Im}(z^2) = 0$$

Figure 13.6

a point

$$0 = \text{Re}[p(z)] = \text{Im}[p(z)] = p(z)$$

For $|z|$ large, the curves are close to the curves $\text{Re}(a_0 z^n) = 0$ and $\text{Im}(a_0 z^n) = 0$, which are families of straight lines through the origin. Moreover, the lines where $\text{Re}(a_0 z^n) = 0$ *alternate* with those where $\text{Im}(a_0 z^n) = 0$ as one makes a circuit around the origin. For example, Figure 13.6 shows $\text{Re}(z^2) = 0$ and $\text{Im}(z^2) = 0$ as alternate solid and dashed lines. It follows that the curves $\text{Re}[p(z)] = 0$ and $\text{Im}[p(z)] = 0$ meet a sufficiently large circle $|z| = R$ *alternately.* Up to this point the argument is comparable to d'Alembert's lemma, and it can be made just as rigorous.

To complete the proof we only have to show that the curves meet inside the circle, and this is the step Gauss thought nobody could doubt. He assumed that the separate pieces of the algebraic curve $\text{Re}[p(z)] = 0$ outside the circle $|z| = R$ would join up inside the circle, as would the separate pieces of $\text{Im}[p(z)] = 0$. Then, since the pieces of $\text{Re}[p(z)] = 0$ alternate with those of

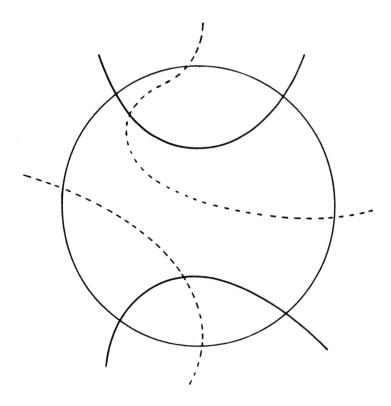

Figure 13.7

$\mathrm{Im}[p(z)]$ on $|z| = R$, it would be "patently absurd" for their connecting pieces inside the circle not to meet. One has only to visualize a situation like that seen in Figure 13.7 to feel sure that Gauss was right. However, the existence of the connecting pieces is extremely hard to prove (and proving that they meet is not trivial either, being at least as hard as the intermediate value theorem). The first proof was given by Ostrowski [1920].

From our present perspective, d'Alembert's route to the fundamental theorem of algebra seems basically easy because it proceeds through general properties of continuous functions. The route of Gauss, although appearing equally easy from a distance, goes through the still-unfamiliar territory of real algebraic curves. The intersections of real algebraic curves are harder to understand than the intersections of complex algebraic curves, and in retrospect they are harder to understand than the fundamental theorem of algebra. Indeed, as we shall see in the next chapter, the fundamental theorem gives us Bézout's theorem, which in turn settles the problem of counting the intersections of complex algebraic curves.

13.8. Biographical Notes: d'Alembert

Jean Le Rond d'Alembert (Fig. 13.8) was born in Paris in 1717 and died there in 1783. He was the illegitimate son of the Chevalier Destouches-Canon, a cavalry officer, and salon hostess Madame de Tencin. His mother abandoned him at birth near the church of St. Jean-le-Rond in the cloisters of Notre

Figure 13.8. Jean Baptiste le Rond d'Alembert.

Dame, and so he was christened Jean Le Rond, following the custom for foundlings. He was subsequently located by his father, who found a home for him with a glazier named Rousseau and his wife. The name d'Alembert came later, for reasons which are unclear.

The Rousseaus must have been devoted foster parents, as d'Alembert lived with them until 1765. He received an annuity from his father, who also arranged for him to be educated at the Jansenist Collège de Quatre-Nations in Paris. There he received a good grounding in mathematics and developed a permanent distaste for theology. After brief studies in law and medicine he turned to mathematics in 1739.

In that year he began sending communications to the Académie des Sciences, and his ambition and talent rapidly carried him to fame. He became a member of the Académie in 1741 and published his best known work, the *Traité de dynamique* in 1743. Having struggled to the top from humble beginnings, d'Alembert did not want to lose his position. Once in the Academy, his struggle was to stay ahead of his rivals. Whether by accident or inborn competitiveness, d'Alembert always seemed to be working on the same problems as other top mathematicians—initially Clairaut, later Daniel Bernoulli and Euler. He was always fearful of losing priority and fell into a cycle of hasty publication followed by controversy over the meaning and significance of his work. Despite the fact that he was an excellent writer (elected to the Académie Française in 1754), his mathematics was almost always poorly presented. Many of his best ideas were not understood until Euler overhauled them and gave them masterly expositions. Since Euler often did this without giving credit, d'Alembert was understandably furious, but he squandered his energy in quarreling instead of giving his own work the exposition it deserved.

Another reason for d'Alembert's lack of attention to his mathematics was his involvement in the broader intellectual life of his time. When d'Alembert came on the scene in the 1740s, mathematics was enjoying great prestige in philosophical circles, largely because of Newton's success in explaining the motions of the planets. Mathematics was a model of rational inquiry which, it was hoped, would allow the proper organization of all knowledge and the proper conduct of all human affairs. The movement to reorganize knowledge and conduct along rational lines became known as the Enlightenment, and it was particularly strong in France, where philosophers saw it as a means to overthrow existing institutions, particularly the church. Around 1745 d'Alembert became immersed in the ferment of the Enlightenment, then bubbling in the salons and cafés of Paris. He made friends with the leading lights—Diderot, Condillac, Rousseau—and was in demand at the most fashionable salons for his wit and gift for mimicry.

The Enlightenment was not all talk, however, and one of its most solid achievements was the 17-volume *Encyclopédie*, edited by Diderot between 1745 and 1772. d'Alembert wrote the introduction to the *Encyclopédie*, the *Discours préliminaire*, and in it summed up his views on the unity of all knowledge. It contributed greatly to the success of the *Encyclopédie*, and was the main reason for his election to the Académie Française. D'Alembert was also scientific editor and wrote many of the mathematics articles. Eventually a split developed among the encyclopedists between the extreme materialists, led by Diderot, and the more moderate faction of Voltaire. Diderot leaned toward biology, for which he conjectured an absurd pseudomathematical basis, while deploring the "impracticality" of ordinary mathematics. D'Alembert sided with Voltaire and cut his ties with the *Encyclopédie* in 1758.

Nevertheless, intellectual fashion was moving away from mathematics, and in the 1760s d'Alembert found himself with only one philosopher friend still interested in it, the probability theorist Condorcet. At about this time d'Alembert met the one love of his life, Julie de Lespinasse. Julie was the cousin of Madame du Deffand, whose salon d'Alembert attended. After a quarrel over poaching the salon's members, Julie set up a salon of her own, with d'Alembert's help. When Julie became ill with smallpox, d'Alembert nursed her back to health; when he himself fell sick, she persuaded him to move in with her. This was in 1765, when he finally left his foster home. For the next 10 years his life revolved around Julie's salon, and her death in 1776 came as a cruel blow. Humiliation was added to sorrow when he discovered from her letters that she had been passionately involved with other men throughout their time together.

D'Alembert spent his last seven years in a small apartment in the Louvre, to which he was entitled as permanent secretary of the Académie Française. He found himself unable to work in mathematics, although it was the only thing that still interested him, and he became gloomy about the future of mathematics itself. Despite his gloom, he did what he could to support and encourage young mathematicians. Perhaps the finest achievement of d'Alembert's later years was to launch the careers of Lagrange and Laplace, whose work in mechanics ultimately completed much of his own. It must have given him some satisfaction to anticipate the future successes of his gifted protégés, even though they effectively ended the theory of mechanics as d'Alembert knew it. What he could not have anticipated was that a minor element of his work, the use of complex numbers, would blossom in the next century (see Sections 15.1 and 15.2), and that mathematics would break out of the bounds set by eighteenth-century thinking.

CHAPTER 14

Complex Numbers and Curves

14.1. Roots and Intersections

There is a close connection between intersections of algebraic curves and roots of polynomial equations, going back as far as Menaechmus' construction of $\sqrt[3]{2}$ (a root of the equation $x^3 = 2$) by intersecting a parabola and a hyperbola (Section 2.4). The most direct connection, of course, occurs in the case of a polynomial curve

$$y = p(x) \tag{1}$$

whose intersections with the axis $y = 0$ are just the real roots of the equation

$$p(x) = 0 \tag{2}$$

If (2) has k real roots, then the curve (1) has k intersections with the axis $y = 0$. Here we must count intersections the same way we count roots, according to *multiplicity*. A root r of (2) has multiplicity μ if the factor $(x - r)$ occurs μ times in $p(x)$, and the root r is then counted μ times. This way of counting is also geometrically natural because if, for example, the curve $y = p(x)$ meets the axis $y = 0$ with multiplicity 2 at 0, then a line $y = \varepsilon x$ "close" to the axis meets the curve twice: once near the intersection with the axis and once precisely there. The intersection with $y = 0$ (Fig. 14.1) can therefore be considered as two coincident points to which the distinct intersections with $y = \varepsilon x$ tend as $\varepsilon \to 0$. Likewise, an intersection of multiplicity 3 can be explained as the limit of three distinct intersections, for example, of $y = \varepsilon x$ with $y = x^3$ (Fig. 14.2).

At first glance this idea seems to break down with multiplicity 4, since $y = \varepsilon x$ meets $y = x^4$ at only two points, $x = 0$ and $x = \sqrt[3]{\varepsilon}$. The explanation is that there are also two complex roots in this case (the two complex values of $\sqrt[3]{\varepsilon}$), hence we cannot neglect complex roots if we want to get the geometrically "correct" number of intersections.

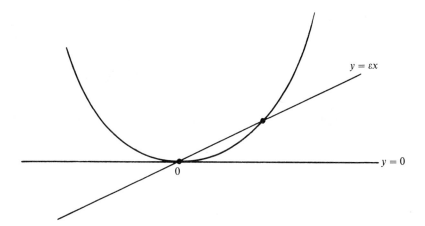

Figure 14.1

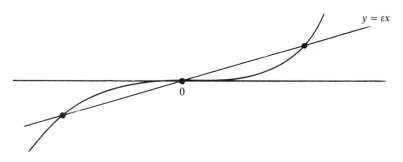

Figure 14.2

The fundamental theorem of algebra (Section 13.6) gives us n roots of an nth-degree equation (2) and consequently n intersections of the polynomial curve (1) with the axis $y = 0$. To get n roots, however, we have to admit complex values of x, hence we have to consider "curves" for which x and y are complex in order to obtain n intersections. This, and other tidy consequences of the fundamental theorem of algebra (e.g., the "coincident point" interpretation of multiplicity; see exercise 1), persuaded eighteenth-century mathematicians to admit complex numbers into the theory of curves before complex numbers themselves were understood—and even before the fundamental theorem of algebra was proved.

The most elegant consequence was Bézout's theorem that a curve C_m of degree m meets a curve C_n of degree n at mn points. As we saw in Section 7.6, if homogeneous coordinates are used to take account of points at infinity, then the intersections of C_m and C_n correspond to the solutions of an equation $r_{mn}(x, y) = 0$ which is homogeneous of degree mn. We can now use the fundamental theorem of algebra to show that $r_{mn}(x, y)$ is the product of mn linear factors as follows:

$$r_{mn}(x, y) = y^{mn} r_{mn}\left(\frac{x}{y}, 1\right)$$

$$= y^{mn} \prod_{i=1}^{p}\left(b_i \frac{x}{y} - a_i\right) \qquad \text{for some } p \leqslant mn$$

by the fundamental theorem, since $r_{mn}(x/y, 1)$ is a polynomial of degree $p \leqslant mn$ in the single variable x/y. But then

$$r_{mn}(x, y) = y^{mn-p} \prod_{i=1}^{p}(b_i x - a_i y)$$

$$= \prod_{i=1}^{mn}(b_i x - a_i y)$$

since each factor y in front (if any) is trivially of the form $b_i x - a_i y$.

It follows that the equation $r_{mn}(x, y) = 0$ has mn solutions, and hence there are mn intersections of C_m and C_n, counting multiplicities.

EXERCISE

1. Show that $y = \varepsilon x$ meets $y = x^n$ in n distinct points.

14.2. The Complex Projective Line

We saw in Section 7.5 that adding a point at infinity to the real line $\mathbb{R}$ in $\mathbb{R} \times \mathbb{R}$ forms a closed curve which is qualitatively like a circle. Indeed, a real projective line in Klein's model of the real projective plane $P_2(\mathbb{R})$ has much the same geometric properties as a great circle on a sphere, after one allows for the fact that antipodal points on the sphere are the same point on $P_2(\mathbb{R})$. The situation with the complex "line" $\mathbb{C}$ is similar but more difficult to visualize. $\mathbb{C}$ is already two-dimensional, as we saw in Gauss' proof of the fundamental theorem of algebra, hence the complex "plane" $\mathbb{C} \times \mathbb{C}$ is four-dimensional and virtually impossible to visualize.

To avoid an excursion into four-dimensional space, we first revise our approach to the real projective line. In Section 7.5 we considered ordinary lines L in a horizontal plane, not passing through the origin, and extended each to a projective line whose "points" were lines through the origin and in the plane containing L. The nonhorizontal lines in this family corresponded to points of L, and the horizontal line in the family to the point at infinity of L. We now use this construction again to demonstrate directly the qualitative, or more precisely *topological*, equivalence between a projective line and a circle (Fig. 14.3).

The origin N is taken to be the top point of a circle which, at its bottom point, touches our line $L = \mathbb{R}$. There is a continuous one-to-one correspondence between lines through N and points of the circle. Each nonhorizontal

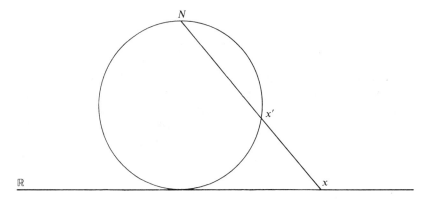

Figure 14.3

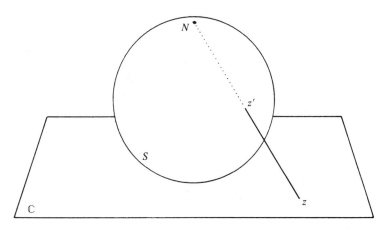

Figure 14.4

line corresponds to its intersection $x' \neq N$ with the circle, while the horizontal line corresponds to N itself. Thus the projective completion of $\mathbb{R}$, which we now call $P_1(\mathbb{R})$, is *topologically the same* as the circle, in the sense that there is a continuous one-to-one correspondence between them. Moreover, we can understand projective completion of $\mathbb{R}$ topologically as a process of adding one "point" which is "approached" as one tends to infinity, in either direction, along $\mathbb{R}$, for as x tends to infinity in either direction, x' tends to the same point, N, on the circle.

We can now view projective completion of $\mathbb{C}$ in the same way using Figure 14.4, which shows the so-called *stereographic projection* of the plane $\mathbb{C}$ into a sphere. Each point $z \in \mathbb{C}$ is projected to a point z' on the tangential sphere S by the ray through z and the north pole N of S. This establishes a continuous one-to-one correspondence between points z of $\mathbb{C}$ and points $z' \neq N$ on S. Moreover, as z tends to infinity in any direction, z' tends to N, hence the projective completion $P_1(\mathbb{C})$ of $\mathbb{C}$ is topologically the same as the complete sphere S, with the point at ∞ of $\mathbb{C}$ corresponding to N.

Since one also wants to complete $\mathbb{C}$ by a point ∞ in this way for complex analysis, geometry and analysis are both served by passing from $\mathbb{C}$ to $P_1(\mathbb{C})$. Gauss seems to have been the first to appreciate the advantages of $\mathbb{C} \cup \{\infty\}$ over $\mathbb{C}$, hence one often calls $P_1(\mathbb{C})$ the *Gauss sphere* in analysis. (Unfortunately, only a few unpublished, undated fragments of Gauss' work on this topic seem to have survived; see Gauss [c. 1819].) Algebraic geometers call $P_1(\mathbb{C})$ the (complex) projective line, since it is the formal equivalent of a real line, even though it is topologically a surface. Similarly, complex curves are topologically surfaces, known to analysts as *Riemann surfaces*, though algebraic geometers prefer to call them "curves."

The "surface" viewpoint is helpful when studying intrinsic properties of complex curves. For example, *genus* (introduced in connection with parametrization in Sections 10.3 to 10.5) turns out to have a very simple meaning in the topology of surfaces (see Section 14.4). On the other hand, the "curve" viewpoint is helpful when studying intersections of curves and their embedding in $\mathbb{C} \times \mathbb{C}$ or $P_2(\mathbb{C})$. Instead of trying to imagine the intersection of two planes in a single point of $\mathbb{C} \times \mathbb{C}$, for example, it is better to imagine the intersection as analogous to that of real lines in a real plane. After all, we are working with $\mathbb{C}$ to remove anomalies that occur with $\mathbb{R}$, not for the sake of doing something different, and we expect that much of the behavior of real curves will recur with complex ones.

EXERCISES

1. Show that the projective completion of the curve $Y = X^2$ is topologically a sphere by considering its parametrization

$$X = t$$
$$Y = t^2$$

where t ranges over the sphere $\mathbb{C} \cup \{\infty\}$. Namely, show that the mapping $t \mapsto (t, t^2)$ is one-to-one and continuous.

2. Similarly show that the projective completion of $Y^2 = X^3$ is topologically a sphere by considering its parametrization

$$X = t^2$$
$$Y = t^3$$

Interpret the mapping $t \mapsto (t^2, t^3)$ geometrically by considering the line $Y = tX$ and its intersection with the curve.

3. Consider the mapping of the t sphere onto the projective completion of $Y^2 = X^2(X + 1)$ defined by $t \mapsto P(t)$ where $P(t)$ is the third intersection of the curve with the line $Y = tX$ through the double point (Fig. 14.5). Show that this mapping is continuous and that it is one-to-one except at the points $t = \pm 1$, which are both mapped to the point 0 on the curve. Conclude that the curve is topologically the same as a sphere with two points identified (Fig. 14.6).

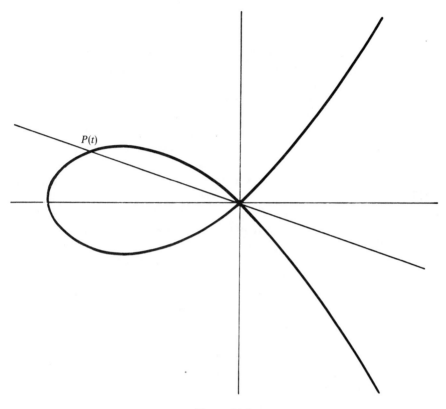

Figure 14.5

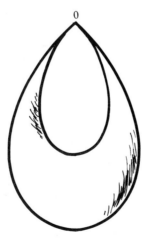

Figure 14.6

14.3. Branch Points

The key to the topological form of a complex curve $p(x, y) = 0$ lies in its *branch points*, the points α where the Newton–Puiseux expansion of y begins with a fractional power of $(x - \alpha)$ (cf. Section 9.5). The nature of branch points was first described by Riemann [1851] as part of a revolutionary new geometric theory of complex functions. Riemann's idea, one of the most illuminating in the history of mathematics, was to represent a relation $p(x, y) = 0$ between complex x and complex y by covering a plane (or sphere) representing the x variable by a surface representing the y variable, the point or points of the y surface over a given point $x = \alpha$ being those values of y that satisfy $p(\alpha, y) = 0$.

If the equation $p(\alpha, y) = 0$ is of degree n in y, there will in general be n distinct y values for a given α, consequently n "sheets" of the y surface lying over the x plane in the neighborhood of $x = \alpha$. At finitely many exceptional values of x, sheets merge due to concidence of roots, and the Newton–Puiseux theory says that at such a point y behaves like a fractional power of x at 0. Our main problem, therefore, is to understand the behavior of the Riemann surface for $y = x^{m/n}$ in the neighborhood of 0.

The idea can be grasped sufficiently well from seeing the special case $y = x^{1/2}$. If we consider the unit disk in the y plane and try to deform it so that the points $y = \pm\sqrt{x}$ lie above the point x in the unit disk of the x plane, then the result is something like Figure 14.7. The angles θ on the disk boundaries are the arguments of the corresponding points $e^{i\theta}$.

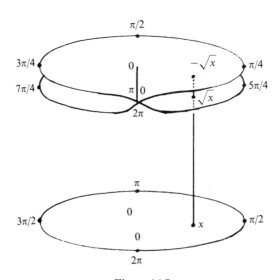

Figure 14.7

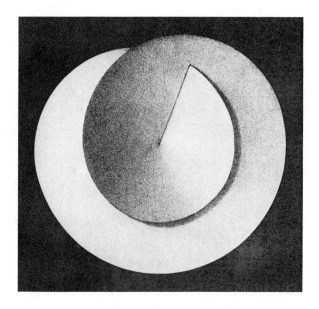

Figure 14.8

If

$$x = e^{i\theta} = e^{i(\theta + 2\pi)}$$

then

$$y = e^{i\theta/2}, \qquad e^{i(\theta/2 + \pi)}$$

giving the values shown. A more graphic depiction of the branch point is seen in Figure 14.8, taken from an early textbook on Riemann's theory (Neumann [1865], endpaper).

It should be mentioned that the awkward appearance of the branch point, in particular the line of self-intersection, is a consequence of representing the relation $y^2 = x$ in fewer dimensions than the four it really requires. If we similarly attempt to represent the relation $y^2 = x$ between real x and y by laying the y axis along the x axis so that $y = \pm\sqrt{x}$ are on top of x, then the result is an awkward folded "branch point" at 0 (Fig. 14.9). This is a consequence of trying to represent the relation in one dimension. In reality, as the second part of the figure shows, when viewed as a curve in the plane the relation is just as smooth at 0 as anywhere else. (Notice, incidentally, that the folded line in Fig. 14.9, the real y axis, corresponds to the self-intersection line in Fig. 14.8.)

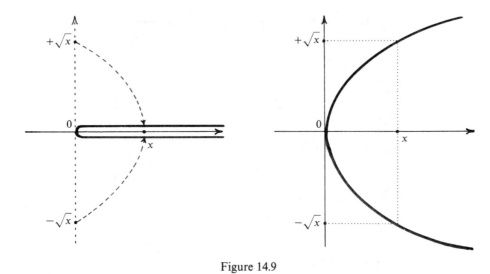

Figure 14.9

14.4. Topology of Complex Projective Curves

To understand the complete structure of the complex projective curve defined by $y^2 = x$ we need to know its behavior at infinity. At ∞ there is another branch point like the one at 0 (just replace x by $1/u$ and y and $1/v$ and notice that we are looking at $v^2 = u$ near $u = 0, v = 0$—the same situation as before). The topological nature of the relation between x and y can then be captured by the model seen in Figure 14.10. A sphere (the x sphere) is covered by two spheres (like skins of an onion), slit along a line from 0 to ∞ and cross-joined. The slit from 0 to ∞ is arbitrary but the cross-joining is necessary to produce the branch point structure at 0 and ∞.

The covering of the x sphere by this two sheeted surface expresses the "covering projection map" $(x, y) \mapsto x$ from a general point on the curve $y^2 = x$ to its x coordinate and shows that it is 2-to-1 except at the branch points 0, ∞. The two-sheeted surface itself captures the intrinsic topological structure of the curve, and this structure can be more readily seen by separating the two skins from the x sphere and each other, then joining the required edges (Fig. 14.11). Edges to be joined are labeled by the same letters, and we see that the resulting surface is topologically a sphere.

This result could have been obtained more directly by projecting each point (x, y) on the curve to y, since this is a 1-to-1 continuous map between the curve and the y axis, which we know to be topologically a sphere (when ∞ is included). The curve here was modeled by cutting and joining sheets on the sphere because this method extends to all algebraic curves. The Newton–Puiseux theorem implies that any algebraic relation $p(x, y) = 0$ can be modeled by a finite-sheeted covering of the sphere, with finitely many branch points.

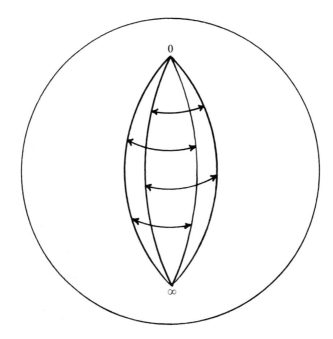

Figure 14.10

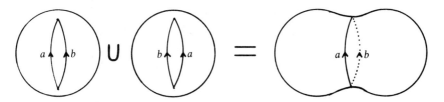

Figure 14.11

The most general branch point structure is given by a prescription for cross-joining (permuting) the sheets, and by slitting the sheets between branch points (or, if necessary, to an auxiliary point) they can be rejoined to produce the prescribed branching behavior.

The most interesting case of this method is the cubic curve

$$y^2 = x(x - \alpha)(x - \beta)$$

This relation defines a covering of the x sphere which is two sheeted, since for each x there are $+$ and $-$ values for y, with branch points at 0, α, β, and ∞. (The branch point at ∞ is explained in the exercises below.) Thus if we slit the sheets from 0 to α and from β to ∞, the required joining is like that shown in Figure 14.12. We find, as Riemann did, that the surface is a torus, and hence *not* topologically the same as a sphere. This discovery proved to be a revelation

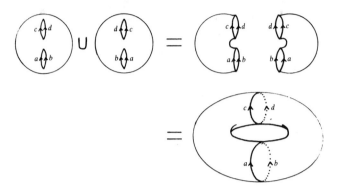

Figure 14.12

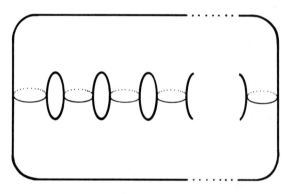

Figure 14.13

for the understanding of cubic curves and elliptic functions, as we shall see in the next chapter.

One quickly sees that by considering relations of the form

$$y^2 = (x - \alpha_1)(x - \alpha_2)\ldots(x - \alpha_{2n})$$

it is possible to obtain Riemann surfaces of the form shown in Figure 14.13. These surfaces are distinguished topologically from each other by the number of "holes": 0 for the sphere, 1 for the torus, and so on. This simple topological invariant turns out to be the *genus* which also determines the type of functions that can parametrize the corresponding complex curve. Other geometric and analytic properties of genus will unfold over the next few chapters. The topological importance of genus was established by Möbius [1863], when he showed that any closed surface in ordinary space is topologically equivalent to one of the form seen in Figure 14.13.

EXERCISES

1. Use the substitution $x = 1/u$, $y = 1/v$ to show that the curve

$$y^2 = x(x - \alpha)(x - \beta)$$

behaves at infinity like the curve

$$v^2 = u^3(1 - u\alpha)^{-1}(1 - u\beta)^{-1}$$

does at 0, which is turn is qualitatively like the behavior of

$$v = u^{3/2}$$

2. Show that $v = u^{3/2}$ has a branch point at 0 like that of $v = u^{1/2}$.

14.5. Biographical Notes: Riemann

Bernhard Riemann (Fig. 14.14) was born in the village of Breselenz, near Hannover, in 1826, and died at Selasca in Italy in 1866. He was the second of six children of Friedrich Riemann, a Protestant minister, and Charlotte Ebell. Up to the age of 13 he was taught by his father, with the help of the village schoolmaster, but he showed such a grasp of mathematics that sometimes they were unable to follow him. In 1840 Riemann went to live with his grandmother in Hannover in order to attend secondary school. After her death in 1842 he continued his studies at a school in Lüneberg, which was nearer to home, his father having moved to a new parish in the village of Quickborn. In Lüneberg it was his good fortune to have a headmaster who recognized his talent and gave him books by Euler and Legendre to read. The story goes that he mastered Legendre's 800-page *Théorie des Nombres* in six days.

The bright side of Riemann's life, which we have seen so far, was not unlike Abel's. But, as with Abel, there was a darker side as well. Riemann's family was also poor and suffered from tuberculosis. His mother, three sisters, and Riemann himself eventually died from the disease. At least Riemann was spared the family discord and very early death which made Abel's life so tragic. At all times he maintained a close and loving relationship with his family, he lived long enough to marry and to be a father, and he also had time to develop his major ideas to maturity and to gain a significant following. Riemann's

Figure 14.14. Bernhard Riemann.

published work—just a single volume—is in fact less copious than that of any important mathematician who lived to his fortieth year. But no other single volume has had such an impact on modern mathematics.

Riemann's career as a mathematician began soon after he entered the University of Göttingen in 1846. He intended to follow in his father's footsteps by studying theology but, like Euler and the Bernoullis before him, he found the call of mathematics too strong and obtained his father's permission to switch fields. The switch to mathematics was in recognition of where his greatest talent lay, not because of disdain for theology or philosophy. In fact, Riemann was deeply pious and well read in philosophy—so much so that readers ever since have lamented the influence of German philosophical writing on his style.

Göttingen in 1846 was not the Mecca for mathematicians one would have expected it to be with the great Gauss in the chair of mathematics. The professors kept aloof from the students and did not encourage original thinking or lecture on current research. Even Gauss himself taught only elementary courses. After a year, Riemann transferred to the University of Berlin, where the atmosphere was more democratic and where Jacobi, Dirichlet, Steiner, and Eisenstein shared their latest ideas. Riemann was too shy to immerse himself fully in this radically different environment, but he became friendly with Eisenstein, who was just three years his senior, and learned a great deal from Dirichlet. Riemann's later work made highly original use of some of Dirichlet's ideas, in particular a quasi-physical principle (actually first stated by Kelvin) Riemann called *Dirichlet's principle*. Among the remarkable conclusions he drew from this principle was the theorem that curves of genus 0 are precisely those that can be parametrized by rational functions.

Dirichlet's forté was the use of analysis in pure mathematics, particularly in number theory, and Riemann too has been broadly classified as an analyst. However, he was not a specialist as analysts usually are today. His field was all of mathematics, seen from the analytic viewpoint. He saw where analysis could be used to illuminate mathematics from number theory to geometry, but he also saw where analysis itself was in need of illumination from outside. The concept of a Riemann surface, and the topological concept of genus in particular, made many previously hard-won results of analysis almost obvious. A vivid example of the illumination of analysis by topology is Riemann's explanation of the double periodicity of elliptic functions, which we shall see in Section 15.4.

Riemann surfaces were introduced in Riemann's doctoral thesis [1851]. He had returned to Göttingen in 1849 and, after gaining his doctorate, began working to qualify for a Privatdozent (lecturing) position. One of the requirements was an essay, which he met with a memoir on Fourier series in which he introduced the "Riemann integral" concept. The Riemann integral is not really one of Riemann's best ideas—although it is the one best known to students today—since the integral later introduced by Lebesgue is far better suited to the subject (see Chapter 20). The other requirement was a lecture, for which he had to submit three titles to the university faculty. He had only prepared the first two and was dismayed when Gauss chose the third: on the foundations of geometry. However, Riemann rose brilliantly to the occasion, and his lecture *Über die Hypothesen, welche der Geometrie zu Grunde liegen*, became one of the classics of mathematics (Riemann [1854]). In it he introduced the main ideas of modern differential geometry: n-dimensional spaces, metrics and curvature, and the way in which curvature controls global geometric properties of a space. In the special case of two dimensions, these ideas had already been grasped by Gauss (see Chapter 16), so it was a joy and a revelation to Gauss, then in the last year of his life, to see how much further Riemann had carried them.

Riemann succeeded in becoming a lecturer and had the satisfaction of attracting an unexpectedly large class (eight students!). During the next few years he developed the material for perhaps his greatest work (Riemann [1857]), which did for algebraic geometry what he had earlier [1854] done for differential geometry. One of his students at this time was Dedekind, who later recast Riemann's theory into the more algebraic form that is used today.

Dedekind also coedited Riemann's collected works and wrote an essay on Riemann's life (Dedekind [1876]), which is the main biographical source for this section. The lecturer's position was very productive mathematically, but it brought in only voluntary fees from students, and Riemann was close to starvation. Other setbacks he suffered were the death of his father and sister Clara and a nervous breakdown brought on by overwork.

When Gauss died in 1855 and was succeeded by Dirichlet there was an unsuccessful move to appoint Riemann as extraordinary professor. This move failed, but Riemann was granted a regular salary, and when Dirichlet died in 1859 Riemann succeeded him. In 1862 he married Elise Koch, a friend of his sisters, and their daughter, Ida, was born in Pisa in 1863. Riemann had begun traveling to Italy for the sake of his health in 1862, and he spent much time there during his remaining years. He loved Italy and its art treasures and also received a warm reception from Italian mathematicians. Two of his friends in Pisa, Enrico Betti and Eugenio Beltrami, were inspired by Riemann's ideas to make important contributions to topology and differential geometry. Beltrami saw how Riemann's concept of curved space could be used as a basis for noneuclidean geometry, a revolutionary discovery which even Riemann may not have anticipated (see Chapter 17).

Riemann's sojourn in Italy was all too short. He died at Selasca on Lake Maggiore in the summer of 1866, with his wife beside him. Dedekind described his last days as follows:

> On the day before his death he lay beneath a fig tree, filled with joy at the glorious landscape, writing his last work, unfortunately left incomplete. His end came gently, without struggle or death agony; it seemed as though he followed with interest the parting of the soul from the body; his wife had to give him bread and wine, he asked her to convey his love to those at home, saying "Kiss our child." She said the Lord's prayer with him, he could no longer speak; at the words "Forgive us our trespasses" he raised his eyes devoutly, she felt his hand in hers becoming colder, and after a few more breaths his pure, noble heart ceased to beat. The gentle mind implanted in him in his father's house stayed with him all his life, and he remained true to his God as his father had, though not in the same form. (Dedekind [1876])

It was said of Abel that he left enough to keep mathematicians busy for 500 years, and the same might be said of Riemann. Today, more than 120 years after Riemann's death, the major unsolved problem in pure mathematics is the so-called *Riemann hypothesis*, a conjecture made casually by Riemann [1859] in his paper on the distribution of prime numbers. Riemann considered Euler's function (discussed in Section 9.7),

$$\zeta(s) = 1 + \frac{1}{2^s} + \frac{1}{3^s} + \cdots$$

introducing the zeta notation for it, and extended it to complex values of s. He observed that if $\zeta(s) = 0$, then $0 \leqslant \text{Re}(s) \leqslant 1$, and added that it was quite likely that all zeros of $\zeta(s)$ had real part 1/2. He did not pursue the matter further, since his initial observation was enough for his purpose, which was to derive an infinite series for $F(x)$, the number of primes less than a positive integer x. Mathematicians later realized that Riemann's hypothesis governs the distribution of prime numbers to an extraordinary extent, which is why its proof is so eagerly sought. Since all the efforts of the best mathematicians have failed so far, perhaps only another Riemann will succeed.

CHAPTER 15

Complex Numbers and Functions

15.1. Complex Functions

When Bombelli [1572] introduced complex numbers, he implicitly introduced complex functions as well. The solution y of the cubic equation $y^3 = py + q$,

$$y = \sqrt[3]{\frac{q}{2} + \sqrt{\left(\frac{q}{2}\right)^2 - \left(\frac{p}{3}\right)^3}} + \sqrt[3]{\frac{q}{2} - \sqrt{\left(\frac{q}{2}\right)^2 - \left(\frac{p}{3}\right)^3}}$$

involves the cube root function of a complex argument when $(q/2)^2 < (p/3)^3$. However, in this context complex functions were no more (or less) problematic than complex numbers. It was sometimes surprising to find that functions turned out to be equal, as when Leibniz and de Moivre showed (Section 5.6) that

$$x = \frac{1}{2}\sqrt[n]{y + \sqrt{y^2 - 1}} + \frac{1}{2}\sqrt[n]{y - \sqrt{y^2 - 1}}$$

where y is the polynomial in $x = \sin\theta$ that equals $\sin n\theta$, but one did not worry about the meaning of the functions as long as their equations could be checked by algebra.

Things became more puzzling with transcendental functions, in particular those defined by integration. A key example was the logarithm function, which comes from integrating $dz/(1 + z)$. Once this function was understood, the reason for algebraic miracles like the Leibniz–de Moivre theorem became much clearer.

The story of the complex logarithm began when John Bernoulli [1702] noted that

$$\frac{dz}{1 + z^2} = \frac{dz}{2(1 + z\sqrt{-1})} + \frac{dz}{2(1 - z\sqrt{-1})}$$

and drew the conclusion that "imaginary logarithms express real circular sectors." He did not actually perform the integration, but he could have got

$$\tan^{-1}z = \frac{1}{2i}\log\frac{i-z}{i+z}$$

since Euler gives him credit for a similar formula when writing to him in [1728]. However, this may have been the young Euler's deference to his former teacher, because John Bernoulli showed poor understanding of logarithms as the correspondence continued. He persistently claimed that $\log(-x) = \log(x)$ on the grounds that

$$d\log(-x) = \frac{1}{x} = d\log(x)$$

despite a remainder from Euler [1728] that equality of differentials does not imply equality of integrals. Euler went on to suggest that the complex logarithm had infinitely many values.

In the meantime, Cotes [1714] had also discovered a relation between complex logarithms and circular functions:

$$\log(\cos x + i\sin x) = ix$$

Recognizing the importance of this result, he entitled his work *Harmonia Mensurarum* (Harmony of measures). The "measures" in question were the logarithm and inverse tangent functions, which "measured" the hyperbola and the circle, respectively, via the integrals $\int dx/(1+x)$ and $\int dx/(1+x^2)$. A wide class of integrals had been reduced to these two types, but it was not understood why two apparently unrelated "measures" should be required. Cotes' result was the first (apart from the near-miss of John Bernoulli [1702]) to relate the two, showing that in the wider domain of complex functions the logarithm and inverse circular functions are essentially the same.

The most compact statement of their relationship was attained around 1740, when Euler shifted attention from the logarithm function to its inverse, the exponential function. The definitive formula

$$e^{ix} = \cos x + i\sin x$$

was first published by Euler [1748], who derived it by comparing series expansions of both sides. Euler's formulation in terms of the single-valued function e^{ix} gave a simple explanation of the many values of the logarithm (which Cotes had missed) as a consequence of the periodicity of cos and sin. A direct explanation, based on the definition of log as an integral, was not possible until Gauss [1811] clarified the meaning of complex integrals and pointed out their dependence on the path of integration (see Section 15.3).

Euler's formula also shows

$$(\cos x + i\sin x)^n = e^{inx} = \cos nx + i\sin nx$$

and hence gives a deeper explanation of the Leibniz–de Moivre formula. More generally, the addition theorems for cos and sin (Section 11.4) could be seen as consequences of the much simpler addition formula for the exponential function

$$e^{u+v} = e^u \cdot e^v$$

The imaginary function e^{ix} was so much more coherent than its real constituents $\cos x$ and $\sin x$ that it was difficult to do without it, and Euler's formula gave mathematicians a strong push toward the eventual acceptance of complex numbers. A more detailed account of the role of the logarithm and exponential functions in the development of complex numbers may be found in Cajori [1913].

Almost simultaneously with Euler's elucidation of cos and sin, d'Alembert's work in hydrodynamics revealed that many real functions occur naturally in pairs which are real and imaginary parts of complex functions. As mentioned in Section 12.5, d'Alembert [1752] discovered the equations

$$\frac{\partial P}{\partial y} - \frac{\partial Q}{\partial x} = 0 \tag{1}$$

$$\frac{\partial P}{\partial x} + \frac{\partial Q}{\partial y} = 0 \tag{2}$$

relating the velocity components P, Q in two-dimensional steady irrotational fluid flow. Equations (1) and (2) come from the requirements that $Q\,dx + P\,dy$ and $P\,dx - Q\,dy$ be complete differentials, in which case another complete differential is

$$Q\,dx + P\,dy + i(P\,dx - Q\,dy) = (Q + iP)\left(dx + \frac{dy}{i}\right) = (Q + iP)d\left(x + \frac{y}{i}\right)$$

D'Alembert concluded that this means $Q + iP$ is a function f of $x + y/i$, so that $Q = \text{Re}(f)$ and $P = \text{Im}(f)$.

To feel the force of this result, one has to forget the modern definition of function, under which $u(x, y) + iv(x, y)$ is a function of $x + iy$ for *any* functions u, v. In the eighteenth-century context, a "function" $f(x + iy)$ of $x + iy$ was calculable from $x + iy$ by elementary operations; at worst, $f(x + iy)$ was a power series in $x + iy$. This imposes a strong constraint on u, v, namely, that

$$\frac{\partial u}{\partial x} = \frac{\partial v}{\partial y}$$

$$\frac{\partial u}{\partial y} = -\frac{\partial v}{\partial x}$$

These were just the equations found by d'Alembert in his hydrodynamical investigations, but they came to be named the Cauchy–Riemann equations, because these mathematicians stressed their key role in the study of complex functions. The concept of complex function solidified when Cauchy [1837] showed that a function $f(z)$, where $z = x + iy$, only had to be differentiable in order to be expressible as a power series in z. Thus it suffices to define a complex function $f(z)$ to be one which is differentiable with respect to z in order to guarantee that f is defined with eighteenth-century strictness. It follows, in particular, that the first derivative of f entails derivatives of all orders, and that the values of f in any neighborhood determine its values everywhere. This "rigidity" in the notion of complex function is enough of a constraint to enable nontrivial properties to be proved, but at the same time it leaves enough flexibility—one might say "fluidity"—to cover important general situations.

EXERCISES

1. Show that i^i has a real value (Euler [1746]).

2. Give a formula for all values of i^i (Euler [1746]), using the formula for e^{ix}.

15.2. Conformal Mapping

Another important general situation clarified by complex functions is the problem of conformal mapping. Mapping a sphere (the earth's surface) onto a plane is a practical problem that has attracted the attention of mathematicians since ancient times. Before the eighteenth century the most notable mathematical contributions to mapping were stereographic projection (Section 14.2), due to Ptolemy c. A.D. 150, and the Mercator projection used by G. Mercator in 1569 (this Mercator was Gerard, not the Nicholas who discovered the series for $\log(1 + x)$). Both these projections are conformal, that is, angle-preserving, or what eighteenth-century mathematicians preferred to call "similar in the small." This means that the image $f(R)$ of any region R tends toward an exact scale map of R as the size of R tends to 0. Since "similarity in the large" is clearly impossible—for example, a great circle cannot be mapped to a closed curve which divides the plane into two equal parts—conformality is the best one can do to preserve the appearance of regions on the sphere. Preservation of angles was intentional in the Mercator projection, whose purpose was to assist navigation, and in the case of stereographic projection conformality was first noticed by Harriot around 1590 (see Lohne [1979]).

   Advances in the theory of conformal mapping were made by Lambert [1772], Euler [1777] (sphere onto plane), and Lagrange [1779] (general surface of revolution onto plane). All these authors used complex numbers, but the presentation of Lagrange is the clearest and most general. Using the method of d'Alembert [1752], he combined a pair of differential equations in two real variables into a single equation in one complex variable and arrived at the result that any two conformal maps of a surface of revolution onto the (x, y) plane were related via a complex function $f(x + iy)$ mapping the plane onto itself. These results were crowned by the result of Gauss [1822], generalizing Lagrange's theorem to conformal maps of an arbitrary surface onto the plane.

Conversely, a complex function $f(z)$ defines a map of the z plane onto itself, and it is easy to see that this map is conformal. In fact, this is a consequence of the differentiability of f. To say that the limit

$$\lim_{\delta z \to 0} \frac{f(z_0 + \delta z) - f(z_0)}{\delta z}$$

exists is to say that the mapping of the disk $\{z : |z - z_0| < |\delta z|\}$ around z_0 to the region around $f(z_0)$ tends to a scale mapping as the radius $|\delta z|$ tends to 0. If the derivative is expressed in polar form as

$$f'(z_0) = re^{i\alpha}$$

then r is the scale factor of this limit mapping and α is the angle of rotation. Riemann [1851] seems to have been the first to take the conformal mapping property as a basis for the theory of complex functions. His deepest result in this direction was the *Riemann mapping theorem*, which states that any region of the plane bounded by a simple closed curve can be mapped onto the unit disk conformally, and hence by a complex function. The proof of this theorem in Riemann [1851] depended on properties of potential functions, which Riemann justified partly by appeal to physical intuition—the so-called *Dirichlet's principle*. Such reasoning went against the growing tendency toward rigor in nineteenth-century analysis, and stricter proofs were given by Schwarz [1870] and Neumann [1870]. However, Riemann's faith in the physical roots of complex function theory was eventually justified when Hilbert [1899] put Dirichlet's principle on a sound basis.

15.3. Cauchy's Theorem

We have seen that interesting complex functions arise from integration. For example, the elliptic functions come from inversion of elliptic integrals (Section 11.3). However, it is not at first clear what the integral $\int_{z_0}^{z} f(t) dt$ means when z_0, z are complex numbers. It is natural, and not technically difficult, to define $\int_{z_0}^{z} f(t) dt$ as $\int_{\mathscr{C}} f(t) dt$, the integral of f along a curve $\mathscr{C}$ from z_0 to z; the problem is that $\int_{\mathscr{C}} f(t) dt$ appears to depend on $\mathscr{C}$ and hence may not be anything like a function of z, as one would wish.

The first to recognize and resolve this problem seems to have been Gauss. In a letter to Bessel, Gauss [1811] raised the problem and claimed its resolution as follows:

Now how is one to think of $\int \phi(z)\,dz$ for $z = a + ib$? Evidently, if one wishes to start from clear concepts, one must assume that z changes by infinitely small increments (each of the form $\alpha + i\beta$) from that value for which the integral is to be 0 to $c = a + ib$, and then *sum* all the $\phi(z)\,dz$. ... But now ... continuous transition from one value of z to another $a + ib$ takes place along a curve and hence is possible in infinitely many ways. I now conjecture that the integral $\int_0^c \phi(z)\,dz$ will always have the same value after two different transitions if z never becomes infinite within the region enclosed by the two curves representing the transitions. (Translation of Gauss [1811] in Birkhoff [1973])

In the same letter, Gauss also observed that if $\phi(z)$ *does* become infinite in the region, then in general $\int_0^c \phi(z)\,dz$ *will* take different values when integrated along different curves. He saw in particular that the infinitely many values of $\log c$ corresponded to the different ways a path from 1 to c could wind around $z = 0$, the point where $\phi(z) = 1/z$ becomes infinite.

The theorem that $\int_{z_0}^z f(t)\,dt$ is independent of the path throughout a region where f is finite (and differentiable, which went without saying for Gauss) is now known as *Cauchy's theorem*, since Cauchy was the first to offer a proof and to develop the consequences of the theorem. An equivalent and more convenient statement is that $\int_{\mathscr{C}} f(t)\,dt = 0$ for any closed curve $\mathscr{C}$ in a region where f is differentiable. Cauchy presented a proof to the Paris Academy in 1814 but first published it later [1825]. In his *Sur les intégrales* [1846] he presented a more transparent proof, based on the Cauchy–Riemann equations and the theorem of Green [1828] and Ostrogradsky [1828] which relates a line integral to a surface integral. The latter theorem, usually known as *Green's theorem*, is a generalization of the fundamental theorem of calculus to real functions $f(x, y)$ of two variables and can be stated as follows: if $\mathscr{C}$ is a simple closed curve bounding a region $\mathscr{R}$ and f is suitably smooth, then

$$\int_{\mathscr{C}} f\,dx = \iint_{\mathscr{R}} \frac{\partial f}{\partial y}\,dx\,dy$$

$$\int_{\mathscr{C}} f\,dy = -\iint_{\mathscr{R}} \frac{\partial f}{\partial x}\,dx\,dy$$

where $\iint_{\mathscr{R}}$ denotes the surface integral over $\mathscr{R}$ and $\int_{\mathscr{C}}$ denotes the line integral around $\mathscr{C}$ in the counterclockwise sense. (The difference in sign in the two formulas reflects the different sense of $\mathscr{C}$ when x and y are interchanged.)

Cauchy's theorem follows from Green's theorem by an easy calculation. If

$$f(t) = u(t) + iv(t)$$

is the decomposition of f into real and imaginary parts, and if we write

$$dt = dx + i\,dy$$

then

$$\int_{\mathscr{C}} f(t)\, dt = \int_{\mathscr{C}} (u + iv)(dx + i\, dy)$$

$$= \int_{\mathscr{C}} (u\, dx - v\, dy) + i \int_{\mathscr{C}} (v\, dx + u\, dy)$$

$$= \int\!\!\int_{\mathscr{R}} \left(\frac{\partial u}{\partial y} + \frac{\partial v}{\partial x} \right) dx\, dy + i \int\!\!\int_{\mathscr{R}} \left(\frac{\partial v}{\partial y} - \frac{\partial u}{\partial x} \right) dx\, dy$$

$$= 0$$

since

$$\frac{\partial u}{\partial y} + \frac{\partial v}{\partial x} = 0 \qquad \text{and} \qquad \frac{\partial v}{\partial y} - \frac{\partial u}{\partial x} = 0$$

by the Cauchy–Riemann equations. This proof requires f to have a continuous first derivative in order to be able to apply Green's theorem. The restriction of continuity of $f'(t)$ in the proof was removed by Goursat [1900]. As it happens, if f' exists, it will have not only continuity but derivatives of all orders. This follows from one of the remarkable consequences Cauchy [1837] drew from the assumption $\int_{\mathscr{C}} f(t)\, dt = 0$, namely, that f has a power series expansion. By Goursat [1900], then, differentiability of a complex function is enough to guarantee a power series expansion. A generalization of this result to f which become infinite at isolated points was made by Laurent [1843] (f then has an expansion including negative powers; that is the *Laurent expansion*) and to "many-valued" f with branch points by Puiseux [1850] (f then has an expansion in fractional powers; the *Newton–Puiseux expansion*).

15.4. The Double Periodicity of Elliptic Functions

The clear view of complex integration provided by Cauchy's theorem is one step toward an understanding of elliptic integrals such as $\int_0^z dt / \sqrt{t(t - \alpha)(t - \beta)}$. The other important step is the idea of a Riemann surface (Section 14.4) which enables us to visualize the possible paths of integration from 0 to z. The "function" $1/\sqrt{t(t - \alpha)(t - \beta)}$ is of course two-valued and, by an argument like that in Section 14.4, is represented by a two-sheeted covering of the t sphere, with branch points at 0, α, β, ∞. Thus the paths of integration, correctly viewed, are curves on this surface, which is topologically a torus (again, as in Section 14.4).

Now a torus contains certain closed curves that do not bound a piece of the surface, such as the curves $\mathscr{C}_1$ and $\mathscr{C}_2$ shown in Figure 15.1. There is no region $\mathscr{R}$ bounded by $\mathscr{C}_1$ or $\mathscr{C}_2$; hence Green's theorem does not apply, and we in fact obtain nonzero values

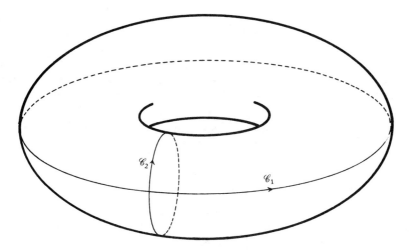

Figure 15.1

$$\omega_1 = \int_{\mathscr{C}_1} \frac{dt}{\sqrt{t(t-\alpha)(t-\beta)}}$$

$$\omega_2 = \int_{\mathscr{C}_2} \frac{dt}{\sqrt{t(t-\alpha)(t-\beta)}}$$

Consequently, the integral

$$\Phi^{-1}(z) = \int_0^z \frac{dt}{\sqrt{t(t-\alpha)(t-\beta)}}$$

will be ambiguous: for each value $\Phi^{-1}(z) = w$ obtained for a certain path $\mathscr{C}$ from 0 to z we also obtain the values $w + m\omega_1 + n\omega_2$ by adding to $\mathscr{C}$ a detour which winds m times around $\mathscr{C}_1$ and n times around $\mathscr{C}_2$. (For topological reasons, this is essentially the most general path of integration.)

It follows that the inverse relation $\Phi(w) = z$, the elliptic function corresponding to the integral, satisfies

$$\Phi(w) = \Phi(w + m\omega_1 + n\omega_2)$$

for any integers m, n. That is, Φ is doubly periodic, with periods ω_1, ω_2. This intuitive explanation of double periodicity is due to Riemann [1851], who later [1858] developed the theory of elliptic functions from this standpoint.

Remarkable series expansions of elliptic functions, which exhibit the double periodicity analytically, were discovered by Eisenstein [1847]. The precedents for Eisenstein's series, as Eisenstein himself pointed out, were partial fraction expansions of circular functions discovered by Euler, for example,

$$\pi \cot \pi x = \sum_{n=-\infty}^{\infty} \frac{1}{x+n}$$

(Euler [1748], p. 191). It is obvious (at least formally, though one has to be a little careful about the meaning of this summation to ensure convergence) that the sum is unchanged when x is replaced by $x + 1$, hence the period 1 of $\pi \cot \pi x$ is exhibited directly by its series expansion. Eisenstein showed that doubly periodic functions could be obtained by analogous expressions, such as

$$\sum_{m,n=-\infty}^{\infty} \frac{1}{(z + m\omega_1 + n\omega_2)^2}$$

which again (with suitable interpretation to ensure convergence) are obviously unchanged when z is replaced by $z + \omega_1$ or $z + \omega_2$. Hence we obtain a function with periods ω_1, ω_2. The function above is in fact identical (up to a constant) with the Weierstrass $\wp$-function, mentioned in Section 11.5 as the inverse to the integral $\int dt/\sqrt{4t^3 - g_2 t - g_3}$. Weierstrass [1863], p. 121, found the relations between g_2, g_3 and the periods ω_1, ω_2:

$$g_2 = 60 \sum \frac{1}{(m\omega_1 + n\omega_2)^4}$$

$$g_3 = 140 \sum \frac{1}{(m\omega_1 + n\omega_2)^6}$$

where the sums are over all pairs $(m, n) \neq (0, 0)$. Elegant modern accounts of the Eisenstein and Weierstrass theories may be found in Weil [1976] and Robert [1973].

15.5. Elliptic Curves

We have seen that nonsingular cubic curves of the form

$$y^2 = ax^3 + bx^2 + cx + d \tag{1}$$

are important not only among the cubic curves themselves (cf. Newton's classification, Sections 6.4 and 7.4), but also in number theory (Section 10.6) and the theory of elliptic functions (Section 11.2). One of the great achievements of nineteenth-century mathematics was the synthesis of a unified view of all these manifestations of cubic curves. The view was first glimpsed by Jacobi [1834], and it came more clearly into focus with the development of complex analysis between Riemann [1851] and Poincaré [1901]. The theory of elliptic curves, as the unified view has come to be known, continues to inspire researchers today, as it seems to encompass some of the most fascinating problems of number theory.

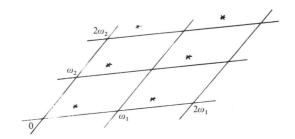

Figure 15.2

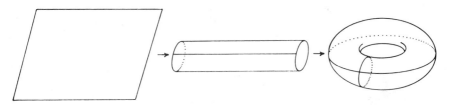

Figure 15.3

Jacobi saw, at least implicitly, that the curve (1) could be parametrized as

$$x = f(z), \qquad y = f'(z) \tag{2}$$

where f and its derivative f' are elliptic functions. Knowing that f and f' were doubly periodic, with the same periods ω_1, ω_2, say, he would have seen that this gave a map of the z plane $\mathbb{C}$ onto the curve (1) for which the preimage of a given point on (1) is a set of points in $\mathbb{C}$ of the form

$$z + \Lambda = \{z + m\omega_1 + n\omega_2 | m, n \in \mathbb{Z}\}$$

where

$$\Lambda = \{m\omega_1 + n\omega_2 | m, n \in \mathbb{Z}\}$$

is called the *lattice of periods* of f. The complex numbers $z + m\omega_1 + n\omega_2$ in $z + \Lambda$ are also called "equivalent with respect to Λ." One such equivalence class is shown by asterisks in Figure 15.2.

The parametrization (2) means that there is a one-to-one correspondence between the points $(f(z), f'(z))$ of the curve and the equivalence classes $z + \Lambda$. Nowadays we express this relation by saying that the curve is *isomorphic to the space* $\mathbb{C}/\Lambda$ of these equivalence classes. Jacobi might have seen, though it was probably not of interest to him, that $\mathbb{C}/\Lambda$ is a torus. One sees this by taking one parallelogram in $\mathbb{C}$, which includes a representative of each equivalence class, and identifying the equivalent points on its boundary (i.e., pasting opposite sides together; Fig. 15.3). Of course, the torus form of (1) eventually came to light through the Riemann surface construction given in Section 14.4.

An elegant way of demonstrating both the double periodicity of elliptic functions and the parametrization of cubic curves was given by Weierstrass [1863]. Beginning with formula

$$\wp(z) = \sum_{m,n=-\infty}^{\infty} \frac{1}{(z + m\omega_1 + n\omega_2)^2}$$

which, as mentioned in Section 15.4, makes the double periodicity evident, Weierstrass showed by simple computations with series that

$$[\wp'(z)]^2 = 4[\wp(z)]^3 - g_2\wp(z) - g_3$$

where g_2, g_3 are the constants depending on ω_1, ω_2, which were defined in Section 15.4. It follows that the point $[\wp(z), \wp'(z)]$ lies on the curve

$$y^2 = 4x^3 - g_2 x - g_3 \tag{3}$$

and a little further checking shows that (3) is in fact isomorphic to $\mathbb{C}/\Lambda$, where Λ is the lattice of periods of $\wp$. The parametrization of all curves (1) by elliptic functions follows by making a linear transformation.

Once the curve (1) is parametrized as

$$x = f(z), \qquad y = f'(z)$$

one sees a natural "addition" of points on the curve induced by adding their parameter values. Because of the double periodicity of f and f' this "addition" is simply ordinary addition in $\mathbb{C}$, modulo Λ. In particular, it is immediate that "addition of points" has some properties of ordinary addition, such as commutativity and associativity. However, as mentioned in Section 10.6, addition of parameter values z is also reflected in the geometry of the curve. The most concise statement of the relationship, due to Clebsch [1864], is that if z_1, z_2, z_3 are parameter values of three collinear points, then

$$z_1 + z_2 + z_3 = 0 \bmod(\omega_1, \omega_2)$$

(or $z_1 + z_2 + z_3 \in \Lambda$). This means that "addition of points" also has an elementary geometric interpretation, for which, incidentally, the algebraic properties are far less obvious.

On the other hand, the straight-line interpretation of "addition" gives the simplest explanation of the addition theorems for elliptic functions. As we saw in Section 10.6, the value of $f(z_3)$ is easy to compute as a rational function of $f(z_1), f'(z_1), f(z_2), f'(z_2)$ when z_1, z_2, z_3 are the parameter values of collinear points. Originally, of course, the formula was obtained by Euler, with great difficulty, by manipulating the integral inverse to f (see Section 11.5).

The view of the curve as $\mathbb{C}/\Lambda$ seems overwhelmingly "right" when one asks for a projective classification of cubic curves. Recall from Section 7.4 that Newton had reduced cubics to the cusp type, the double-point type, and three nonsingular types using real projective transformations. All cubics with a cusp are in fact equivalent to $y^2 = x^3$, and all with a double point are equivalent to $y^2 = x^2(x + 1)$, while the distinction between the nonsingular types disappears over the complex numbers where, as we now know, all are equivalent to tori $\mathbb{C}/\Lambda$. The problem that remains is to decide projective equivalence among the nonsingular cubics. Salmon [1851] showed that this was determined by a certain complex number τ, which could be computed from the equation of the curve. He defined τ geometrically, so that its projective invariance was obvious, with no thought of elliptic functions. But τ turned out to be nothing but ω_1/ω_2, which means that two nonsingular cubics are projectively equivalent if and only if their period lattices Λ have the same shape.

15.6. Uniformization

The characteristic of nonsingular cubics that allows their parametrization by elliptic functions is their topological form. The two periods correspond to the two essentially different circuits around the torus (Fig. 15.4). Other curves of the same genus can be similarly parametrized by elliptic functions and are also called elliptic curves.

A representation of the x and y values on a curve by simultaneous functions of a single parameter z is sometimes called a *uniform* representation, and so the problem of parametrizing all algebraic curves in this way came to be known as the *uniformization* problem. Once the elliptic case was understood, it became clear that a solution of the uniformization problem for arbitrary algebraic curves would depend on a better understanding of surfaces: their

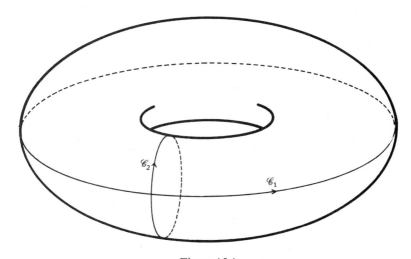

Figure 15.4

topology, the periodicities associated with their closed curves, and the way these periodicities could be reflected in $\mathbb{C}$. These problems were first attacked by Poincaré and Klein in the 1880s, and their work led to the eventual positive solution of the uniformization problem by Poincaré [1907] and Koebe [1907].

Even more important than the solution of this single problem, however, was the amazing convergence of ideas in the preliminary work of Poincaré and Klein. They discovered that multiple periodicities were reflected in $\mathbb{C}$ by groups of transformations and that the transformations in question were of the simple type $z \mapsto (az + b)/(cz + d)$, called *linear fractional*. Linear fractional transformations generalize the linear transformations $z \mapsto z + \omega_1, z \mapsto z + \omega_2$ naturally associated with the periods of elliptic functions. However, while the transformations $z \mapsto z + \omega_1$, $z \mapsto z + \omega_2$ are algebraically and geometrically transparent—they commute, and they generate the general transformations $z \mapsto z + m\omega_1 + n\omega_2$, which are simply rigid translations of the plane—the more general linear fractional transformations are not as easily understood. Linear fractional transformations do not normally commute, and their mastery requires a simultaneous grasp of their algebraic, geometric, and topological aspects.

The simultaneous view proved to be enormously fruitful in the development of group theory and topology, as we shall see in Chapters 18 and 19. Geometry was also given a new lease of life when Poincaré [1882] discovered that linear fractional transformations gave a natural interpretation of noneuclidean geometry, a field which until then had been a curiosity on the fringes of mathematics. In the next two chapters we shall look at the origins of non-euclidean geometry and see how the subject was transformed by Poincaré's discovery.

15.7. Biographical Notes: Lagrange and Cauchy

Joseph Louis Lagrange (Fig. 15.5) was born in Turin in 1736 and died in Paris in 1813. He was the oldest of 11 children of Giuseppe Lagrangia, treasurer of the Office of Public Works in Turin, and Teresa Grosso, the daughter of a physician, and a member of the wealthy Conti family. Despite this background, Lagrange's family was not well off, as his father had made some unwise financial speculations. Lagrange eventually came to appreciate the loss of his chance to become a wealthy idler, saying "If I had inherited a fortune I should probably not have cast my lot with mathematics."

His prowess in mathematics developed with amazing speed after he first encountered calculus in 1753, at the age of 17. By 1754 he was writing to Euler about his discoveries, and in 1755 he was made professor at the Royal Artillery School in Turin. As early as 1756 he was offered a superior position in Prussia, but he was too shy, or too reluctant to leave home, to accept it. As his

Figure 15.5. Joseph Louis Lagrange.

reputation grew, he also won the support of d'Alembert. When Euler left Berlin in 1766, d'Alembert arranged for Lagrange to take Euler's place. In 1767, perhaps missing the company of his family in Turin, Lagrange married his cousin Vittoria Conti. In a letter to d'Alembert in 1769 he said he had chosen a wife "who is one of my cousins and who even lived for a long time with my family, is a very good housewife and has no pretensions at all," adding that they had no children and did not want any.

Notwithstanding this lackluster beginning and the ill health of both Lagrange and his wife, the marriage strengthened over the years. Lagrange nursed Vittoria as her health worsened, and he was heartbroken when she died in 1783. He became deeply depressed about his work and the future of mathematics itself, writing to d'Alembert: "I cannot say that I shall still be doing mathematics 10 years from now. It also seems to me that the mine is already too deep, and that unless new veins are discovered it will have to be abandoned." Not long before this, Lagrange had completed one of his greatest works, the *Mécanique analytique*, but when a copy of the book reached him from the printer, he left it unopened on his desk.

Frederick II died in 1786, and Lagrange's position in Berlin became less secure. After receiving several offers from Italy and France, he accepted a position in the Paris Academy in 1787. The change of scene did not appreciably revive his general spirits or enthusiasm for mathematics. Though always welcome at social and scientific gatherings, he was always politely detached, sympathetic but uninvolved. At least it can be said of his detachment that it enabled him to survive the 1789 revolution, which took the lives of his more committed friends Condorcet and Lavoisier. The revolution did in fact stir some activity in Lagrange. In 1790 he became a member of the commission on weights and measures, which introduced the metric system now used universally in science. An interesting glimpse of mathematics during the revolution, in the form of a "panel discussion" between Lagrange, Laplace, and members of a student audience, may be found in Dedron and Itard [1973], pp. 302–310.

In 1792 Lagrange married Renée-Françoise-Adelaïde Le Monnier, the teenaged daughter of an astronomer colleague. His interest in life and mathematics revived, and even in his seventies he made some brilliant contributions to celestial mechanics, which he incorporated in a second edition of his *Mécanique analytique*. When he died in 1813 he was buried in the Pantheon in Paris.

Lagrange is known for an uncompromisingly formal approach to analysis and mechanics. He viewed all functions as power series and attempted to reduce all mechanics to the analysis of such functions, without use of geometry. He was proud of the fact that the *Mécanique analytique* contained no diagrams. His fear that mathematics would have to be abandoned "if new veins were not discovered" was of course unfounded, but understandable as an admission of the limitations of his own approach. The great advances of nineteenth-century analysis were due, more than anything else, to a revival of geometry. In particular, Lagrange's own view of functions as power series became intelligible only in the domain of complex functions, when it re-emerged from the geometric theory of complex integration discovered by Gauss and Cauchy.

Augustin-Louis Cauchy (Fig. 15.6) was born in Paris in 1789, only weeks after the storming of the Bastille, but he was anything but a child of the revolution. His father, Louis-François, was a lawyer and government official who fled to Paris with his wife, Marie-Madeleine Desestre, during the Terror. Augustin-Louis was the first of their six children. Throughout his life Cauchy was to hold extreme antirevolutionary and proroyalist views. The family settled in the village of Arcueil, and Cauchy received his early education from his father. He also had the benefit of contact with famous scientists who came to visit Laplace, who was a neighbor. Lagrange is said to have predicted that Cauchy would become a scientific genius but advised his father not to show him a mathematics book before he was 17.

As Napoleon rose to power at the end of the eighteenth century, Cauchy's father returned to government service, and the family moved back to Paris. Cauchy concentrated on classics in secondary school, which he completed in 1804, but then gravitated toward a scientific career. He entered the École Polytechnique in 1805, transferred to the École des Ponts in 1807, and began working as an engineer around 1809. In 1810 he went to Cherbourg to help

Figure 15.6. Augustin-Louis Cauchy.

in the construction of Napoleon's naval base, carrying with him, so it was said, Laplace's *Mécanique céleste* and Lagrange's *Traité des fonctions analytiques.*

His first important mathematical work was the solution of a problem posed to him by Lagrange: to show that any convex polyhedron is rigid. (More precisely, to show that the dihedral angles of a convex polyhedron are uniquely determined by its faces.) An accessible proof of his result, which deserves to be better known, is in Lyusternik [1966]. Cauchy's theorem partially settled a conjecture of Euler that any closed surface is rigid, and was in fact the best positive result obtainable, as Connelly [1978] has found a nonconvex polyhedron which is *not* rigid. Cauchy's second major discovery was his proof, in 1812, of Fermat's conjecture that every integer is the sum of n-agonal numbers (cf. Section 3.2).

The Cauchy integral theorem, which he submitted to the French Academy in 1814, carried him into the mathematical mainstream. He also managed to catch the political tide, which was turning royalist again, and became a member of the Academy on the expulsion of some republican members in 1816. At the same time he became professor at the École Polytechnique, where the 1820s saw the publication of his classic analysis texts and also one of his most important creations, the theory of elasticity. He also gained additional chairs at the Sorbonne and the Collège de France. He and Aloïse de Bure were married in 1818, and they had two daughters.

The mild revolution of 1830, which replaced the Bourbon King Charles X with the Orléans King Louis-Phillipe, was a catastrophe in Cauchy's view. From principles which were curious, though certainly firmly held, Cauchy refused to take the oath of allegiance to the new king. This meant he had to resign his chairs, but Cauchy went further than this—he left his family and followed the old king into exile. He did not return to Paris until 1838, and it was another 10 years before he regained one of his former chairs. Ironically, he had the revolution of 1848 to thank for this, because it abolished the oath of allegiance. He returned to the Sorbonne and kept up a steady flow of mathematical papers until his death in 1857.

CHAPTER 16

Differential Geometry

16.1. Transcendental Curves

We saw in Chapter 8 that the development of calculus in the seventeenth century was greatly stimulated by problems in the geometry of curves. Differentiation grew out of methods for the construction of tangents, and integration grew out of attempts to find areas and arc lengths. Calculus not only unlocked the secrets of the classical curves and of the algebraic curves defined by Descartes; it also extended the concept of curve itself. Once it became possible to handle slopes, lengths, and areas with precision, it also became possible to use these quantities to define new, nonalgebraic curves. These were the curves called "mechanical" by Descartes (Sections 6.3 and 12.3) and "transcendental" by Leibniz. In contrast to algebraic curves, which could be studied in some depth by purely algebraic methods, transcendental curves were inseparable from the methods of calculus. Hence it is not surprising that a new set of geometric ideas, the ideas of "infinitesimal" of *differential* geometry, first emerged from the investigation of transcendental curves.

A more surprising by-product of the investigation of transcendental curves was the first solution of the ancient problem of arc length. The problem was first posed for an algebraic curve, the circle, by the Greeks and in this case is equivalent to an area problem ("squaring the circle"), since both area and arc length of the circle depend on the evaluation of π. As we now know, π is a transcendental number (Section 2.3), so the arc length problem for the circle has no solution by the elementary means allowed by the Greeks. The first curve whose arc length could be found by elementary means was discovered by Harriot around 1590. It is the curve defined by the polar equation

$$r = e^{k\theta}$$

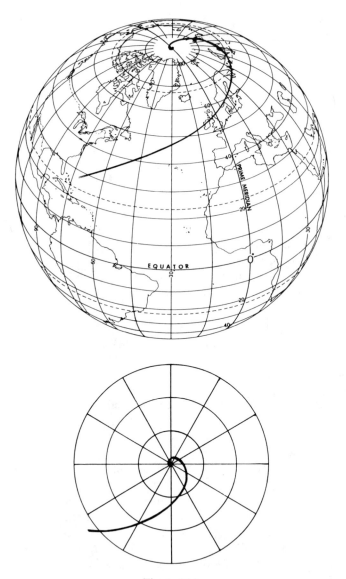

Figure 16.1

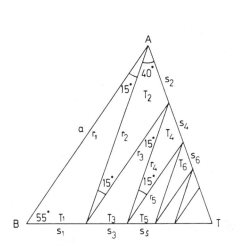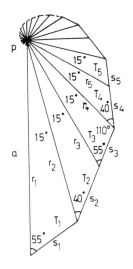

Figure 16.2

known as the *logarithmic* or *equiangular* spiral. Harriot did not have the exponential function and knew the curve only by its equiangular property, which is that the tangent makes a constant angle ϕ (depending on k) with the radius vector. (The spiral turned up in his researches on navigation and map projections (Section 15.2) as the projection of a *rhumb line* on the sphere (Fig. 16.1). A rhumb line is a curve that meets the meridians at a constant angle; in practical terms, it represents the course of a ship sailing in a fixed compass direction.)

Not having the tools of calculus, Harriot had to rely on ingenious geometry and a simple limit argument. His construction is illustrated in Figure 16.2 (from Lohne [1979, p. 273]). The spiral of angle $55°$ is approximated by a polygon with sides $s_1, s_2, s_3, \ldots$, which yield triangles $T_1, T_2, T_3, \ldots$, when connected to the origin p. $T_1, T_2, T_3, \ldots$, can be reassembled to form triangle ABT, whose area therefore equals that of the spiral (when the areas of overlapping turns are added together). Also

$$BT + TA = s_1 + s_2 + s_3 + \cdots = \text{length of the spiral}$$

When the approximation is made with shorter sides $s_1', s_2', s_3', \ldots$, but otherwise in the same way, the *same triangle ABT* results: the isosceles triangle with base a and base angles $55°$. Hence the length and area of the polygonal spiral remain the same. Yet, plainly, the polygonal spiral tends to the smooth equiangular spiral as $s_1' \to 0$, hence we have also found length and area of the smooth curve.

Harriot's work was not published, and the arc length of the equiangular spiral was rediscovered by Torricelli [1645]. Gradually the problem of arc length became understood more systematically as a problem of integration,

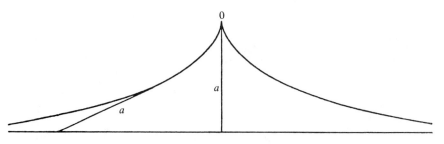

Figure 16.3

though usually a rather intractable one. The first solution for an algebraic curve was for $y^2 = x^3$, by Neil and Heuraet in 1657. Soon after this Wren solved the problem for the cycloid, and his solution was given by Wallis [1659]. A remarkable feature of Wren's result is that the length of one arch of the cycloid is a rational multiple (namely, 4) of the diameter of the generating circle.

As mentioned in Section 12.3, other extraordinary properties of the cycloid related to mechanics, and one of these will be reinterpreted geometrically in the next section. One transcendental curve that we did not discuss in connection with mechanics is the *tractrix* of Newton [1676']. Newton defined this curve by the property that the length of its tangent from point of contact to the x axis is constant (Fig. 16.3). It follows that the curve satisfies

$$\frac{dy}{ds} = \frac{y}{a}$$

where s denotes arc length. By using $ds = \sqrt{dx^2 + dy^2}$, this differential equation can be solved to give

$$x = -\sqrt{a^2 - y^2} + a\log\frac{a + \sqrt{a^2 - y^2}}{y}$$

the equation for the curve given, in more geometric language, by Huygens [1693]. Huygens pointed out that the curve could be interpreted as the path of a stone pulled by a string of length a (hence the name "tractrix"). Thus the tractrix, too, has some mechanical significance. In fact it can be constructed from the famous mechanical curve, the catenary, by a method we shall see in the next section. However, its most important role was in the generation of the *psuedosphere*, a surface discussed in Section 16.4.

EXERCISES

1. Deduce the equiangular property of the logarithmic spiral from its polar equation.

2. Also deduce that the logarithmic spiral is *self-similar*. That is, magnifying $r = e^{k\theta}$ by a factor m to $r = me^{k\theta}$ gives a curve which is congruent to the original. (James

Bernoulli was to impressed by this property of the logarithmic spiral that he arranged to have the spiral engraved on his tombstone, with a motto: *Eadem mutato resurgo* ("Though changed, I arise again the same"). (See James Bernoulli [1692], p. 213.)

16.2. Curvature of Plane Curves

One of the most important ideas in differential geometry is that of *curvature*. The development of this idea from curves to surfaces then to higher dimensional spaces has had many important consequences for mathematics and physics, among them clarification of both the mathematical and physical meaning of "space," "space-time," and "gravitation." In this section we shall look at the beginnings of the theory of curvature in the seventeenth-century theory of curves. As discussed here, this theory concerns *plane* curves only; space curves involve an additional consideration of *torsion* (twisting), which will not concern us.

Just as the direction of a curve C at a point P is determined by its straight-line approximation, that is, tangent, at P, the curvature of C at P is determined by an approximating circle. Newton [1665″] was the first to single out the circle that defines the curvature: the circle through P whose center R is the limiting position of the intersection of the normal through P and the normal through a nearby point Q on the curve (Fig. 16.4). R is called the *center of curvature*, $RP = \rho$ the *radius of a curvature*, and $1/\rho = \kappa$ the *curvature*. It

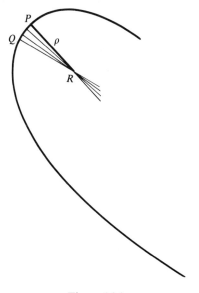

Figure 16.4

follows that the circle of radius r has constant curvature $1/r$. The only other curve of constant curvature is the straight line, which has curvature 0. This is a consequence of the formula for curvature discovered by Newton [1671]:

$$\rho = \frac{[1 + (dy/dx)^2]^{3/2}}{d^2y/dx^2}$$

There is an interesting relationship between a curve C and the locus C' of the center of curvature of C. C is the so-called *involute* of C', which, intuitively speaking, is the path of the end of a piece of string as it is unwound from C' (Fig. 16.5). It is intuitively clear that P, the end of the string, is instantaneously moving in a circle with center at R, the point where the string is tangential to C'.

The geometric property of the cycloid which Huygens [1673] used to design the cycloidal pendulum (Section 12.3) can now be seen as simply this: the involute of a cycloid is another cycloid. Two other stunning results on involutes were obtained by the Bernoulli brothers. James Bernoulli [1692] found that the involute of the logarithmic spiral is another logarithmic spiral, and John Bernoulli [1691/2′] found that the tractrix is the involute of the catenary.

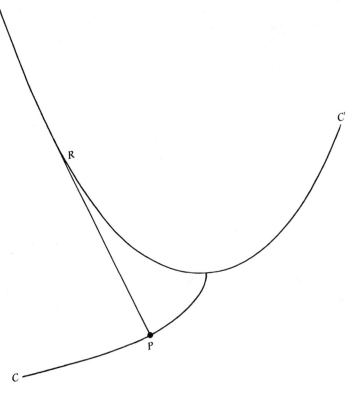

Figure 16.5

Another useful and intuitive definition of curvature, which turns out to be equivalent to the preceding one, was given by Kaestner [1761]. He defined curvature as the rate at which the tangent turns, that is, $d\theta/ds = \lim_{\Delta s \to 0} (\Delta\theta/\Delta s)$, where $\Delta\theta$ is the angle between the tangents at points separated by an arc of the curve of length Δs. It follows from this definition that $\int_{\mathscr{C}} \kappa\, ds = 2\pi$ for a simple closed curve $\mathscr{C}$, since the tangent makes one complete turn on a circuit around $\mathscr{C}$. We shall see in Section 16.6 that this result has a very interesting generalization for curves on nonplanar surfaces.

EXERCISES

1. Use Newton's formula for curvature to show that $\kappa = 0$ implies y is a linear function of x.

2. Show that $d\theta/ds = 1/r$ for the circle of radius r, and deduce that $d\theta/ds = \kappa$ for any curve.

16.3. Curvature of Surfaces

The first approach to defining curvature at a point P of a surface S in three-dimensional space was to express it in terms of the curvature of plane curves, by considering sections of S by planes through the normal at P. Of course, different planes normal to the surface at P may cut the surface in quite different curves, with different curvatures, as the example of the cylinder shows (Fig. 16.6). However, among these curves there will be one of maximum curvature and one of minimum curvature (which may be negative, since we give a sign to curvature according to the side on which the center of curvature

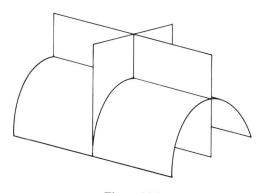

Figure 16.6

lies). Euler [1760] showed that these two curvatures κ_1 and κ_2, called the *principal curvatures*, occur in perpendicular sections, and that together they determine the curvature κ in a section at angle α to one of the principal sections by

$$\kappa = \kappa_1 \cos^2 \alpha + \kappa_2 \sin^2 \alpha$$

This is as far as one can go as long as the curvature of surfaces is subordinated to the curvature of plane curves. A deeper idea occurred to Gauss in the course of his work in geodesy (surveying and mapmaking): curvature of a surface may be detectable *intrinsically*, that is, by measurements which take place entirely on the surface. The curvature of earth, for example, was known on the basis of measurements made by explorers and surveyors, *not* (in the time of Gauss) by viewing it from space. Gauss [1827] made the extraordinary discovery that the quality $\kappa_1 \kappa_2$ can be defined intrinsically and hence can serve as an intrinsic measure of curvature. He was so proud of this result that he called it the *theorema egregium* (excellent theorem). It follows in particular that $\kappa_1 \kappa_2$, which is called the *Gaussian curvature*, is unaffected by bending (without creasing or stretching).

The plane, for example, has $\kappa_1 = \kappa_2 = 0$ and thus zero Gaussian curvature. Hence so has any surface obtained by bending a plane, such as a cyclinder. We can verify the theorema egregium in this case, because one of the principal curvatures of a cylinder is obviously zero.

Surfaces S_1, S_2 obtained from each other by bending are said to be *isometric*. More precisely, S_1 and S_2 are isometric if there is a one-to-one correspondence between points P_1 of S_1 and points P_2 of S_2 such that distance between P_1, P_1' in $S_1 =$ distance between the corresponding P_2, P_2' in S_2, where the distances are measured *within* the respective surfaces. A more precise statement of the theorema egregium then is: if S_1, S_2 are isometric, then S_1, S_2 have the same Gaussian curvature at corresponding points. The converse statement is not true: there are surfaces S_1, S_2 which are not isometric even though there is a one-to-one (and continuous) correspondence between them for which Gaussian curvature is the same at corresponding points. An example is given in Strubecker [1964], Vol. 3, p. 121, involving surfaces of nonconstant Gaussian curvature.

For surfaces of constant Gaussian curvature there is better agreement between isometry and curvature, as we shall see in the next section. From now on, unless otherwise qualified, "curvature" will mean Gaussian curvature.

16.4. Surfaces of Constant Curvature

The simplest surface of constant positive curvature is the sphere of radius r, which has curvature $1/r^2$ at all points. Other surfaces of curvature $1/r^2$ may be obtained by bending portions of the sphere; however, all such surfaces have either edges or points where they are not smooth, as was proved by Hilbert

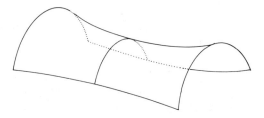

Figure 16.7

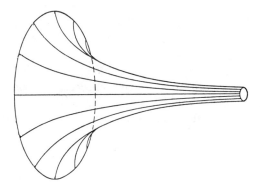

Figure 16.8

[1901]. The plane, as we have observed, has zero curvature, and so have all surfaces obtained by bending the plane or portions of it.

It remains to investigate whether there are surfaces of constant *negative* curvature. In ordinary space, such a surface has principal curvatures of opposite sign at each point, giving it the appearance of a saddle (Fig. 16.7). A number of surfaces of constant negative curvature were given by Minding [1839]. The most famous of them is the *pseudosphere*, the surface of revolution obtain by rotating a tractrix about the x axis (Fig. 16.8). This surface was investigated as early as 1693 by Huygens, who found its surface area, which is finite, and the volume and center of mass of the solid it encloses, which are also finite (Huygens [1693']).

The pseudosphere is in some ways the negative-curvature counterpart of the cylinder, and hence one may wonder whether there is a surface of constant negative curvature that is more like a plane. It was proved by Hilbert [1901] that there is no smooth unbounded surface of constant negative curvature in ordinary space, so this rules out planelike surfaces and also accounts for the "edge" on the pseudosphere. One can, however, obtain a "plane" of negative curvature by introducing a nonstandard notion of length into the euclidean plane. This discovery of Beltrami [1868] will be discussed in the next chapter, along with other implications of negative curvature for noneuclidean geometry.

These geometric implications can also be glimpsed if we return to the question of whether surfaces S_1, S_2 of equal curvature are isometric. Even with constant curvature this is still not true, since a plane is not isometric to a cylinder. What *is* true, though, is that any sufficiently small portion of the plane can be mapped isometrically into any part of the cylinder. Minding [1839] showed that the analogous result is true for any two surfaces S_1, S_2 of the same constant curvature. Taking $S_1 = S_2$, this result can be interpreted as saying *rigid motion* is possible within S_1; a body within S_1 can be moved, without any shrinking or stretching, to any part of S_1 large enough to contain it. The latter restriction is necessary, for example, for the pseudosphere since it becomes indefinitely narrow as $x \to \infty$.

The possibility of rigid motion was fundamental to Euclid's geometry of the plane, and with the discovery of curved surfaces which supported rigid motion, Euclid's geometry could be seen as a special case—the zero curvature case—of something broader. The broader notion of geometry on a surface begins to take shape once one has an appropriate notion of "straight line." This will be developed in the next section.

16.5. Geodesics

A "straight line," or *geodesic* as it is called, can be defined equivalently by a shortest distance property or a zero curvature property. The shortest distance definition was historically first, even though it is mathematically deeper and subject to the inconvenience that a geodesic is *not* necessarily the shortest way between two points. On a sphere, for example, there are two geodesics between two nearby points P_1, P_2: the short portion and the long portion of the great circle through P_1, P_2. We can cover both by saying that a geodesic gives the shortest distance between any two of its points which are sufficiently close together. In talking about shortest distance, even between nearby points P_i, P_j, one still has the calculus of variations problem of finding which curve from P_i to P_j has minimum length. Nevertheless, this is how geodesics were first defined, by James and John Bernoulli, and Euler [1728'] found a differential equation for geodesics from this approach.

A more elementary approach is to define the *geodesic curvature* κ_g at P of a curve C on a surface S as the ordinary curvature of the orthogonal projection of C in the tangent plane to S at P. As one might expect, geodesic curvature can also be defined intrinsically, and κ_g was introduced in this way by Gauss [1825]. A geodesic is then a curve of zero geodesic curvature. This is the definition of Bonnet [1848].

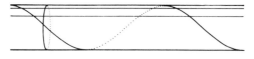

Figure 16.9

The latter definition shows the great circles on the sphere to be geodesics immediately, since their projections onto tangent planes are straight lines. Other examples are the horizontal lines, vertical circles, and helices on the cylinder (Fig. 16.9). These all come from straight lines on the plane that is rolled up to form the cylinder. Geodesics on the pseudosphere, and other surfaces of negative curvature, are not all so simple to describe. However, we shall see in the next chapter that they become simple when one maps the surface of constant negative curvature onto a plane with a suitable distance function.

16.6. The Gauss–Bonnet Theorem

In Section 16.2 we observed that

$$\int_{\mathscr{C}} \kappa \, ds = 2\pi$$

for a simple closed curve $\mathscr{C}$ in the plane. This result has a profound generalization to curved surfaces known as the *Gauss–Bonnet* theorem. On a curved surface κ must be replaced by the geodesic curvature κ_g, and the theorem states

$$\int_{\mathscr{C}} \kappa_g \, ds = 2\pi - \iint_{\mathscr{R}} \kappa_1 \kappa_2 \, dA$$

where A denotes area and $\mathscr{R}$ is the region enclosed by $\mathscr{C}$ (Bonnet [1848]). Gauss himself published only a special case, or rather the limit of a special case, in which $\mathscr{C}$ is a geodesic triangle. In this case, of course, $\kappa_g = 0$ along the sides of $\mathscr{C}$, and κ_g becomes infinite at the corners. By rounding off the corners by small arcs ds one sees (Fig. 16.10) that

$$\int_{\mathscr{C}^*} \kappa_g \, ds \cong \alpha' + \beta' + \gamma'$$

where α', β', γ' are the external angles of the triangle and $\mathscr{C}^*$ is the rounded approximation to the triangle $\mathscr{C}$. Then by letting the size of the roundoffs tend to zero one gets

$$\int_{\mathscr{C}} \kappa_g \, ds = \alpha' + \beta' + \gamma'$$
$$= 3\pi - (\alpha + \beta + \gamma)$$

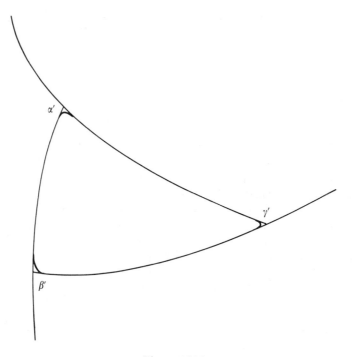

Figure 16.10

where α, β, γ are the internal angles of the triangle. Introducing the quantity

$$(\alpha + \beta + \gamma) - \pi$$

called the *angular excess* of the triangle (because an ordinary triangle has angle sum π), we have

$$\int_{\mathscr{C}} \kappa_g \, ds = 2\pi - \text{angular excess}$$

and the result of Gauss [1827] was that

$$\text{angular excess} = \iint_{\mathscr{R}} \kappa_1 \kappa_2 \, dA$$

We see that the *integral* of the Gaussian curvature has a more elementary geometric meaning than the curvature $\kappa_1 \kappa_2$. It appears, in fact, that Gauss thought about angular excess first, then the curvature integral, and only last about the curvature itself. The decomposition into principal curvatures probably came later, when he reworked his geometric ideas into analytic form, reversing the order of discovery in the process. Dombrowski [1979] made a plausible reconstruction of the original approach, using clues from the unpublished work of Gauss.

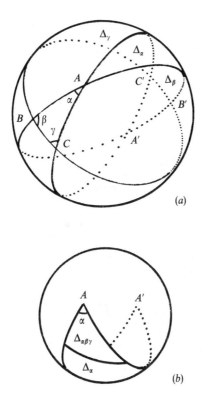

Figure 16.11

The role of angular excess can be seen more plainly in the case of constant curvature $\kappa_1 \kappa_2 = c$. In this case

$$\text{angular excess} = c \times \text{area of } \mathscr{R}$$

so the angular excess gives a measure of area, a result Gauss claimed, in a letter [1846], to have known in 1794. In fact, the special case of this result for the sphere was known to Thomas Harriot in 1603 (see Lohne [1979]). Harriot's elegant proof goes as follows (Fig. 16.11). Prolonging the sides of triangle ABC yields a partition of the sphere into four pairs of congruent, diametrically opposite triangles (Fig. 16.11a). We denote the area of ABC and its diametric opposite $A'B'C'$ by $\Delta_{\alpha\beta\gamma}$. The other three pairs represent areas $\Delta_\alpha, \Delta_\beta, \Delta_\gamma$, which complement $\Delta_{\alpha\beta\gamma}$ in "slices" of the sphere of angles α, β, γ, respectively (Fig. 16.11b). Since the area of a slice is $2r^2$ times its angle, where r is the radius of the sphere, we have

$$\Delta_{\alpha\beta\gamma} + \Delta_\alpha = 2r^2\alpha$$

$$\Delta_{\alpha\beta\gamma} + \Delta_\beta = 2r^2\beta$$

$$\Delta_{\alpha\beta\gamma} + \Delta_\gamma = 2r^2\gamma$$

whence, by addition

$$3\Delta_{\alpha\beta\gamma} + (\Delta_\alpha + \Delta_\beta + \Delta_\gamma) = 2r^2(\alpha + \beta + \gamma) \tag{1}$$

on the other hand,

$$2(\Delta_{\alpha\beta\gamma} + \Delta_\alpha + \Delta_\beta + \Delta_\gamma) = \text{area of sphere}$$
$$= 4\pi r^2$$

and substituting this in (1) gives

$$\Delta_{\alpha\beta\gamma} = r^2(\alpha + \beta + \gamma - \pi)$$

as required, since $1/r^2$ = curvature of the sphere.

Gauss was interested in the counterpart of this result for negative curvature, in which case the angle sum of a triangle is less than π and one has angular *defect* rather than angular excess. His investigations in this case led him not only to Gaussian curvature but also to noneuclidean geometry.

16.7. Biographical Notes: Harriot and Gauss

The discoveries of Thomas Harriot described in this chapter and the last seem to entitle him to a secure place in the history of mathematics, perhaps alongside others who made a few penetrating contributions, such as Desargues and Pascal. Unfortunately, Harriot's place is not yet clear. It was clouded by exaggerated claims made by seventeenth- and eighteenth-century admirers, and until recently the disorder and inaccessibility of his papers have made any claims difficult to verify. In addition, Harriot was a very secretive man, and little is known about his life. He lived in the world of Sir Walter Raleigh, Christopher Marlowe, and Guy Fawkes—a lurid and fascinating world, but a very dangerous one—and probably believed that secrecy was necessary for his survival. As a result, our present understanding of Harriot (see the biography by Shirley [1983]) is based on a meager set of facts about him and a good deal of extrapolation from knowledge of his less discreet contemporaries.

All that we know of Harriot's early life comes from a record of his entry into Oxford University in December 1577, stating that his age was then 17 and his father "plebeian." The only other information about his family comes from his will of 1621, which mentions a sister and a cousin. It seems probable that he never had children and never married. At Oxford, Harriot gained the standard bachelor's degree in classics, but he must have picked up some Euclid and astronomy, which were offered to masters' candidates. He would also have heard Richard Hakluyt, author of the famous *Voyages*, who was then just beginning to lecture on the geography of the New World opened up by the sixteenth-century navigators.

It was probably Hakluyt who inspired Harriot to travel to London in the early 1580s and seek out Sir Walter Raleigh. Raleigh was then about 30 and the most powerful member of Queen Elizabeth's inner circle, with grand dreams of wealth through exploration. Harriot must have impressed Raleigh with his grasp of the mathematical problems of navigation, for around 1583 he joined Raleigh's household as an instructor, with considerable freedom to conduct his own research. Harriot held classes in navigation as part of Raleigh's preparations for the voyage to Virginia in 1585, led by Sir Richard Grenville, which was the first attempt at British settlement in the New World. Although the attempt was unsuccessful, it was the biggest adventure of Harriot's life. He studied Indian languages and customs and wrote a book on the settlement entitled *A Brief and True Report of the New Found Land of Virginia* (1588), the only one of Harriot's works published in his lifetime.

With Raleigh as patron, Harriot was financially secure, and he remained so for the rest of his life. However, he was also at the mercy of Raleigh's political fortunes. By 1592, the 40-year-old Raleigh was finding his role as favorite of the nearly 60-year-old queen increasingly irksome, and he secretly married one of the queen's servants, Elizabeth Throckmorton. He may have married her as early as 1588, but at any rate the secret was out when Lady Raleigh gave birth to a son in 1592, and Raleigh was imprisoned in the Tower of London. Harriot did not suffer for his direct association with Raleigh, but through him he was linked with Christopher Marlowe, at the latter's sensational trial for atheism in 1593.

Marlowe, the dramatist, had a secret life in espionage and other unsavory activities, and any number of accusations could have been made against him, though which ones were true it is now impossible to say. Unfortunately for Harriot, the second of Marlowe's offenses against religion was said to be that "He affirmeth that Moyses was but a Jugler, and that one Heriots being Sir W. Raleighs man can do more than he." As it happened, the proceedings were terminated by the murder of Marlowe in a tavern brawl, and Harriot was not called to testify, but he was left publicly under suspicion.

Harriot did not desert Raleigh, but he was prudent enough to seek another patron, and he found one in Henry Percy, the Ninth Earl of Northumberland. Henry, known as the "Wizard Earl," was a friend of Raleigh and, like him, interested in science and philosophy. In 1593 he gave Harriot a grant, later to become an annual pension, of £80. This sum was twice the salary of the best paid teachers of the time, and it enabled Harriot to maintain a house and servants on the earl's property on the Thames near London. The house, known as Sion House, remained Harriot's home and laboratory for the rest of his life.

But once again Harriot was unlucky in his choice of friends. The earl's cousin, Thomas Percy, was the man who rented the cellars under the Houses of Parliament in the famous plot to blow up King James I with gunpowder on November 5, 1605. Harriot was dragged into the investigation and imprisoned for a short time, on suspicion that he had secretly cast a horoscope of the king. James I was terrified of black magic and indiscriminately viewed all mathematicians as astrologers and magicians. In the end, though, no evidence against Harriot was found, and it was the earl who suffered more, being imprisoned in the Tower from 1605 to 1621.

Meanwhile, Raleigh had fared even worse. After several spells in the Tower, he was released in 1616 to lead an expedition in search of the mythical city of gold, El Dorado. When the expedition returned in a shambles, Raleigh was rearrested and executed on an old treason charge from 1603. One of the few personal documents kept by Harriot is his summary of the speech made by Raleigh at the scene of his execution in 1618 (see Shirley [1983], p. 447).

A month after Raleigh's death, a bright comet appeared in the skies, and Harriot's observations of it were his last major scientific endeavor. He had been suffering for some years from a painful cancer of the nose and finally succumbed to it on a visit to London in 1621. He was buried at St. Christopher's Church in Threadneedle Street, later destroyed in the Great Fire of 1666. The site is now part of the Bank of England, where a replica of Harriot's original monument was installed on July 2, 1971, the three hundred and fiftieth anniversary of his death.

Carl Friedrich Gauss was born in Brunswick in 1777 and died in Göttingen in 1855. He was the only child of Gebhard Gauss and Dorothea Benze, though his father had another son from a previous marriage. Gebhard earned his living mainly from manual labor, but he also did a little accounting, and Gauss is said to have corrected an error in his father's arithmetic at the age of three. (It should be borne in mind here that stories about Gauss' youth were told by Gauss himself in old age, and in a few cases there is evidence that he was prone to exaggerate his own precocity.) Gauss was not close to his father and believed that his genius was inherited from his mother. He started school in 1784 and his teacher, Büttner, soon recognized his ability and obtained advanced books for him. Büttner's assistant Martin Bartels (1769–1836) also gave Gauss special attention. Bartels was himself a beginning mathematician who later became professor at the University of Kazan and the teacher of Lobachevsky (see next chapter).

Gauss entered secondary school in 1788, and in 1791 he won an annual grant from the Duke of Brunswick, something like a government scholarship. He was also selected to enter the Collegium Carolinum, a new scientific academy for outstanding secondary students. In his years there, 1792–1795, Gauss studied the works of Newton, Euler, and Lagrange and began investigations of his own, mainly numerical experiments on such things as the arithmetic-geometric mean. In 1795 he left Brunswick to study at Göttingen in the adjoining state of Hannover, which was then ruled by George III of England. The duke would have preferred Gauss to remain in Brunswick, at the local university of Helmstedt, but continued his financial support nevertheless. Gauss actually chose Göttingen because of its better library and later spoke very disparagingly of its mathematics professor, Kaestner. It is true that the student achievements of Gauss, which began with his construction of the regular 17-gon (Section 2.3) and culminated in his proof of the fundamental theorem of algebra (Section 13.7), dwarfed those of his teachers. Still, one

Figure 16.12. Carl Friedrich Gauss.

wonders whether Kaestner's definition of curvature (Section 16.2) might not
have been useful to Gauss when he took up differential geometry later.

Gauss returned to Brunswick in 1798 and lived there until 1807. Figure
16.12 is a portrait of him from this period, which was the happiest and most
productive of his life. Gauss published his great work on number theory, the
Disquisitiones Arithmeticae, in 1801, made a spectacular entry into astronomy
in the same year by predicting the position of the asteroid Ceres, and married
Johanna Osthoff in 1805. Writing to his friend Wolfgang Bolyai in 1804, Gauss
was uncharacteristically warm and open when it came to Johanna:

> The beautiful face of a madonna, a mirror of peace of mind and health, tender,
> somewhat fanciful eyes, a blameless figure—this is one thing; a bright mind and
> an educated language—this is another; but the quiet, serene, modest and chaste
> soul of an angel who can do no harm to any creature—that is the best.
> (Translation from Bühler [1981], p. 49)

If only Johanna had lived longer, Gauss might have become a very different man. But in 1809 she died, shortly after giving birth to their third child. Gauss was devastated by the blow and never quite recovered his equilibrium.

Less than a year after Johanna's death, he married Minna Waldeck, the daughter of a Göttingen professor. Unlike Johanna, who was a tanner's daughter, Minna had social status and pretensions which caused Gauss uneasiness and embarrassment. Soon after their engagement, for example, he had to tell Minna not to write to his mother, as his mother could not read. Minna also suffered from poor health, and after the couple had had three children between 1811 and 1816 she became virtually a permanent invalid. Gauss found this burden difficult to bear and compounded his problems by unsympathetic treatment of his children. The family drama came to a head in 1830, when their eldest son, Eugen, emigrated to America after a row with his father. The following year Minna died of tuberculosis.

During this unhappy period Gauss was less productive mathematically, but this was not due to family troubles so much as his choice of career. He had become director of the Göttingen observatory in 1807 and in 1817 substituted geodesy for some of his astronomical duties, doing arduous field work every summer from 1818 to 1825 for the geodetic survey of Hannover. Gauss appears to have seldom regretted this choice of career—he disliked teaching and thought that other mathematicians had little to teach him—but it cannot be said that his contributions to astronomy and geodesy were as great as his contributions to mathematics. Indeed, the best things to come out of his geodetic work were his theorem on conformal mapping and complex functions (Section 15.2) and his intrinsic notion of curvature (Section 16.3).

In the 1830s Gauss experienced something of a rebirth with the arrival of the young physicist Wilhelm Weber in Göttingen. The two collaborated enthusiastically in the investigation of magnetism, with Gauss making contributions to both the theory and practice (the electric telegraph). However, their partnership was broken in 1837 when Weber was fired for his courageous refusal to swear an oath of allegiance to the new king of Hannover.

Among the few bright spots of Gauss' later years were his students Eisenstein and Dedekind, as well as Riemann's lecture on the foundations of geometry in 1854. After spending most of his life aloof from other mathematicians, Gauss must have been comforted to find at last that there *were* students capable of understanding his ideas and carrying them further. We can only wonder what might have been if he had made this discovery earlier.

CHAPTER 17

Noneuclidean Geometry

17.1. The Parallel Axiom

Until the nineteenth century, Euclid's geometry enjoyed absolute authority, both as an axiomatic system and as a description of physical space. Euclid's proofs were regarded as models of logical rigor, and his axioms were accepted as correct statements about physical space. Even today, euclidean geometry is the simplest type of geometry, and it furnishes the simplest description of physical space for everyday purposes. Beyond the everyday world, however, lies a vast universe that can be understood only with the help of an expanded geometry. The expansion of geometric concepts initially grew from dissatisfaction with one of Euclid's axioms, the *parallel axiom.*

For our purposes, the most convenient statement of the parallel axiom is as follows:

Axiom P_1. For each straight line L and point P outside L there is exactly one line through P that does not meet L.

There are many other equivalent statement of Axiom P_1, some obviously fairly close to it, for example, Euclid's own:

> That if a straight line falling on two straight lines make the interior angles on the same side less than two right angles, the two straight lines, if produced indefinitely, meet on that side on which are the angles less than the two right angles. (Heath [1925], p. 202)

Other statements of Axiom P_1 are less obviously equivalent. For example,

(i) The angle sum of a triangle $= \pi$ (Euclid).
(ii) The locus of points equidistant from a straight line is a straight line (al-Haytham, c. A.D. 1000).

(iii) Similar triangles of different sizes exist (Wallis [1663]; see Fauvel and
      Gray [1987], p. 510).

Thus a denial of the parallel axiom entails denial of (i), (ii), and (iii). A denial
of (iii) means in particular that scale models are impossible, since three points
in the original object and the three corresponding points of a scale model
would define similar triangles of different sizes.

Such unlikely consequences convinced many people that the parallel axiom
was a logically necessary property of straight lines, already implied by the
other axioms of Euclid, and so efforts were made to prove it outright.

The most tenacious attempt, entitled *Euclides ab omni naevo vindicatus*
(Euclid cleared of every flaw), was made by Saccheri [1733]. Saccheri's plan
of attack began by subdividing the denial of the parallel axiom into two
alternatives:

Axiom P_0. There is no line through P that does not meet L.

Axiom P_2. There are at least two lines through P that do not meet L.

The next step was to destroy each alternative by deducing a contradiction
from it. He succeeded in deducing a contradiction from Axiom P_0, using other
axioms of Euclid, such as the axiom that a straight line can be prolonged
indefinitely. (Such additional assumptions are certainly necessary, since great
circles on the sphere have some properties of straight lines, except that they
are finite in length.)

Saccheri was less successful with Axiom P_2. The consequences he derived
from it, hoping to obtain a contradiction, were as follows: among the lines M
through P which do not meet L there are two extremes, M^+ and M^-, called
parallels or *asymptotic lines* (Fig. 17.1). Any line M strictly between M^+ and
M^- has a common perpendicular with L and, furthermore, the position of this
perpendicular tends to infinity as M tends to M^+ or M^-. Although curious,
these consequences to Axiom P_2 were not contradictory and Saccheri, sensing

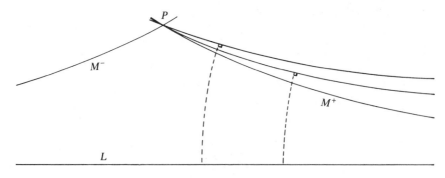

Figure 17.1

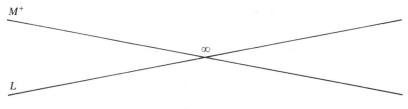

M^+

L

∞

Figure 17.2

that the contradiction was slipping away from him, tried to overtake it by proceeding to infinity.

He claimed that the asymptotic line M^+ would meet L at infinity and have a common perpendicular with it there. This was perhaps plausible, given similar arguments in projective geometry, though it certainly would not have been accepted by Euclid. But it *still* was no contradiction. Saccheri merely claimed that such a conclusion was "repugnant to the nature of the straight line" (Saccheri [1733/1920], p. 173), perhaps visualizing an intersection like Figure 17.2. But why should asymptotic lines not be tangential at infinity? History was to show that this was an appropriate resolution of Saccheri's "contradiction" (see Section 17.5). Thus Saccheri's results were not, as he thought, steps toward a proof of the parallel axiom; they were the first theorems of a *noneuclidean* geometry in which Axiom P_2 replaces the parallel axiom.

17.2. Spherical Geometry and Its Conjectured Hyperbolic Analogue

In rejecting P_0 because of its incompatibility with infinite lines, Saccheri avoided having to consider the most natural geometry in which P_0 holds, that of the sphere with great circles as "lines." Spherical geometry had been cultivated since ancient times to meet the needs of astronomers and navigators, and formulas for the side lengths and areas of spherical triangles were well known. But because it was such a practical subject, and not considered by Euclid, the axiomatic significance of spherical geometry was at first ignored. What did happen, though, was that the first explorations of Axiom P_2 were guided by the analogy of the sphere.

Lambert [1766] made the striking discovery that Axiom P_2 implies that the area of a triangle with angles α, β, γ is proportional to $\pi - (\alpha + \beta + \gamma)$, its angular defect. In other words,

$$\text{area} = -R^2(\alpha + \beta + \gamma - \pi)$$

for some positive constant R^2. Having rediscovered Harriot's theorem that

$$\text{area} = R^2(\alpha + \beta + \gamma - \pi)$$

for a triangle on the sphere of radius R, Lambert mused that one "could almost conclude that the new geometry would be true on a sphere of imaginary radius." What a sphere of radius iR might be was never explained, but the idea of using complex numbers to generate the formulas of a hypothetical geometry proved fruitful.

It was found that formulas derived from Axiom P_2 could also be obtained by replacing R by iR in corresponding formulas of spherical geometry. For example, Gauss [1831] deduced from Axiom P_2 that the circumference of a circle of radius r is $2\pi R \sinh r/R$. The same result follows by replacing R by iR in $2\pi R \sin r/R$, which is the circumference of a circle of radius r on the sphere (where, of course, r is measured *on* the spherical surface. See exercise 1).

The geometry of Axiom P_2 was called *hyperbolic* by Klein [1871]. One reason for this is that its formulas involve hyperbolic functions, whereas those of spherical geometry involve circular functions. Lambert [1766] introduced the hyperbolic functions and noted their analogy with the circular functions, but he did not follow through with a complete translation of spherical formulas into hyperbolic formulas. This was first done by Taurinus [1826], one of a small circle who corresponded with Gauss on geometric questions.

This gave hyperbolic geometry a second leg to stand on, but there was still nothing solid under its feet. Neither Gauss nor Taurinus seemed confident of finding a convincing interpretation of hyperbolic geometry, even though Gauss [1827] came remarkably close with the "Gauss–Bonnet" theorem. As mentioned in Section 16.6, this theorem shows that surfaces of constant negative curvature give a geometry in which angular defect is proportional to area, and Gauss knew that the pseudosphere was such a surface. Gauss' student Minding [1840] even showed that the hyperbolic formulas for triangles hold on the pseudosphere, but no one at that time commented on the likely importance of this result for hyperbolic geometry. Perhaps it was clear that the pseudosphere cannot serve as a "plane," because it is infinite in only one direction. It was not until 1868, when Beltrami extended the pseudosphere to a true *hyperbolic plane*—a surface locally like the pseudosphere but infinite in all directions—that hyperbolic geometry was finally placed on a firm foundation.

EXERCISES

1. Prove that the circumference of the circle C of radius r on the sphere of radius R (Fig. 17.3) is $2\pi R \sin r/R$.

2. Show that both $2\pi R \sin r/R$ and $2\pi R \sinh r/R$ tend to $2\pi r$ as $R \to \infty$.

3. Show directly that angular excess (or defect) is *additive*, that is, if a triangle Δ is the union of triangles Δ_1, Δ_2 with only an edge in common, then

   angular excess (Δ) = angular excess (Δ_1) + angular excess (Δ_2)

   Deduce an analogous result for angular excess of polygons.

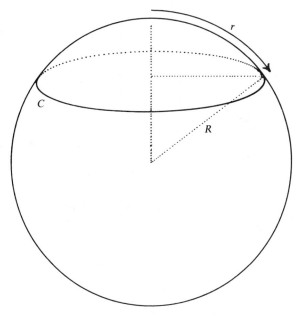

Figure 17.3

4. Conclude, from exercise 3 and suitable assumptions about area, that area is proportional to angular excess (Bonola [1912], p. 46).

17.3. The Hyperbolic Geometry of Bolyai and Lobachevsky

The most important contributors to hyperbolic geometry between Gauss and Beltrami were Lobachevsky and Bolyai, who published independent discoveries of the subject (Lobachevsky [1829]; Bolyai [1832]). Because of their courage in advocating an unconventional geometry, Bolyai and Lobachevsky have won the admiration of many historians. Nevertheless, the historical significance of their work is debatable. The bulk of their results were already known to Gauss and his circle and could have been picked up, at least in hazy form, from existing publications and personal contacts. Lambert [1766] and Taurinus [1826] were in print, and Bolyai's father, W. F. Bolyai, was a lifelong friend of Gauss, as was Lobachevsky's teacher Bartels. In any case their work, though more systematic than previous attempts and expressed with a lot more conviction, attracted very little attention at first. We have seen how the possibility of using differential geometry to justify hyperbolic geometry was overlooked until 1868. Up to that time, there seemed no reason to take hyperbolic geometry seriously.

In retrospect, of course, the theorems of Bolyai and Lobachevsky can be seen to unify the fragmentary results of their predecessors very nicely. They cover the basic relations between sides and angles of triangles (hyperbolic trigonometry), the measure of polygonal areas by angular defect, and formulas for circumference and area of circles. Lobachevsky [1836] broke new ground by finding volumes of polyhedra, which turn out to be far from elementary, involving the function $\int_0^\theta \log 2|\sin t|\, dt$.

Both Bolyai and Lobachevsky considered a three-dimensional space satisfying Axiom P_2 and made extensive use of a surface peculiar to this space, the *horosphere*. A horosphere is a "sphere with center at infinity," and it is *not* a hyperbolic plane. Wachter, a student of Gauss, observed in a letter of 1816 (published in Stäckel [1901]) that the geometry of the horosphere is in fact euclidean. This astonishing result was rediscovered by Bolyai and Lobachevsky, and they anticipated that it would make euclidean geometry subordinate to hyperbolic. We shall see in Section 17.5 how this view was vindicated in the work of Beltrami.

17.4. Beltrami's Projective Model of Hyperbolic Geometry

Interest in hyperbolic geometry was rekindled in the 1860s when unpublished work of Gauss, who had died in 1855, came to light. Learning that Gauss had taken hyperbolic geometry seriously, mathematicians became more receptive to noneuclidean ideas. The works of Bolyai and Lobachevsky were rescued from obscurity and, approaching them from the viewpoint of differential geometry, Beltrami [1868] was able to give them the concrete explanation that had eluded all his predecessors.

Beltrami was interested in the geometry of surfaces and had found the surfaces which could be mapped onto the plane in such a way that their geodesics went to straight lines [1865]. They turned out to be just the surfaces of constant curvature. In the case of positive curvature, the sphere, such a mapping is central projection onto a tangent plane (Fig. 17.4), though of

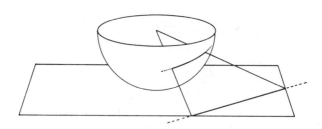

Figure 17.4

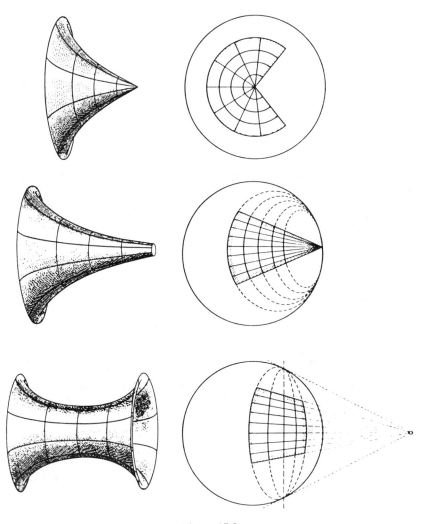

Figure 17.5

course this maps only half the sphere onto the whole plane. The mappings of surfaces of constant negative curvature, on the other hand, take the *whole* surface onto only *part* of the plane. Figure 17.5, adapted from Klein [1928], shows some of these mappings (the middle one being of the pseudosphere).

Each negatively curved surface S is mapped onto a portion of the unit disk. Beltrami [1868] realized that the disk can then be viewed as a natural extension of S to an "infinite plane," thus bypassing the problem of constructing "planelike" surfaces of constant negative curvature in ordinary space. Instead, one takes the disk as the "plane," line segments within it as "lines," and "distance" between two points of the disk as the distance between their preimage points on the surface S. The function $d(P, Q)$ giving "distance"

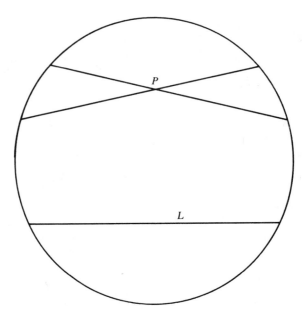

Figure 17.6

between points P, Q of the disk in this way turns out to be meaningful for all points inside the unit circle, so the notion of "distance" extends to the whole open disk. As Q approaches the unit circle, $d(P,Q)$ tends to infinity, so the "plane," and hence the "lines" in it, are indeed infinite with respect to this nonstandard "distance."

It follows that all the axioms of Euclid, except the parallel axiom, are satisfied with this new interpretation of "plane," "line," and "distance." Instead of the parallel axiom, one has of course Axiom P_2, since there is more than one "line" through a point P outside a given "line" L (Fig. 17.6). Beltrami also observed that the rigid motions of the "plane," since they preserve straight lines, are necessarily projective transformations. They are precisely those projective transformations of the plane that map the unit circle onto itself. Consequently, this model of the hyperbolic plane is often called the *projective model*. Cayley [1859] had already observed that these projective transformations could be used to define a "distance" $d(P,Q)$ in the unit disk—by saying $d(P,Q) = d(P',Q')$ if a transformation preserving the unit circle sends P to P' and Q to Q'—but he had not realized that the geometry obtained was that of Bolyai and Lobachevsky.

The pseudosphere is not entirely superseded by the projective model, since it remains the source of "real" distances and angles, whereas those in the projective model are necessarily distorted. One of the distinctive curves of the hyperbolic plane, the *horocycle*, or circle with center at infinity, is shown particularly clearly on the pseudosphere. If one imagines, following Beltrami

[1868], the pseudosphere wrapped by infinitely many turns of an infinitely thin covering, then the edge of this covering (along the rim of the pseudosphere) is a horocycle. The middle picture of Figure 17.5 shows the image of one turn of the covering, drawn solidly, and horocycles resulting from continued unwrapping are shown as dashed lines.

EXERCISE

1. Similarly interpret the top and bottom pictures in Figure 17.5. Pick out *hyperbolic circles* (a hyperbolic circle is the locus of a point moving at constant "distance" from a fixed point) and *equidistant curves* (an equidistant curve is the locus of a point moving at constant "distance" from a "line").

17.5. Beltrami's Conformal Models of Hyperbolic Geometry

The projective model of the hyperbolic plane distorts angles as well as lengths. One can see this with the asymptotic geodesics on the pseudosphere, which clearly tend to tangency at infinity yet are mapped onto lines meeting at a nonzero angle at the boundary of the unit disk (Fig. 17.5). Beltrami [1868'] found that models with true angles—the so-called *conformal models*—could be obtained by sacrificing straightness of "lines." His basic conformal model is not in fact part of the plane but part of a hemisphere. It is erected over the projective model and its "lines" are vertical sections of the hemisphere (hence semicircles) over the "lines" of the projective model (Fig. 17.7).

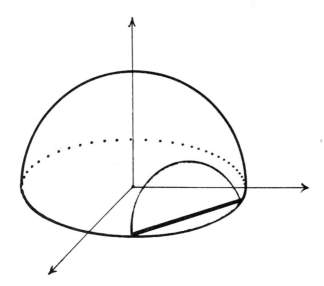

Figure 17.7

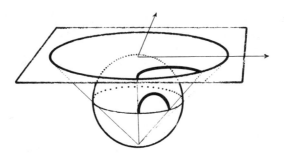

Figure 17.8

The "distance" between points on the hemisphere is equal to the "distance" between the points beneath them in the projective model. Later we shall see that "distance" on the hemisphere also has a simple direct definition.

Two planar conformal models are obtained from the hemisphere model by stereographic projection, which, as we know from Section 14.2, preserves angles and sends circles to circles. The first of these is a disk (Fig. 17.8) which, by change of scale, can again be taken as the unit disk. The second (Fig. 17.9) is a half plane, which we take to be the upper half plane, $y > 0$. Since the "lines" in the hemisphere model are circular and orthogonal to the equator, "lines" in the planar conformal models are again circular, orthogonal to the boundary of the disk and half plane, respectively, or straight lines in exceptional cases. To avoid continual mention of these exceptional cases—namely, line segments through the disk center and lines $x =$ constant in the half plane—we consider lines to be circles of infinite radius.

One of the beauties of the conformal models is that other important curves—hyperbolic "circles," horocycles, and equidistant curves—are also real circles. Each curve equidistant from a given "line" L is a circle through the end points of L on the boundary. Horocycles are circles tangential to the boundary and also, in the half plane model, the lines $y =$ constant. A circle that does not meet the boundary is a hyperbolic "circle" but its "center", at equal "distance" from all its points, is not at the euclidean center. Figure 17.10 shows some of these curves. Note also that asymptotic "lines" are tangential at "infinity" (the boundary), and that the boundary is their common perpendicular, thus resolving the situation that Saccheri (Section 17.1) thought to be contradictory.

"Distance" is particularly easy to express in the half plane model. The

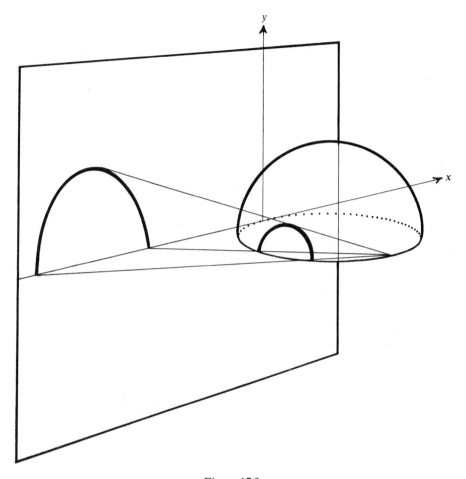

Figure 17.9

"distance" ds between infinitesimally close points (x, y) and $(x + dx, y + dy)$ is

$$ds = \frac{\sqrt{dx^2 + dy^2}}{y}$$

that is, the euclidean distance divided by y. Thus "distance" $\rightarrow \infty$ as a point approaches the boundary $y = 0$ of the half plane, as expected. Keeping x constant, we find by integration that "distance" along a vertical line increases exponentially with respect to euclidean distance as y decreases. For example, the "distances" between the successive points at which $x = 0$ and $y = 1, \frac{1}{2}, \frac{1}{4}, \ldots$, are equal. The formula for ds was first obtained by Liouville [1850] by directly mapping the pseudosphere into the half plane, making simplifying

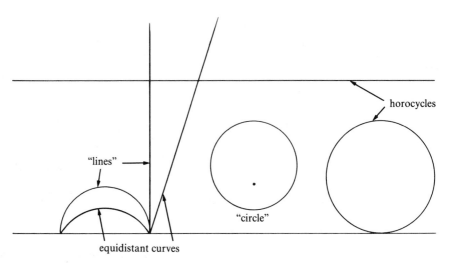

Figure 17.10

changes of variable. However, Liouville did not realize that the half plane with his "distance" formula was a model of hyperbolic geometry. The "distance" formula for the conformal disk had also been obtained before Beltrami, by Riemann [1854], but again without noticing the hyperbolic geometry.

Beltrami [1868'] not only obtained these models, in a unified way, but he also extended the idea to n dimensions. For example, he gave a model of the three-dimensional space considered by Bolyai and Lobachevsky as the upper half, $z > 0$, of ordinary (x, y, z)-space, with "distance"

$$ds = \frac{\sqrt{dx^2 + dy^2 + dz^2}}{z}$$

"Lines" are then semicircles orthogonal to $z = 0$ and "planes" are hemispheres orthogonal to $z = 0$. Restricting the "distance" function to such a hemisphere turns out to give Beltrami's hemisphere model. Thus the hemisphere model can be viewed as a hyperbolic plane lying in hyperbolic 3-space. The horospheres of the half space model are spheres tangential to $z = 0$, together with the planes $z = $ constant. Beltrami [1868'] pointed out that on $z = $ constant we have

$$ds = \frac{\sqrt{dx^2 + dy^2 + dz^2}}{\text{constant}}$$

that is, "distance" is proportional to euclidean distance. Thus he had an immediate proof of Wachter's wonderful theorem that the geometry of the horosphere is euclidean.

17.6. The Complex Interpretation of Two-dimensional Geometries

One of the characteristics of the euclidean plane is the existence of *regular tessellations*: tilings of the plane by regular polygons. There are of course three such tilings, based on the square, equilateral triangle, and regular hexagon (Fig. 17.11). Associated with each tiling is a *group of rigid motions* of the plane which maps the tiling pattern onto itself. For example, the unit square pattern is mapped onto itself by unit translations parallel to the x and y axes and by the rotation of $\pi/2$ about the origin, and these three motions generate all motions of the tessellation onto itself. If we write $z = x + iy$, then these generating motions are given by the transformations

$$z \mapsto z + 1, \qquad z \mapsto z + i, \qquad z \mapsto zi$$

The triangle and hexagon tessellations have a similar group of motions, generated by

$$z \mapsto z + 1, \qquad z \mapsto z + \tau, \qquad z \mapsto z\tau$$

where $\tau = e^{i\pi/3}$ is the third vertex of the equilateral triangle whose other vertices are at 0, 1 (Fig. 17.12). More generally, any motion of the euclidean plane can be composed from translations $z \mapsto z + a$ and rotations $z \mapsto ze^{i\theta}$.

The sphere also admits a finite number of regular tessellations, obtained by central projection of the regular polyhedra (Section 2.2). Figure 17.13 shows the spherical tessellation corresponding to the icosahedron. (Each face has

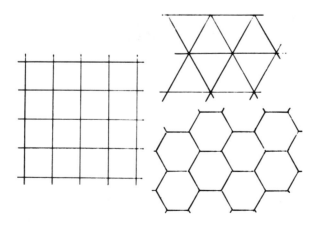

Figure 17.11

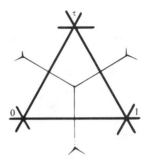

Figure 17.12

Figure 17.13

been further subdivided into six congruent triangles.) The motions that map such a tessellation onto itself can also be expressed as complex transformations by interpreting the sphere as $\mathbb{C} \cup \{\infty\}$ via stereographic projection (Section 14.2). Gauss [c. 1819] found that any motion of the sphere can be expressed by a transformation of the form

$$z \longmapsto \frac{az + b}{-\bar{b}z + \bar{a}}$$

where $a, b \in \mathbb{C}$ and an overbar denotes the complex conjugate.

The conformal models of the hyperbolic plane can be regarded as parts of $\mathbb{C}$, namely, the unit disk $\{z \mid |z| < 1\}$ and the half plane $\{z \mid \text{Im}(z) > 0\}$. Their rigid motions, being conformal transformations, are then complex functions, and Poincaré [1882] made the beautiful discovery that they are of the form

$$z \longmapsto \frac{az + b}{\bar{b}z + \bar{a}}$$

(for the disk) and

$$z \longmapsto \frac{\alpha z + \beta}{\gamma z + \delta}$$

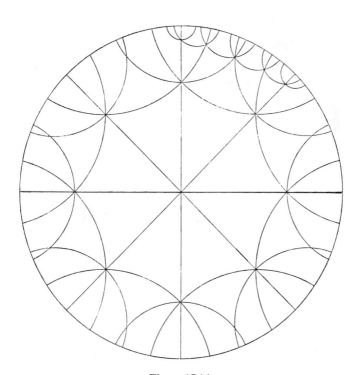

Figure 17.14

where α, β, γ, $\delta \in \mathbb{R}$ (for the half plane). Infinitely many regular tessellations are possible, since the angles of a regular n-gon can be made arbitrarily small by increasing its area. For example, there are tessellations by equilateral triangles in which n triangles meet at each vertex, for each $n \geqslant 7$, and similar variety is possible for other polygons (see exercises).

Some of these tessellations were known *before* Poincaré [1882] gave the complex interpretation of hyperbolic geometry, and even before any model of hyperbolic geometry was known at all. Figure 17.14 shows a tessellation by equilateral triangles of angle $\pi/4$ found in unpublished, and unfortunately undated, work of Gauss (*Werke*, vol. VIII, p. 104). Such tessellations arise from the so-called hypergeometric differential equation and were rediscovered in the same context by Riemann [1858/9] and Schwarz [1872] (the first published example, Fig. 17.15). By explaining these tessellations in terms of hyperbolic geometry, Poincaré [1882] showed for the first time that hyperbolic geom-

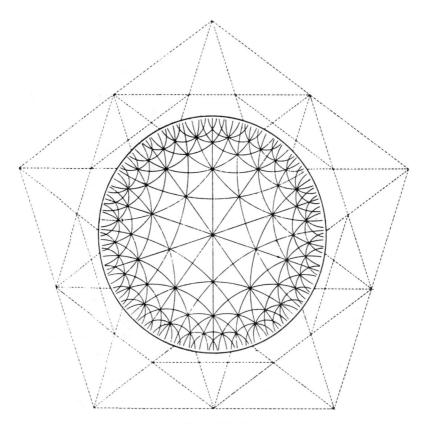

Figure 17.15

etry was part of preexisting mathematics, whose geometric nature had not previously been understood.

In a subsequent paper, Poincaré [1883] explained the geometric nature of *linear fractional transformations,*

$$z \mapsto \frac{az + b}{cz + d}$$

special cases of which, as we have seen, express the rigid motions of the two-dimensional euclidean, spherical, and hyperbolic geometries. He showed that each linear fractional transformation of the plane $\mathbb{C}$ is induced by hyperbolic motion of the three-dimensional half space with boundary plane $\mathbb{C}$; Thus Poincaré's theorem embraces those of Wachter and Beltrami on the representation of two-dimensional euclidean, spherical, and hyperbolic geometry within three-dimensional hyperbolic geometry.

EXERCISES

1. Show that a triangle in the hyperbolic plane can have any angle sum $< \pi$.

2. Deduce that there are equilateral triangles with angle $2\pi/n$ for each $n \geq 7$.

3. Find corresponding results for regular n-gons.

17.7. Biographical Notes: Bolyai and Lobachevsky

János Bolyai was born in 1802 in Kolosvár, then in the Transylvanian part of Hungary (and now Cluj, Romania) and died in Marosvásárhely in Hungary (now Târgu-Mureş, Romania) in 1860. His father, Farkas (also known by his German name, Wolfgang), was professor of mathematics, physics, and chemistry, and his mother, Susanna von Árkos, was the daughter of a surgeon. János received his early education from his father and also studied at the Evangelic-Reformed College, where his father taught, from 1815 to 1818. Farkas had been a fellow student of Gauss at Göttingen, and he hoped that János would follow him there, but instead the younger Bolyai opted for a military career. He studied at the Vienna engineering academy from 1818 to 1822 and then entered the army.

In the army, János became known as an invincible duelist, but he suffered from bouts of fever and was eventually pensioned off in 1833. He returned to Marosvásárhely to live with his father, but the two did not get along, and in 1834 he moved to a small family estate. He set up house with his mistress, Rosalie von Orbán; they had three children. This could have been the start of a mathematical career, in the style of Descartes, as a leisured country gentleman. But, sad to say, Bolyai's mathematical career was already over in 1833, and it was not until after his death that the world knew he had accomplished anything.

János had inherited a passion for the foundations of geometry from his father, so much so that in 1820 Farkas tried almost desperately to steer him away from the problem of parallels: "You should not tempt the parallels in this way, I know this way until its end—I also have measured this bottomless night, I have lost in it every light, every joy of my life" (Stäckel [1913], pp. 76–77). Of course János ignored this warning, but eventually he found the way out that Farkas had missed. After unsuccessful attempts to prove the euclidean parallel axiom, he discarded it and proceeded to derive consequences of Axiom P_2. By 1823 his results seemed so complete and elegant they somehow had to be real, and he wrote triumphantly to his father: "From nothing I have created another entirely new world."

Farkas was unwilling to accept the new geometry, but in June 1831 he agreed to send his son's results to Gauss, who did not answer for over six months (admittedly, this was the time of his wife's death). When Gauss did answer it was in the most self-serving way imaginable:

> Now something about the work of your son. You will probably be shocked for a moment when I begin by saying *that I cannot praise it*, but I cannot do anything else, since to praise it would be to praise myself. The whole content of the paper, the path that your son has taken, and the results to which he has been led, agree almost everywhere with my own meditations, which have occupied me in part already for 30–35 years. (Gauss [1832])

Later in the letter, Gauss offered Bolyai the same backhanded thanks that he had offered to Abel (cf. Section 11.6) for "saving him the trouble" of writing up the results himself, and he raised the question of the volume of the tetrahedron as a problem for further research.

As we now know, Gauss *did* have many of the results of noneuclidean geometry by this time, including the answer to the volume problem he had raised to test his young rival (see Gauss [1832′]). Nevertheless, Gauss was almost certainly wrong to imply that his understanding of noneuclidean geometry went back 35 years. As late as 1804, when Farkas Bolyai wrote to him about the problem of parallels, Gauss could offer no help except the hope that the problem would be settled one day (see Bühler [1981], p. 100).

János Bolyai was disillusioned and embittered by Gauss' reply but did not give up immediately. He published his work as an appendix to his father's book the *Tentamen* (W. Bolyai [1832]). However, when there was no response from other mathematicians he became discouraged and never published again. He was also troubled by the possibility that there might, after all, be contradictions in his geometry. As we know, this possibility was not ruled out until 1868, and by then Gauss, Bolyai, and Lobachevsky were all dead.

Nikolai Ivanovich Lobachevsky (Fig. 17.16) was born in Novgorod (now Gorki) in 1792 and died in Kazan in 1856. He was the son of Ivan Maksimovich Lobachevsky and Praskovia Aleksandrovna. His father died when Nikolai was five years old, and his mother moved with her three sons to Kazan. By persistent efforts, she was able to secure scholarships for their education, and in 1807 Nikolai entered Kazan University, which had been founded just two

Figure 17.16. Nikolai Ivanovich Lobachevsky.

years earlier. He was supervised by Martin Bartels, the former teacher of Gauss, but a link to Gauss' geometric ideas seems less likely than in the case of Bolyai, since Bartels had little contact with Gauss after his school days.

Lobachevsky stayed at Kazan for the rest of his life, becoming professor in 1814 and making many contributions to the growth of the university. He married the wealthy Lady Varvara Aleksivna Moisieva in 1832 and was himself raised to the nobility in 1837, in recognition of his services to education. The couple had seven children.

Lobachevsky's investigation of parallels began in 1816, when he lectured on geometry, and he at first thought he could prove the euclidean axiom. Gradually he became aware of the way in which parallels regulate other geometric properties, such as areas, and in 1823 he wrote *Geometriya*, which separated theorems not requiring an assumption about parallels from those that did. He still believed in the euclidean axiom, however, so Bolyai was ahead of him at this stage. Lobachevsky's publications in noneuclidean geometry began in 1829, but at first they attracted no attention, since they were in Russian and Kazan University was little known. He did gain a wider audience with an article in French in Crelle's journal in 1837, but Gauss seems to have been the only one to recognize its importance. Gauss was in fact so impressed that he collected Lobachevsky's obscure Kazan publications and taught himself Russian in order to read them, but once again he was unwilling to admit to others how impressed he was. It seems that he never contacted Lobachevsky at all, and it is only through a letter [1846'] published after his death that his opinion became public. As usual, Gauss first thought was to guard his own priority, and his memory of when he discovered noneuclidean geometry seems to have improved with age:

> Lobachevsky calls it imaginary geometry. You know that I have had the same conviction for 54 years (since 1792), with a certain later extension which I do not want to go into here. There was nothing materially new for me in Lobachevsky's paper, but he explains his theory in a way which is different from mine, and does this in a masterful way, in a truly geometric spirit. (Bühler [1981], p. 150)

Perhaps Lobachevsky was lucky *not* to hear the kind of praise Gauss bestowed on his competitors, although he was certainly less easily discouraged than Bolyai. Despite the silence of foreign mathematicians, opposition from mathematicians in Russia, and the handicap of blindness in his later years, he continued to refine and expand his theory. The final version of his work, *Pangéométrie*, was published in 1855–56, the last year of his life.

CHAPTER 18

Group Theory

18.1. The Group Concept

The notion of group is one of the most important unifying ideas in mathematics. It draws together a wide variety of mathematical objects for which a notion of combination, or "product," exists. Such products include the ordinary arithmetical product of numbers, but a more typical example is the product, or composition, of functions. If f, g are functions, then gf is the function whose value for argument x is $f[g(x)]$. (The reason for writing $f[g(x)]$ as gf is that its meaning is "apply g, then f." We have to pay attention to order because in general $gf \neq fg$.)

A group G is defined formally to be a set with an operation, called *product* and denoted by juxtaposition, a specific element called the *identity* and written 1 and, for each $g \in G$, an element called the *inverse* of g and written g^{-1}, with the following properties:

(i) $g_1(g_2 g_3) = (g_1 g_2) g_3$ for all $g_1, g_2, g_3 \in G$.
(ii) $g1 = 1g = g$ for all $g \in G$.
(iii) $gg^{-1} = g^{-1}g = 1$ for all $g \in G$.

These axioms evolved over more than a century of work with particular groups, during which their essential features emerged only gradually. We shall look at some of the groups that played an important role in this process in the other sections of this chapter. In practice, the properties (i) (associativity) and (ii) (identity) are usually evident, and it is more important to ensure that the product operation is in fact *defined* for all elements of G. Many mathematical concepts have been created in response to the desire, at first unconscious, for products to exist.

For example, we saw in Section 7.2 that a perspective view of a perspective view is not in general a perspective view. Thus if we take the "product" of a perspective transformation g and a perspective transformation f to be the result of performing g then f, then gf does not always belong to the set of perspective transformations. The set of *projective* transformations is the simplest possible extension of the set of perspective transformations to a set on which the product is always defined, namely, the set of finite products of perspective transformations.

In other instances concepts have arisen from the desire to have inverses. Negative numbers, for example, can be regarded as the result of extending the set $\{0, 1, 2, 3, \ldots\}$ to one in which each element has an inverse under the $+$ operation. Another example is the enlargement of the plane by points at infinity, which ensures that each projective transformation has an inverse, because it enables points that are projected to infinity to be projected back again.

Perhaps the earliest nontrivial use of an inverse occurs with the operation of "multiplication modulo p," which Euler [1758] (and possibly Fermat before him) used to give an essentially group theoretic proof of Fermat's little theorem. Recalling from Section 10.2 that integers m and n are called congruent modulo p if they differ by an integer multiple of p, we call b an *inverse of a* with respect to multiplication mod p if ab is congruent to 1 modulo p, that is, if

$$ab + kp = 1$$

for some integer k. If p is prime and a is not a multiple of p, then such a b exists by application of the euclidean algorithm to the relatively prime numbers a, p (Section 3.3). Euler did not define a group in his proof, but it is easy for us to do so (and to rephrase his proof accordingly; see exercises). The group elements are the *nonzero residue classes* mod p:

$$1(\bmod p) = \{\ldots, -p + 1, 1, p + 1, 2p + 1, \ldots\}$$
$$2(\bmod p) = \{\ldots, -p + 2, 2, p + 2, 2p + 2, \ldots\}$$
$$3(\bmod p) = \{\ldots, -p + 3, 3, p + 3, 2p + 3, \ldots\}$$
$$\vdots$$
$$(p - 1)(\bmod p) = \{\ldots, -1, p - 1, 2p - 1, 3p - 1, \ldots\}$$

with product defined by

$$a(\bmod p)b(\bmod p) = (a \cdot b)(\bmod p)$$

where $a \cdot b$ is the ordinary arithmetic product. Group properties (i) and (ii) follow from ordinary arithmetic; (iii), as we have seen, follows from the euclidean algorithm.

The preceding examples illustrate the influence of geometry and number theory on the group concept. An even more decisive influence was the theory of equations, which we shall look at briefly in the next section. A more detailed account of the development of the group concept may be found in Wussing [1969].

1. Show that if $a(\bmod)p$ is a nonzero residue class, then

$$\{a(\bmod p)1(\bmod p), a(\bmod p)2(\bmod p), \ldots, a(\bmod p)(p-1)(\bmod p)\}$$

   is the same set as

$$\{1(\bmod p), 2(\bmod p), \ldots, (p-1)(\bmod p)\}$$

2. Deduce that

$$a^{p-1} \cdot 1 \cdot 2 \cdot \cdots \cdot (p-1)(\bmod p) = 1 \cdot 2 \cdot \cdots \cdot (p-1)(\bmod p)$$

   and hence that

$$a^{p-1}(\bmod p) = 1(\bmod p)$$

   that is,

$$a^{p-1} \equiv 1(\bmod p)$$

   (Fermat's little theorem).

18.2. Permutations and the Theory of Equations

We saw in Section 10.1 that, as early as 1321, Levi ben Gershon found that there are $n!$ permutations of n things. These permutations are invertible functions which form a group S_n under composition, though their behavior under composition was not considered until the eighteenth century. It was when the idea of permutation was applied to the roots of polynomial equations, by Vandermonde [1771] and Lagrange [1771], that the first truly group theoretic properties of permutations were discovered. At the same time, Vandermonde and Lagrange discovered the key to the understanding of solution of equations by radicals.

They began with the observation that if an equation

$$x^n + a_1 x^{n-1} + \cdots + a_{n-1}x + a_n = 0 \tag{1}$$

has roots $x_1, x_2, \ldots, x_n$, then

$$x^n + a_1 x^{n-1} + \cdots + a_{n-1}x + a_n = (x - x_1)(x - x_2)\ldots(x - x_n) \tag{2}$$

and by multiplying out the right-hand side and comparing coefficients one finds that the a_i are certain functions of $x_1, x_2, \ldots, x_n$. For example,

$$a_n = (-1)^n x_1 x_2 \ldots x_n$$

$$a_1 = -(x_1 + x_2 + \cdots + x_n)$$

These functions are *symmetric*, that is, unaltered by any permutation of $x_1, x_2, \ldots, x_n$, since the right-hand side of (2) is unaltered by such permutations. Consequently, any rational function of $a_1, a_2, \ldots, a_n$ is symmetric as a function of $x_1, x_2, \ldots, x_n$. Now the object of solution by radicals is to apply rational operations *and radicals* to $a_1, a_2, \ldots, a_n$ so as to obtain the roots, that is, the completely *asymmetrical* functions x_i.

Radicals must therefore reduce symmetry in some way, and one can see that they do in the quadratic case. The roots of

$$x^2 + a_1 x + a_2 = (x - x_1)(x - x_2) = 0$$

are

$$x_1, x_2 = \frac{-a_1 \pm \sqrt{a_1^2 - 4a_2}}{2} = \frac{(x_1 + x_2) \pm \sqrt{x_1^2 - 2x_1 x_2 + x_2^2}}{2}$$

and we notice that the symmetric functions $x_1 + x_2$ and $x_1^2 - 2x_1 x_2 + x_2^2$ yield the two asymmetric functions x_1, x_2 when the two-valued radical $\sqrt{}$ is introduced. In general, introduction of radicals $\sqrt[p]{}$ multiplies the number of values of the function by p and divides symmetry by p, in the sense that the group of permutations leaving the function unaltered is reduced to $1/p$ of its previous size.

Vandermonde and Lagrange found that they could explain the previous solutions of cubic and quartic equations in terms of such symmetry reduction in the corresponding permutation groups, S_3 and S_4. They also found some properties of subgroups. For example, Lagrange essentially found the result now known as "Lagrange's theorem": the order of a subgroup divides the order of the group. However, they were unable to obtain sufficient understanding of the relation between radicals and subgroups of S_n to settle the equations of degree $\geqslant 5$. Ruffini [1799] and Abel [1826] made enough progress with S_5 to be able to prove the unsolvability of the quintic, but none of these authors had a firm enough grip on the relation between radicals and permutations to handle arbitrary equations. They were not in fact conscious of the group concept, and it is only with hindsight that we can interpret their results in group theoretic terms.

The concept, and indeed the word "group," first appears in Galois [1831]. Along with it is the concept of *normal subgroup*, which finally unlocks the secret of solvability by radicals. A subgroup $H = \{h_1, h_2, \ldots, h_k\}$ of a group G is called normal if

$$\{gh_1, gh_2, \ldots, gh_k\} = \{h_1 g, h_2 g, \ldots, h_k g\}$$

for each $g \in G$. Galois showed that each equation E has a group G_E consisting of the permutations of the roots which leave rational functions of the roots unaltered, and that the reduction of symmetry accompanied by introduction of a radical corresponds to formation of a normal subgroup. Then solution of E by radicals is possible only if G_E can be reduced to the identity permutation by a chain of normal subgroups (nested in a certain way). If E is the general equation of degree n, then $G_E = S_n$ and the theorem of Ruffini and Abel is recovered by showing that S_n has no such chain of normal subgroups (see, e.g., Dickson [1903]).

This brief sketch of Galois' ideas covers only a part of his theory. Another part is his theory of *fields*, which is needed to clarify the notion of rational function. The group theory and the field theory make up what is currently known as "Galois theory" (e.g., Edwards [1984]). What one might consider to be the summit of Galois' theory, rising above the confines of algebra, is currently neglected. This is the solution of equations by elliptic and related functions, for which one must consult earlier books such as Jordan [1870] and Klein [1884]. The greatest triumph of this theory was the solution of the general quintic equation by elliptic modular functions in Hermite [1858], following a hint in Galois [1831'] (cf. Section 5.5).

EXERCISES

1. Define an *even permutation* f of $\{1, 2, \ldots, n\}$ to be one with an even number of *inversions*, that is, pairs (i, j) for which $i < j$ and $f(i) > f(j)$ (Cramer [1750], p. 658). Show that the product of even permutations is even, and hence that the even permutations of $\{1, 2, \ldots, n\}$ form a group A_n.

2. If g is an odd permutation, that is, $g \in S_n - A_n$, show that $gA_n = \{gf \mid f \in A_n\}$ is the set of all odd permutations in S_n, hence A_n contains exactly half the members of S_n.

18.3. The Group Concept

Galois understood "group" to mean a group of permutations of a finite set, so his definition stated only that the product of two permutations in the group must again be a member of the group. Associativity, identity, and inverses were consequences of his assumptions, and indeed too obvious to be considered important from his point of view. Galois' work was published only in 1846, and by that time the theory of finite permutation groups had been taken up and systematized by Cauchy [1844]. Cauchy likewise required only closure under product in his definition of group, but he recognized the importance of identity and inverses by introducing the notation of 1 for the identity and f^{-1} for the inverse of f.

Cayley [1854] was the first to consider the possibility of more abstract group elements, and with it the necessity of postulating associativity. (Incidentally, one of the few groups for which associativity is not obvious is that defined by the chord construction on a cubic curve: cf. Sections 10.6 and 15.5) He took group elements to be simply "symbols," with a symbolic product of A and B written $A \cdot B$ and subject to the law $A \cdot (B \cdot C) = (A \cdot B) \cdot C$, and a unique element 1 subject to the laws $A \cdot 1 = 1 \cdot A = A$. He still assumed that each group was finite, however; this meant that the existence of inverses did not have to be postulated.

The existence of inverses in a finite group G, as defined by Cayley, follows from an argument used by Cauchy [1815] and developed more fully in Cauchy [1844]. If $A \in G$, then the powers A^2, A^3, ..., all belong to G and hence they eventually include a recurrence of the same element:

$$A^m = A^n \qquad \text{where} \quad m < n$$

Then A^{n-m} is the identity element 1 and A^{n-m-1} is the inverse of A.

The need to postulate inverses first arises with infinite groups, where this argument no longer holds. Geometry was historically the most important source of infinite groups, as we shall see in Section 18.5. It was in extending Cayley's abstract group theory to cover the symmetry groups of infinite tessellations that Dyck [1883] made the first mention of inverses in the definition of group. We shall return to Dyck's concept of group in Section 18.6.

A theorem of Cayley [1878] shows that abstraction of the group concept is, in a sense, empty, because every group is essentially the same as a group of permutations. Cayley proved the theorem for finite groups only, where it is more valuable, but the proof easily extends to arbitrary groups (exercise 1).

EXERCISE

1. Given a group G, which consists of elements f_i, associate any $f_j \in G$ with the permutation that sends each f_i to $f_i f_j$. Show that this group of permutations is isomorphic to G.

18.4. Polyhedral Groups

A beautiful illustration of Cayley's theorem that every group is a permutation group is provided by the regular polyhedra, whose symmetry groups turn out to be important subgroups of S_4 and S_5. The regular polyhedra also show us the more literal, geometric, meaning of "symmetry." If we imagine a polyhedron P occupying a region R in space, the symmetries of P can be viewed as the different ways of fitting P into R. Each symmetry is obtained by a rotation from the initial position, and the product of symmetries is the product of rotations.

To begin with the symmetries of the tetrahedron T: T has four vertices, V_1, V_2, V_3, V_4, so each symmetry of T is determined by a permutation of the four things V_1, V_2, V_3, V_4. There are $4 \times 3 = 12$ symmetries, because V_1 can be put at any of the four vertices of R, after which three choices remain for the remaining triangle of vertices V_2, V_3, V_4. One can check, using the fact that a permutation that leaves one element fixed and rotates the other three

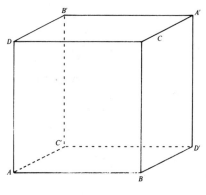

Figure 18.1

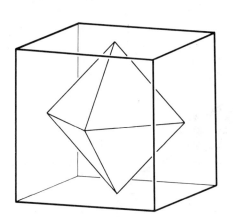

Figure 18.2

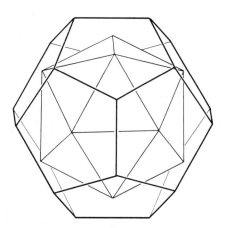

Figure 18.3

is even and that all the symmetries of T are even permutations of V_1, V_2, V_3, V_4. But the subgroup A_4 of *all* even permutations in S_4 has $\frac{1}{2} \times 4! = 12$ elements by exercise 18.2, hence the symmetry group of T is precisely A_4.

The full permutation group S_4 can be realized by the symmetries of the cube. The four elements of the cube that are permuted are the long diagonals AA', BB', CC', DD' (Fig. 18.1). One has to check, first, that each permutation of the diagonals is actually realizable. While doing this, it will become apparent that the position of the diagonals (bearing in mind that end points could be swapped) really determines the position of the cube (exercise 1). S_4 is also the symmetry group of the octahedron, because of the dual relationship between cube and octahedron seen in Figure 18.2. Each symmetry of the cube is clearly a symmetry of its dual octahedron, and conversely.

Likewise, the dual relationship between dodecahedron and icosahedron (Fig. 18.3) shows that they have the same symmetry group. This group turns out to be A_5, the subgroup of even permutations in S_5. The five elements of the dodecahedron whose even permutations determine these symmetries are tetrahedra formed from sets of four diagonals (see Fig. 18.4, which is from Coxeter and Moser [1980], p. 35).

For more information on the polyhedral groups, see Klein [1884]. This book relates the theory of equations to the symmetries of the regular polyhedra and functions of a complex variable. The complex variable makes its appearance when the regular polyhedra are replaced by regular tessellations of the sphere $\mathbb{C} \cup \{\infty\}$, and their symmetries by linear fractional transformations, as in Section 17.6. Klein [1876] showed that, with trivial exceptions, *all* finite groups of linear fractional transformations come from the symmetries of the regular polyhedra in this way.

Figure 18.4

The regular polyhedra were also the source of another approach to groups: *presentation by generators and relations.* Hamilton [1856] showed that the icosahedral group can be generated by three elements ι, χ, λ subject to the relations

$$\iota^2 = \chi^3 = \lambda^5 = 1, \qquad \lambda = \iota\chi \tag{1}$$

This means that any element of the icosahedral group is a product (possibly with repetitions) of ι, χ, λ, and that any relation between ι, χ, λ is a consequence of the relations (1). Dyck [1882] gave similar presentations of the cube and tetrahedron groups, and for the groups of certain infinite tessellations, as part of the first general discussion of generators and relations. We shall return to this in Section 18.6.

EXERCISE

1. Show that each permutation of the diagonals of a cube is realizable and that such a permutation uniquely determines the position of the cube.

18.5. Groups and Geometries

As the regular polyhedra show, geometric symmetry is fundamentally a group theoretic notion. More generally, many notions of "equivalence" in geometry can be explained as properties that are preserved by certain groups of transformations. However, some revision of classical notions was necessary before geometry could benefit from group theoretic ideas.

The oldest notion of geometric equivalence is that of *congruence*. The Greeks understood figures F_1 and F_2 to be congruent if there was a rigid motion of F_1 which carried it into F_2. The disadvantage of this idea was that motion had meaning only for the individual figure. The "product" of motions of different figures was meaningless, and hence one did not have a group of motions.

The step that paved the way for the introduction of group theory into geometry was the extension of the idea of motion to the whole plane by Möbius [1827], which gave a meaning to the product of motions. Möbius in fact considered all continuous transformations of the plane that preserve straightness of lines and gave separate attention to several subclasses of these transformations: those that preserve length (congruences), shape (similarities), and parallelism (affinities). He showed that the most general continuous transformations preserving straightness were just the projective transformations. Thus in one stroke Möbius defined the notions of congruence, similarity, affinity, and projective equivalence as properties that were invariant under certain classes of transformations of the plane. That the classes in question were groups was obvious as soon as one recognized the concept of group. It is an indication of the slowness with which the group concept *was* recognized that the restatement of Möbius' ideas in terms of groups occurred only with Klein [1872].

Klein's formulation became known as the *Erlanger Programm* because he announced it at the University of Erlangen. His idea is to associate each geometry with a group of transformations which preserve the characteristic properties of that geometry. For example, plane euclidean geometry is associated with the group of transformations of the plane that preserve the euclidean distance $\sqrt{(x_2 - x_1)^2 + (y_2 - y_1)^2}$ between arbitrary points (x_1, y_1) and (x_2, y_2). Plane projective geometry is associated with the group of projective transformations. Plane hyperbolic geometry, in view of the projective model, can be associated with the group of projective transformations that map the unit circle onto itself. An important influence on the Erlanger Programm was indeed Cayley [1859], where this group was first shown to determine a geometry, and the subsequent realization of Klein [1871] that the elements of this group are the rigid motions of hyperbolic geometry.

When geometry is reformulated in this way, certain geometric questions become questions about groups. A regular tessellation, for example, corresponds to a subgroup of the full group of motions, consisting of those motions that map the tessellation onto itself. In the case of hyperbolic geometry, where the problem of classifying tessellations is formidable, the interplay between geometric and group theoretic ideas proved to be very fruitful. In the work of Poincaré [1882, 1883] and Klein [1882], group theory is the catalyst for a new synthesis of geometric, topological, and combinatorial ideas, which will be described in Sections 18.6 and 19.7.

18.6. Combinatorial Group Theory

As mentioned in Section 18.4, the groups of the regular polyhedra were the first to be defined in terms of generators and relations. With finite groups such as these, however, one is concerned mainly with the simplicity and elegance of a presentation; the question of *existence* does not arise. For any finite group G one can trivially obtain a finite set of generators (namely, *all* the elements $g_1, \ldots, g_n$ of G) and defining relations (namely, all equations $g_i g_j = g_k$ holding among the generators). Of course the same argument gives an infinite set of generators and defining relations for an infinite group, but this is also not interesting. The real problem is to find finite sets of generators and defining relations for infinite groups where possible.

This problem was first solved for the symmetry groups of certain regular tessellations, and such examples were the basis of the first systematic study of generators and relations, by Klein's student Dyck. Dyck's papers [1882, 1883] laid the foundations of this approach to group theory, now called *combinatorial*.

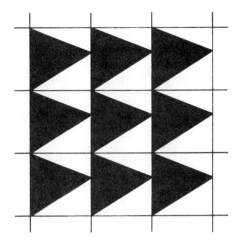

Figure 18.5

For more technical information, as well as detailed history of the development of combinatorial group theory, see Chandler and Magnus [1982].

Figure 18.5 illustrates how generators and relations arise naturally from tessellations. This tessellation is based on the regular tessellation of the euclidean plane by unit squares, but each square has been subdivided into black and white triangles to eliminate symmetries by rotation and reflection. The symmetries that remain are generated by

(a), horizontal translation of length 1
(b), vertical translation of length 1

These generators are subject to the obvious relation

$$ab = ba$$

which implies that any element of the group can be written in the form $a^m b^n$. If $g = a^{m_1} n^{n_1}$ and $h = a^{m_2} b^{n_2}$, then $g = h$ only if $m_1 = m_2$ and $n_1 = n_2$, that is, only if $g = h$ is a *consequence* of the relation $ab = ba$. Thus all relations $g = h$ in the group follow from $ab = ba$, which means that the latter relation is a defining relation of the group.

The obviousness of the defining relation in this case blinds us to a fact that becomes more evident with tessellations of the hyperbolic plane: *the generators and relations can be read off from the tessellation.* Group elements correspond to cells in the tessellation, squares in the present example. If we fix the square corresponding to the identity element 1, then the square to which square 1 is sent by the group element g may be called square g. The generators $a^{\pm 1}, b^{\pm 1}$ are the elements that send square 1 to adjacent squares. They generate the group because square 1 can be sent to any other square by a series of moves

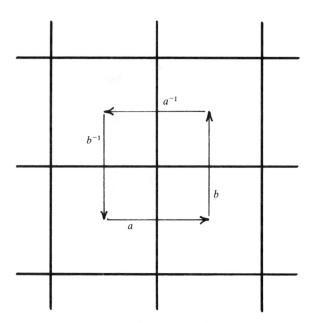

Figure 18.6

from square to adjacent square. Relations correspond to equal sequences of moves or, what comes to the same thing, to sequences of moves that return square 1 to its starting position. These sequences can all be derived from a circuit around a vertex (Fig. 18.6), that is, the sequence $aba^{-1}b^{-1}$. Thus all relations are derived from $aba^{-1}b^{-1} = 1$, or, equivalently, $ab = ba$.

Extending these ideas to more general tessellations, Poincaré [1882] showed that the symmetry groups of all regular tessellations, whether of the sphere, euclidean plane, or hyperbolic plane, can be presented by finitely many generators and relations. Generators correspond to moves of the basic cell to adjacent cells, and hence to the sides of the basic cell; defining relations correspond to its vertices. These results are also important for topology, as we shall see in the next chapter.

The notion of group abstracted from such examples was expressed in a somewhat technical way, involving normal subgroups, by Dyck [1882]. The following, simpler, approach was worked out by Dehn, and used by Dehn's student Magnus [1930]. A group G is defined by a set $\{a_1, a_2, \ldots,\}$ of *generators* and a set $\{W_1 = W_1', W_2 = W_2', \ldots,\}$ of *defining relations*. Each generator a_i is called a *letter*; a_i has an *inverse* a_i^{-1}, and arbitrary finite sequences ("products") of letters and inverse letters are called *words*.

Words W, W' are called *equivalent* if $W = W'$ is a consequence of the defining relations, that is, if W may be converted to W' by a sequence of replacements of subwords W_i by W_i' (or vice versa) and cancellation (or insertion) of subwords $a_i a_i^{-1}$, $a_i^{-1} a_i$. The elements of G are the equivalence classes

$$[W] = \{W' \,|\, W' \text{ is equivalent to } W\}$$

and the product of elements $[U]$, $[V]$ is defined by

$$[U][V] = [UV]$$

where UV denotes the result of concatenating the words U, V. It has to be checked that this product is well defined, but once this is done, group properties (i), (ii), and (iii) follow easily.

EXERCISES

1. If U is equivalent to U', show that UV is equivalent to $U'V$. Conclude, using this and a similar result for V', that the product $[U][V]$ is independent of the choice of representatives for $[U]$, $[V]$.

2. $[U]([V][W]) = ([U][V])[W]$ is trivial. Why?

3. Show that $1 = $ equivalence class of the empty word.

4. Show that $[W]^{-1} = [W^{-1}]$, where W^{-1} is the result of writing W backward and changing the sign of each exponent.

18.7. Biographical Notes: Galois

Evariste Galois (Fig. 18.7) was born in the town of Bourg-la-Reine, near Paris, in 1811, and died, from wounds received in a duel, in Paris in 1832. The tragedy and mystery of his short life make him the most romantic figure in mathematics, and several biographers have been tempted to cast Galois in the role of misunderstood genius and victim of the Establishment. However, it has been amply documented by Rothman [1982] that Galois does not easily fit this role. Though the known facts of his life should satisfy anyone's appetite for drama, his tragedy is of the more classical kind, whose seeds lie in the character of the victim himself.

Galois was the second of three children of Nicholas-Gabriel Galois, the director of a boarding school and later mayor of Bourg-la-Reine, and Adelaïde-Marie Demante, who came from a family of jurists. Both parents were well educated, and Galois seems to have had a happy, if unconventional, childhood. Up to the age of 12, he was educated entirely by his mother, a severe classicist who instilled in him a knowledge of Latin and Greek and a respect for Stoic morality. His father was far less of a stoic, but unconventional in a different way, being a republican at a time when France was returning to the monarchy.

In October 1823 Evariste entered the Lycée of Louis-le-Grand, a well-known school whose pupils had included Robespierre and Victor Hugo and

Figure 18.7. Evariste Galois at the age of 15.

would later include the mathematician Charles Hermite, who found the transcendental solution to the quintic equation. There does not seem to have been any mathematics in Galois family background, and he did not begin studying it at school until February 1827. To make the progress he did, he must have devoured mathematics at a greater rate than anyone in history, except perhaps Newton in 1665/6, so it is no wonder that for the first time his school reports noted unsatisfactory progress in other subjects. Comments about his character being "singular" and "closed" but "original" also began to be made. At this stage Galois was studying Legendre's *Geometry* and Lagrange's works on the theory of equations and analytic functions. He believed he was ready to enter the École Polytechnique but, due to his lack of preparation in the standard syllabus, failed the entrance examination.

In 1828 he had the good fortune to study mathematics under a teacher who recognized his genius, Louis-Paul-Emile Richard. This led to Galois' first publication, a paper on continued fractions that appeared in the *Annales* of Gergonne in March 1829. Thanks to Richard, many pieces of Galois' early work still exist and have been published in Bourgne and Azra [1962]. They include class papers saved by Richard and later preserved for posterity by Hermite. One fragment from 1828 shows that Galois, like Abel, initially believed he could solve the quintic equation.

One might think that publication in a respected journal was reasonable encouragement for a 17-year-old mathematician, but it was not enough for

Galois. He nursed a grudge against the examiners of the École Polytechnique for failing him, and he was supported by Richard, who declared that he should be admitted without examination. Needless to say, this did not happen, but worse disappointments were to follow.

Galois had already begun working on his theory of equations and submitted his first paper on the subject to the Paris Academy in May 1829. Cauchy was referee and even seemed to be favorably impressed (see Rothman [1982], p. 89), but months went by and the paper failed to appear. Then, in July 1829, Galois' father committed suicide. The cause was trivial, even childish—a spiteful attack on him by the priest of Bourg-la-Reine—but it unleashed political passions with which Galois senior could not cope. Nor could Evariste cope with the loss of his father. His distrust of the political and educational establishment deepened into paranoia, and the sacrifice of his own life must suddenly have seemed a real possibility. It was almost the last straw when, a few days after his father's death, he failed the entrance examination for the École Polytechnique a second time.

Despite these crushing blows, Galois persevered with examinations and succeeded in entering the less prestigious École Normale in November 1829. In early 1830 he got his theory of equations into print (though not through the Academy) with the publication of three papers. The more decisive event of 1830, however, was the July revolution against the Bourbon monarchy. It gave Galois the ideal focus for his rage over the death of his father and his own humiliations, and he emerged as a republican firebrand. He made friends with the republican leaders Blanqui and Raspail and began political agitation at the École Normale—until he was expelled in December 1830 for an article he wrote against its director. In the same month, the Bourbons fled France and, as mentioned in Section 15.7, Cauchy fled with them.

Immediately after leaving the École Normale, Galois joined the Artillery of the National Guard, a republican stronghold, to concentrate on revolutionary activity. At a republican banquet on May 9, 1831, he proposed a toast with a dagger in his hand, implying a threat against the life of the new king, Louis-Philippe. Galois was arrested the following day and held until June 15 in Sainte-Pélagie prison. He was then tried for threatening the life of the king, but he was acquitted almost immediately, evidently on the grounds that he was young and foolish. The acquittal was an act of considerable leniency, as Galois gave full vent to his opinions during the trial. He admitted that he still intended to kill the king "if he betrays" and added his view that the king "will soon turn traitor if he has not done so already."

Galois was arrested a second time on Bastille Day 1831, for illegal possession of weapons and for wearing the uniform of the Artillery Guard (which had been disbanded at the end of 1830). He was held in Sainte-Pélagie prison until October and then sentenced to a further six months. Galois became very despondent and once, thinking of his father, attempted suicide. Thus he was not in a receptive mood when he finally heard from the Academy—that they were returning his manuscript—even though he was invited to submit a more complete account of his theory. Galois did in fact begin to revise his work, but he poured most of his energy into the preface, a scorching condemnation of the scientific establishment and Academicians in particular "who already have the death of Abel on their consciences." The last six weeks of his imprisonment were spent in a nursing home. Some prisoners were transferred there as a measure against cholera, which was then epidemic in Paris. In these relatively pleasant surroundings, Galois resumed his research and managed to write a few philosophical essays.

He was released on April 29, 1832. Frustratingly little is known about the next, final, month of his life. He wrote to his friend Chevalier on May 25, expressing his complete disenchantment with life and hinting that a broken love affair was the reason. It appears that the woman was Stéphanie Dumotel, daughter of the resident physician at the nursing home. Two letters from her to Galois exist, though they are defaced (presumably by Galois himself) so as to be only partly readable. One, dated May 14, says "Please let us break up this affair." The other mentions sorrows someone else had caused her, in such a way that Galois might have felt obliged to come to her defense. Whether this was the cause of the fatal duel we do not know. It is also possible that Galois felt the duel had been hanging over his head for a long time. When he first went to prison in 1831, one of his comrades was Raspail, who, in a letter from prison on July 25 that year, quoted Galois as follows: "And I tell you, I will die in a duel over some low class coquette. Why? Because she will invite me to avenge her honour which another has compromised" (Raspail [1839], p. 89). In letters he wrote to friends on the night before the duel, Galois again spoke of an "infamous coquette."

He also wrote: "Forgive those who kill me for they are of good faith." His opponent was in fact a fellow republican, Pescheux d'Herbinville. Authors who like conspiracy theories have since conjectured that d'Herbinville was really a police agent, but no evidence exists for this. His revolutionary credentials were as good as those of Galois. The police agent theory seems rather to reflect twentieth-century bafflement over dueling, something which we no longer understand or sympathize with (though we still applaud successful duelists, such as Bolyai and Weierstrass). There may be *no* rational explanation for the duel, but no doubt the suicide of his father and Galois' own self-destructive tendencies were among the conditions that made it possible. Galois was convinced he was going to die over something small and contemptible, and the tragedy is that he let it happen.

The tragedy for mathematics was the incompleteness of Galois' work at the time of his death. The night before the duel, he wrote a long letter to Chevalier outlining his discoveries and hoping "some men will find it profitable to sort out this mess." Chevalier and Alfred Galois (Evariste's younger brother) later copied out the mathematical papers and sent them to Gauss and Jacobi, but there was no response. The first to study them conscientiously was Liouville, who became convinced of their importance in 1843 and arranged to have them published. They finally appeared in 1846, and by the 1850s the algebraic part of the theory began to creep into textbooks. But, as mentioned in Section 18.2, there was more. Galois also talked of connections between algebraic equations and transcendental functions and made a cryptic reference to a "theory of ambiguity." The latter probably concerned the many-valuedness of algebraic functions, and we may be fairly sure that whatever Galois did was later superseded by Riemann. As for the transcendental functions, we also know that Hermite [1858] successfully completed one of Galois' investigations in solving the quintic equation by means of elliptic modular functions, and that Jordan [1870] exposed the group theory governing the behavior of such functions. However, these results only scratch the surface, and it is still possible that a bigger "Galois theory" remains to be discovered.

CHAPTER 19

Topology

19.1. Geometry and Topology

Topology is concerned with those properties that remain invariant under continuous transformations. In the context of Klein's Erlanger Programm (where it receives a brief mention under its old name of *analysis situs*) it is the "geometry" of groups of continuous invertible transformations, or *homeomorphisms*. The "space" to which transformations are applied and indeed the meaning of "continuous" remain somewhat open. When these terms are interpreted in the most general way, as subject only to certain axioms (which we shall not bother to state here), one has *general topology*. The theorems of general topology, important in fields ranging from set theory to analysis, are not very geometric in flavor. *Geometric topology*, which we shall be concerned with in this chapter, is obtained when the transformations are ordinary continuous functions on $\mathbb{R}^n$ or on certain subsets of $\mathbb{R}^n$.

Geometric topology is more recognizably "geometric" than general topology, though the "geometry" is necessarily of a discrete and combinatorial kind. Ordinary geometric quantities—such as length, angle, and curvature—admit continuous variation and hence cannot be invariant under continuous transformations. The kind of quantities that are topologically invariant are such things as the number of "pieces" of a figure or the number of "holes" in it. It turns out, though, that the combinatorial structures of topology can often be reflected by combinatorial structures in ordinary geometry, such as polyhedra and tessellations. In the case of surface topology, this geometric modeling of topological structure is so complete that topology becomes essentially a part of ordinary geometry. "Ordinary" here means geometry with notions of length, angle, and curvature, not necessarily euclidean geometry. In fact, the natural geometric models of most surfaces are hyperbolic.

It remains to be seen whether topology as a whole will ever be subordinate to ordinary geometry. This is conjectured to be the case in three dimensions, though the situation has so far been too complicated to resolve completely (see Thurston [1982] and Scott [1983]). It does appear that here, too, hyperbolic geometry is the most important geometry. In four or more dimensions it would be rash to speculate, though geometric methods have been important in recent breakthroughs (e.g., Donaldson [1983]). In this chapter we make a virtue of a necessity by confining our discussion to the topology of surfaces. This is the only area that is sufficiently understandable and relevant when set against the background of the rest of this book. Fortunately, this area is also rich enough to illustrate some important topological ideas, while still being mathematically tractable and visual.

We shall begin the discussion of surface topology at its historical starting point, the theory of polyhedra.

19.2. The Polyhedron Formulas of Descartes and Euler

The first topological property of polyhedra seems to have been discovered by Descartes around 1630. Descartes' short paper on the subject is lost, but its contents are known from a copy made by Leibniz in 1676, discovered among Leibniz' papers in 1860 and published in Prouhet [1860]. A detailed study of this paper, including a translation and facsimile of the Leibniz manuscript, has been published by Federico [1982].

The same property was rediscovered by Euler [1752], and it is now known as the *Euler characteristic*. If a polyhedron has V vertices, E edges, and F faces, then its Euler characteristic is $V - E + F$. Euler showed that this quantity has a certain invariance by showing

$$V - E + F = 2$$

for all convex polyhedra, a result now known as the *Euler polyhedron formula*. Descartes already had the same result implicitly in the pair of formulas

$$P = 2F + 2V - 4, \qquad P = 2E$$

where P is the number of what Descartes calls "plane angles": corners of faces determined by pairs of adjacent edges. The relation $P = 2E$ then follows from the observation that each edge participates in two corners. It should be stressed that Descartes' "plane angle" has nothing to do with angle measure, and hence is just as topological a concept as Euler's "edges." Thus Descartes' result belongs to topology just as much as Euler's does, even though it fails to isolate the concept of Euler characteristic quite as well. Some rather hairsplitting distinctions have been made between Euler and Descartes in an effort to show that Euler invented topology and Descartes did not (see Federico [1982] for a review of different opinions).

Actually, neither of these mathematicians understood the polyhedron formula in a fully topological way. They both used nontopological concepts, such as angle measure, in their proofs, and they did not realize that "vertices," "edges," and "faces" are meaningful on any surface: edges need not be straight and faces need not be flat. Other early proofs of the Euler polyhedron formula also rely on angle measure and other ordinary geometric quantities. For example, that of Legendre [1794] assumes the polyhedron can be projected onto the sphere, then uses the Harriot relation between angular excess and area for spherical polygons (exercise 1).

Probably the first to achieve a purely topological understanding of $V - E + F$ was Poincaré [1895]. Poincaré in fact generalized the Euler characteristic to n-dimensional figures, but in the case of polyhedra his essential observation was this: a vertex divides an edge into two edges, and an edge divides a face into two faces. It follows that any subdivision of edges or faces of a polyhedron leaves $V - E + F$ unchanged: if a new vertex is introduced on an edge, V and E both increase by 1; if a new edge is introduced across a face, E and F both increase by 1. The reverse processes of amalgamation, where they make sense, likewise leave $V - E + F$ unchanged.

The constancy of $V - E + F$ over, say, the class of convex polyhedra then follows if it can be shown that any polyhedron P_1 in the class can be converted to any other, P_2, by subdivisions and amalgamations. A plausible argument for this (due to Riemann [1851]) is to view P_1 and P_2 as subdivisions of the same surface, say a sphere. Assuming the edges of P_1 and P_2 meet only finitely often, superimposing P_1 on P_2 gives a common subdivision P_3 whose $V - E + F$ value is therefore the same as that of P_1 and P_2. Hence the $V - E + F$ values of P_1 and P_2 are equal. The assumption of only finitely many intersections is hard to justify, however. A different approach, which also yields the value of $V - E + F$ for *nonspherical* surfaces, will be explained in the next section.

EXERCISE

1. (Legendre [1794]) Consider the spherical projection of a convex polyhedron which has spherical polygons for faces. Use the fact that

$$\text{area of a spherical } n\text{-gon} = \text{angle sum} - (n - 2)\pi$$

to conclude that

$$\text{total area} = 4\pi = (\Sigma \text{ all angles}) - \pi(\Sigma \text{ all } n) + 2\pi F$$

Show also that

$$\Sigma \text{ all } n = 2E, \qquad \Sigma \text{ all angles} = 2\pi V$$

whence

$$V - E + F = 2$$

19.3. The Classification of Surfaces

Between the 1850s and the 1880s, several different lines of research led to the demand for a topological classification of surfaces. One line, descending from Euler, was the classification of polyhedra. Another was the Riemann surface representation of algebraic curves, coming from Riemann [1851, 1857]. Related to this was the problem of classifying symmetry groups of tessellations, considered by Poincaré [1882] and Klein [1882] (see Section 19.6). Finally, there was the problem of classifying smooth closed surfaces in ordinary space (Möbius [1863]). These different lines of research converged when it was realized that in each case it was possible to subdivide the surface into faces by edges (not necessarily straight, of course) so that it became a generalized polyhedron. The generalized polyhedra are what were traditionally called *closed* surfaces, now described by topologists as *compact and without boundary*.

The subdivision argument for the invariance of the Euler characteristic $V - E + F$ applies to any such polyhedron, not just those homeomorphic to the sphere and not just those with straight edges and flat faces. Various mathematicians (Riemann [1851], Möbius [1863], Jordan [1866]) came to the conclusion that any closed surface is determined, up to homeomorphism, by its Euler characteristic. It also emerged that the different possible Euler characteristics were represented by the "normal form" surfaces seen in Figure 19.1, which were discovered by Möbius [1863]. It is certainly plausible that these forms are distinct, topologically, because of their different numbers of "holes." The main part of the proof is to show that any closed surface is homeomorphic to one of them.

The assumptions of Riemann (that the surface is a Riemann surface) and Möbius (that the surface is smoothly embedded in $\mathbb{R}^3$) were a little too special to yield a purely topological proof, and in addition they contained a hidden assumption of *orientability* ("two-sidedness"). A rigorous proof, from an axiomatic definition of generalized polyhedron, was given by Dehn and Heegaard [1907]. The closed orientable surfaces indeed turn out to be those pictured in Figure 19.1, but in addition there are *nonorientable* surfaces, which are not homeomorphic to orientable surfaces. A nonorientable surface may be defined as one that contains a *Möbius band*, a nonclosed surface discovered independently by Möbius and Listing in 1858 (Fig. 19.2). Closed nonorientable surfaces cannot occur as Riemann surfaces, nor can they lie in $\mathbb{R}^3$ without

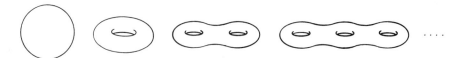

Figure 19.1

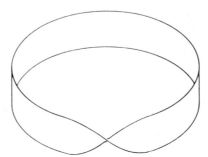

Figure 19.2

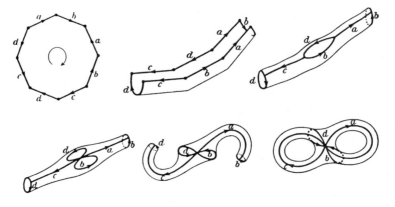

Figure 19.3

crossing themselves; nevertheless, they include some important surfaces, such as the projective plane (exercise 7.5.5). The nonorientable surfaces are also determined, up to homeomorphism, by the Euler characteristic.

The Möbius forms of closed orientable surfaces were given standard polyhedral structures by Klein [1882]. These are "minimal" subdivisions with just one face and, except for the sphere, with just one vertex. When the Klein subdivision of a surface is cut along its edges, one obtains a *fundamental polygon*, from which the surface may be reconstructed by identifying like-labeled edges (Fig. 19.3). It is often more convenient to work with the polygon rather than the surface or its polyhedral structure. For example, since Brahana [1921], most proofs of the classification theorem have used polygons rather than polyhedra, "cutting and pasting" them (instead of subdividing and amalgamating) until Klein's fundamental polygons are obtained. The fundamental polygon gives a very easy calculation of the Euler characteristic χ and shows it to be related to the *genus g* (number of "holes") by

$$\chi = 2 - 2g$$

(exercise 1). The genus of course determines the surface more simply than the Euler characteristic, but we shall see that the Euler characteristic is a better reflection of geometric properties.

EXERCISE

1. Show that the standard polyhedron for a surface of genus $g \geqslant 1$ has $V = 1$, $E = 2g$, $F = 1$, whence $\chi = 2 - 2g$.

19.4. Descartes and the Gauss–Bonnet Theorem

The first theorem in the Descartes manuscript is a remarkable statement about the total "curvature" of a convex polyhedron, not at first appearing to have any topological content. It is a spatial analogue of the obvious theorem that the sum of the external angles of a convex polygon is 2π. The latter theorem can be seen intuitively by considering the total turn of a line that is transported around the polygon (Fig. 19.4). A different proof, which generalizes to polyhedra, is the following (Fig. 19.5). At each vertex, construct a sector of a unit circle, bounded by normals to the two edges at that vertex. Clearly,

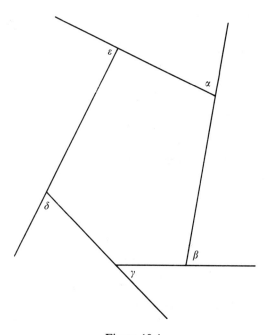

Figure 19.4

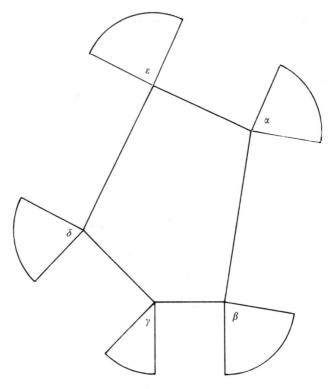

Figure 19.5

the angle of the sector equals the external angle at that vertex. Also, adjacent sides of adjacent sectors are perpendicular to the same edge, hence parallel, so the sectors can be fitted together to form a complete disk, of total angle 2π.

To generalize this to polyhedra, define the *exterior solid angle* at each vertex P to be the sector of a unit ball bounded by planes normal to the edges at P (Fig. 19.6). As before, adjacent sides of adjacent sectors are parallel, hence the sectors can be fitted together to form a complete ball, of total solid angle 4π. Descartes only stated that the total exterior solid angle is 4π, without even defining exterior solid angle. The foregoing proof is based on the reconstruction by Polya [1954].

The theorem about polygons has an analogue for simple closed smooth curves $\mathscr{C}$, namely, $\int_{\mathscr{C}} \kappa \, ds = 2\pi$ (Section 16.2). This leads us to wonder whether the Descartes theorem has an analogue for smooth closed convex surfaces $\mathscr{S}$, say, $\iint_{\mathscr{S}} \kappa_1 \kappa_2 \, dA = 4\pi$, where $\kappa_1 \kappa_2$ is the Gaussian curvature. This is so, and in fact there is a proof like the polyhedron proof using yet another characterization of Gaussian curvature due to Gauss [1827].

If we take a small geodesic polygon $\mathscr{P}$ on the surface $\mathscr{S}$, then the "total curvature" of the portion $\mathscr{P}$ can be represented by an "exterior solid angle" $\mathscr{A}$ bounded by parallels to the normals to $\mathscr{S}$ along the sides of $\mathscr{P}$ (Fig. 19.7).

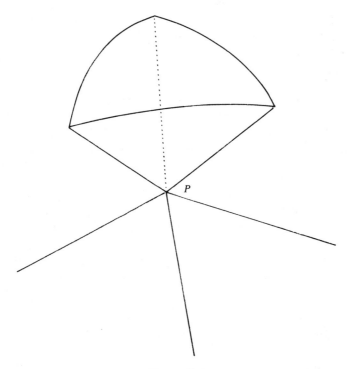

Figure 19.6

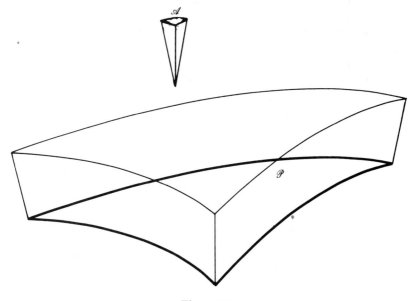

Figure 19.7

Gauss showed that the measure of $\mathscr{A}$—the area it cuts out of the unit sphere—is $\iint_{\mathscr{P}} \kappa_1 \kappa_2 \, dA$. But it is also clear, by the parallelism of adjacent sides of adjacent exterior solid angles $\mathscr{A}$, that the $\mathscr{A}$'s corresponding to a partition of $\mathscr{S}$ by geodesic polygons $\mathscr{P}$ fit together to form a complete ball. Hence $\iint_{\mathscr{S}} \kappa_1 \kappa_2 \, dA = 4\pi$.

This is a "global" form of the Gauss–Bonnet theorem. When Descartes' theorem was first published in 1860, the Gauss–Bonnet theorem was already known, and the analogy between the two was noted by Bertrand [1860]. Bertrand, however, made the qualification that "the beautiful conception of Gauss could not in any manner be considered as corollary to that of Descartes." This may be true in a narrow sense; nevertheless, the Descartes and Gauss–Bonnet theorems can be viewed as limiting cases of each other. Gauss–Bonnet → Descartes by concentrating the curvature of a surface at vertices until it becomes a polyhedron, while Descartes → Gauss–Bonnet by increasing the number of vertices of a polyhedron until it becomes a smooth surface. It is interesting, though probably accidental, that Descartes actually uses the word "curvaturam" to describe the exterior solid angle.

19.5. Euler Characteristic and Curvature

There is another, more "intrinsic" proof of Descartes' theorem which reveals the fact that total exterior solid angle is really $2\pi \times$ Euler characteristic. In fact, knowledge of the total exterior angle yields a proof of the Euler polyhedron formula. This seems to have been the way in which Descartes discovered his version of the formula.

The key step is to show that the exterior solid angle at a vertex $\mathscr{P}$ is expressible intrinsically as $2\pi - (\alpha_1 + \alpha_2 + \cdots + \alpha_n)$, where $\alpha_1, \alpha_2, \ldots, \alpha_n$ are the face angles that meet at $\mathscr{P}$. These are *not* the angles $\alpha'_1, \alpha'_2, \ldots, \alpha'_n$ between the planes that bound the exterior solid angle, but it turns out (exercise 1) that

$$\alpha_i + \alpha'_i = \pi$$

for each i, whence the measure of the exterior solid angle, which comes from $\alpha'_1 + \alpha'_2 + \cdots + \alpha'_n$ by Harriot's theorem (Section 16.6), also comes from $\alpha_1 + \alpha_2 + \cdots + \alpha_n$.

Knowing now that the exterior solid angle at $\mathscr{P}$ equals $2\pi - \Sigma$ face angles at $\mathscr{P}$, we get

$$\text{total exterior solid angle} = 2\pi V - \Sigma \text{ all face angles}$$

where V is the total number of vertices. By grouping the face angles according to faces, we also find (exercise 2) that

$$\Sigma \text{ all face angles} = \pi(-2E + 2F)$$

whence

$$\text{total exterior solid angle} = 2\pi(V - E + F)$$

$$= 2\pi \times \text{Euler characteristic}$$

In the case of convex polyhedra, where we already know that total exterior solid angle $= 4\pi$, this gives Euler characteristic $= 2$. More important, the derivation is valid for polyhedra of arbitrary Euler characteristic, showing that the total exterior solid angle is really the *same* as the Euler characteristic, up to a constant multiple.

There is a similar intrinsic proof of the Gauss–Bonnet theorem, again valid for arbitrary Euler characteristic, which shows that

$$\text{total curvature} = \iint_{\mathscr{S}} \kappa_1 \kappa_2 \, dA = 2\pi \times \text{Euler characteristic}$$

(exercise 3). Legendre's [1794] proof of the Euler polyhedron formula is the special case of the argument for constant positive curvature.

Thus the Euler characteristic regulates the total curvature of a surface. In particular, if the curvature is constant, then it must have the same sign as the Euler characteristic. This in turn has implications for the geometry of the surface. As we saw in Section 16.4, surfaces of constant positive curvature have spherical geometry, those of zero curvature have euclidean geometry, and those of negative curvature have hyperbolic geometry. In the next section we shall see that there is a natural way to impose constant curvature on surfaces of arbitrary Euler characteristic. It will then follow that the natural geometry of a surface is spherical, euclidean, or hyperbolic according to whether its Euler characteristic is positive, zero, or negative. Moreover, if the absolute value of the curvature is taken to be 1, then the Gauss–Bonnet theorem gives

$$\text{area} = |2\pi \times \text{Euler characteristic}|$$

This makes surface topology completely subordinate to geometry, at least for orientable surfaces, because it says that the topology of a surface is completely determined by the sign of its curvature and its area.

These results were implicit in the work of Poincaré and Klein in the 1880s. Perhaps Klein was the first to see clearly how the geometry of a surface determines its topology (see, e.g., Klein [1928], p. 264).

EXERCISES

1. Figure 19.8 shows the region around a vertex P of a polyhedron and the exterior solid angle of P centered at O and bounded by the planes OAB, OBC, OCA perpendicular to the edges through P. Show that there are right angles where indicated, and hence that

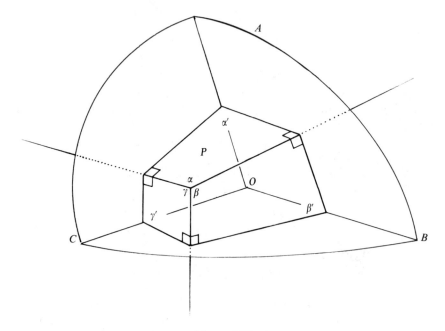

Figure 19.8

$$\alpha + \alpha' = \pi$$
$$\beta + \beta' = \pi$$
$$\gamma + \gamma' = \pi$$

2. Show that in an ordinary polyhedron (i.e., one with flat faces)

$$\Sigma \text{ all face angles} = \pi(-2E + 2F)$$

using the fact the angle sum of an n-gon is $(n - 2)\pi$.

3. Prove the global form of the Gauss–Bonnet theorem,

$$\iint_{\mathscr{S}} \kappa_1 \kappa_2 \, dA = 2\pi \times \text{Euler characteristic}$$

by partitioning the closed surface $\mathscr{S}$ into geodesic polygons and applying the ordinary form of the Gauss–Bonnet theorem (Section 16.6).

19.6. Surfaces and Planes

In Section 15.5 we noticed that an elliptic function defines a mapping of a plane onto a torus. Such mappings are also interesting in the topological context, where they are called *universal coverings*. In general, a mapping

$\varphi\colon \tilde{S} \to S$ of a surface $\tilde{S}$ onto a surface S is called a *covering* if it is a homeomorphism locally, that is, when restricted to sufficiently small pieces of $\tilde{S}$. The mapping of the plane onto the torus in Section 15.5 is a covering because it is a homeomorphism when restricted to any region smaller than a period parallelogram.

Another interesting example of a covering we have already met is the mapping of the sphere onto the projective plane given by Klein [1874] (Section 7.5). This map sends each pair of antipodal points of the sphere to the same point of the projective plane, and hence is a homeomorphism when restricted to any part of the sphere smaller than a hemisphere.

One more example is Beltrami's [1868] covering of the pseudosphere by a horocyclic sector (Section 17.4). Topologically, this covering is the same as the covering of a half-cylinder by a half-plane (Fig. 19.9). All these coverings are *universal* in the sense that the covering surface $\tilde{S}$ (sphere or plane) can be covered only by $\tilde{S}$ itself.

An example of a *nonuniversal* covering is the covering of the torus by the cylinder, intuitively like an infinite snake swallowing its own tail (Fig. 19.10). This is nonuniversal because the cylinder can in turn be covered by the plane, just as the half-cylinder is covered by the half-plane in Figure 19.9. In fact, by composing the coverings plane $\to$ cylinder $\to$ torus, we recover our first example, the plane $\to$ torus covering.

Since the sphere can be covered only by itself, the first interesting examples of coverings are those of orientable surfaces of genus $\geqslant 1$ (i.e., Euler char-

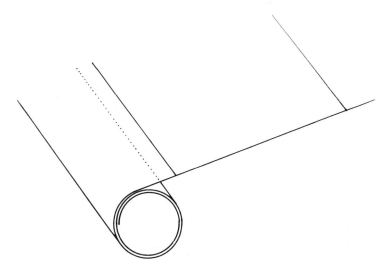

Figure 19.9

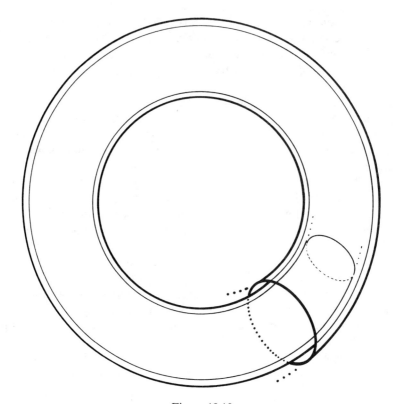

Figure 19.10

acteristic $\leqslant 0$). All of these surfaces can be covered by planes. Moreover, each nonorientable surface can be doubly covered by an orientable surface in the same way that the projective plane is covered by the sphere, so the main thing we need to understand is the universal covering of orientable surfaces of genus $\geqslant 1$ by planes.

The basic idea is due to Schwarz, and it became generally known as a result of a letter from Klein to Poincaré (Klein [1882′]). To construct the universal covering of a surface S one takes infinitely many copies of a fundamental polygon F for S and arranges them in the plane so that *adjacent* copies of F meet in the same way that F meets *itself* on S. For example, the torus T in Figure 19.11a has the fundamental polygon F shown in Figure 19.11b, which meets itself along $\vec{a}$ and $\vec{b}$ in S (where the arrows indicate that edges must agree in direction as well as label). If instead we take infinitely many separate copies of F and join adjacent copies $\vec{a}$ to $\vec{a}$ and $\vec{b}$ to $\vec{b}$, then we obtain a plane $\tilde{T}$, tessellated as in Figure 19.12. The universal covering $\tilde{T} \to T$ is then

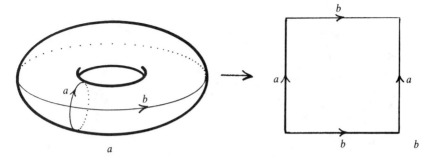

Figure 19.11

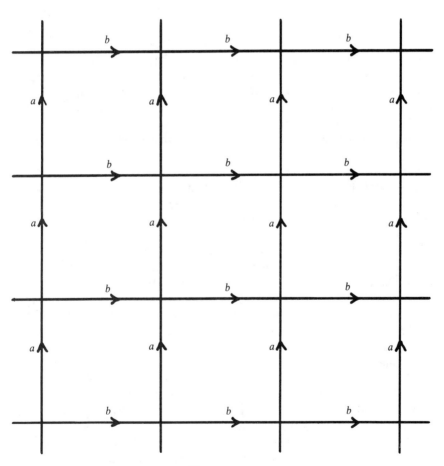

Figure 19.12

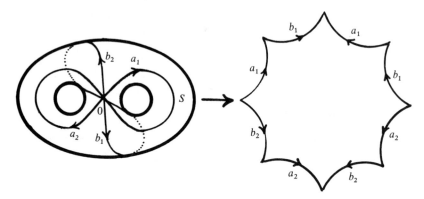

Figure 19.13

defined by mapping each copy of F in $\tilde{T}$ in the natural way onto the F in T.

The tessellation of Figure 19.12 can of course be realized by squares in the euclidean plane. We can therefore impose a euclidean geometry on the torus by defining the distance between (sufficiently close) points on the torus to be the euclidean distance between appropriate preimage points in the plane. In particular, the "straight lines" (geodesics) on the torus are the images of straight lines in the euclidean plane. The torus geometry is not quite the geometry of the plane, of course, since there are closed geodesics, such as the images of the line segments a and b. However, it is euclidean when restricted to sufficiently small regions. For example, the angle sum of each triangle on the torus is π.

For surfaces of genus >1, that is, negative Euler characteristic, the Gauss–Bonnet theorem predicts negative curvature and hence the natural covering plane should be hyperbolic. This can also be seen directly from the combinatorial nature of the tessellation on the universal cover. For example, the fundamental polygon F of the surface S of genus 2 is an octagon (Fig. 19.13). In the universal covering, eight of these octagons have to meet at each vertex, as the eight corners of the single F meet on S. Such a tessellation is impossible, by regular octagons, in the euclidean plane, but it exists in the hyperbolic plane, as Figure 19.14 shows. In fact, this tessellation is obtained by amalgamating triangles in the Gauss tessellation (Fig. 17.14). The tessellations for general genus >1 can similarly be realized geometrically in the hyperbolic plane and were among the hyperbolic tessellations considered by Poincaré [1882] and Klein [1882]. The distance function, hence the curvature and local geometry, can be transported from the covering plane to the surface as we did above for the torus.

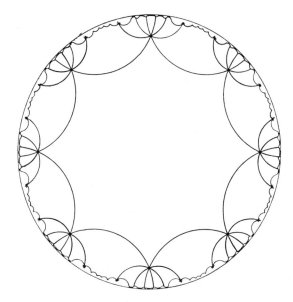

Figure 19.14

EXERCISE

1. Show that the fundamental polygon for an orientable surface of genus p is a $4p$-gon with angle sum 2π. Conclude that its Euler characteristic is proportional to its angular defect and hence to its area.

19.7. The Fundamental Group

Another way to explore the meaning of the universal cover $\tilde{S}$ is to use it to plot paths on the surface S. As a point P moves on S, each preimage $\tilde{P}$ of P moves in an analogous manner on $\tilde{S}$. The only difference is that as P crosses one of the edges of the fundamental polygon on S, $\tilde{P}$ crosses from one fundamental polygon to another on $\tilde{S}$. Thus $\tilde{P}$ will not necessarily return to its starting point, even when P does. In fact, we can see that the displacement of $\tilde{P}$ in some way measures the extent to which P winds around the surface S. Figure 19.15 shows an example. As P winds once round the torus, more or less in the direction of $\vec{a}$, $\tilde{P}$ wanders from one end to the other of a segment $\tilde{a}$ on $\tilde{S}$.

We say that closed paths p, p' with initial point 0 on S "wind in the same way," or are *homotopic*, if p can be deformed into p' with 0 fixed and without

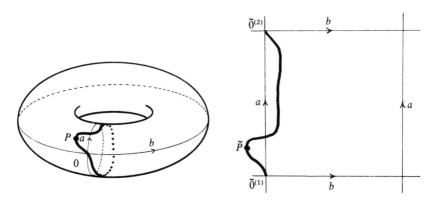

Figure 19.15

leaving the surface. Now if the path p of P is deformed into p', with 0 fixed, then the path $\tilde{p}$ of $\tilde{P}$ is deformed into a $\tilde{p}'$ with the same initial and final points, $\tilde{0}^{(1)}$ and $\tilde{0}^{(2)}$, as $\tilde{p}$. Hence each homotopy class corresponds simply to a *displacement* of the universal cover $\tilde{S}$ which moves $\tilde{0}^{(1)}$ to $\tilde{0}^{(2)}$. The different preimages $\tilde{P}$ will of course start at different preimages $\tilde{0}^{(1)}$ of 0, but a single displacement of $\tilde{S}$ moves them all to their final positions $\tilde{0}^{(2)}$. Moreover, the displacement moves the whole tessellation of $\tilde{S}$ onto itself: it is a rigid motion of the tessellation.

Thus from the topological notion of homotopic closed paths we have arrived once again at ordinary geometry. We have also arrived at a group, called the *fundamental group* of S. Geometrically, it is the group of motions of $\tilde{S}$ which map the tessellation onto itself (which includes mapping each edge to a like-labeled edge). Topologically, it is the group of homotopy classes of closed paths, with a common initial point 0, on S. The product of homotopy classes is defined by successive traversal of representative paths.

The fundamental group was first defined in the topological way by Poincaré [1895]. Poincaré defined it for much more general figures, whose universal covers are not so apparent, so the interpretation as a covering motion group did not emerge until later. As we know, Poincaré had already studied groups of motions of tessellations [1882]. He reconsidered these earlier results from the topological viewpoint [1904], arriving at the interpretation just given. This paper was very influential on the later work of Dehn [1912] and Nielsen [1927] and has been indirectly responsible for a recent surge of interest in hyperbolic geometry.

The more general notion of fundamental group in Poincaré [1895] has also been influential outside topology. It turns out, for example, that for any "reasonably described" figure $\mathscr{F}$ it is possible to compute generators and defining relations for the fundamental group of $\mathscr{F}$. The defining relations of a fundamental group can be quite arbitrary (in fact, *completely* arbitrary, as was shown by Dehn [1910] and Seifert and Threlfall [1934], p. 180), so the question arises: can the properties of a group be determined from its defining relations? One would like to know, for example, when two different sets of relations define the same group. The latter question was raised by Tietze [1908] in the first paper to follow up Poincaré's work. Tietze made the remarkable conjecture—which could not even be precisely formulated at the time—that the problem is unsolvable. The *isomorphism problem for groups*, as it came to be known, was indeed shown to be unsolvable by Adian [1957], in the sense that no algorithm can settle the question for *all* finite sets of defining relations. Adian's result was based on the development of a theory of algorithms that will be outlined in the next chapter. A complete proof of the unsolvability of the isomorphism problem, together with the relevant history, may be found in Stillwell [1982].

By combining Adian's result with some of Tietze [1908] and the result of Seifert and Threlfall mentioned above, Markov [1958] was able to show the unsolvability of the *homeomorphism problem*. This is the problem of deciding, given "reasonably described" figures $\mathscr{F}_1$ and $\mathscr{F}_2$, whether $\mathscr{F}_1$ is homeomorphic to $\mathscr{F}_2$. Thus Poincaré's construction of the fundamental group led in the end to quite an unexpected conclusion: the basic problem of topology is unsolvable.

19.8. Biographical Notes: Poincaré

Henri Poincaré (Fig. 19.16) was born in Nancy in 1854 and died in Paris in 1912. His father, Leon, was a physician and professor of medicine at the University of Nancy, and Henri grew up in a comfortable academic environment. He and his younger sister Aline were at first educated by their mother, and Poincaré later traced his mathematical ability to his maternal grandmother. At the age of five he suffered an attack of diphtheria, which weakened his health and excluded him from the more boisterous childhood games. He made up for this by organizing charades and playlets, and he later became a keen dancer. Many photographs of Poincaré and his family may be seen in the centenary volume [1955] which forms the second half of vol. 11 of Poincaré's *Oeuvres*.

Being excluded from most games, Poincaré had ample time to read and study, and when he began attending school, at the age of eight, he made rapid progress. His ability first showed in French composition, but by the end of his school career his awesome mathematical talent was also apparent. He won first prize in a nationwide mathematics competition and topped the entrance exam to the École Polytechnique in 1873. This, incidentally, was despite the Franco-Prussian War (1870–1871), during which Poincaré's home province of Lorraine bore the brunt of the German invasion. Poincaré accompanied his father on ambulance rounds at this time, becoming a fervent French patriot as a result. However, he never held German mathematicians responsible

Figure 19.16. Henri Poincaré.

for the brutalities of their compatriots. He learned German during the war in order to read the news, and he later put the knowledge to good use in communicating with his German colleagues Fuchs and Klein.

At the École Polytechnique, Poincaré continued to do well, though clumsiness in drawing and experimental work cost him first place. (His marks in drawing, though mediocre, were never zero, despite oft-told tales to that effect. Poincaré's results may be seen in the centenary volume [1955].) Curiously, he planned to become an engineer at this stage and studied at the École des Mines from 1875 to 1879, at the same time writing a doctoral thesis in mathematics. He worked briefly as a mining engineer before becoming an instructor in mathematics at the University of Caen in 1879.

It was at Caen that Poincaré made his first important discovery: the occurrence of noneuclidean geometry in the theory of complex functions. He had been thinking about periodicity with respect to linear fractional transformations, after encountering functions with this property in the work of Lazarus Fuchs. The functions in question arose from differential equations, and Poincaré had been struggling to understand them analytically when he was struck by an unexpected geometric inspiration:

> Just at this time I left Caen, where I was then living, to go on a geological excursion under the auspices of the school of mines. The changes of travel made me forget my mathematical work. Having reached Coutances, we entered an omnibus to go some place or other. At the moment when I put my foot on the step the idea came to me, without anything in my former thoughts seeming to have paved the way for it, that the transformations I had used to define the Fuchsian functions were identical with those of Non-Euclidean geometry. (Poincaré [1918]; translation from Halsted, 1929, p. 387).

The discovery of the underlying geometry (and topology, which soon followed) put Fuchsian functions in a completely new light, rather like the illumination of elliptic functions by Riemann's discovery that they belonged to the torus. For the next few years Poincaré worked feverishly to develop these ideas, in friendly competition with Klein. There were some reservations about his style—undisciplined and lacking in rigor, though very readable—but his brilliance was not contested. He was appointed to a chair at the University of Paris in 1881 and remained there, winning ever-higher honors, until the end of his life. In 1881 he was married to Louise Poulain; they had a son and three daughters.

Poincaré's work on Fuchsian functions led him to topology, as we have seen in Sections 19.6 and 19.7. So did another of his great inventions, the qualitative theory of differential equations. He used this theory, which deals with such questions as the long-term stability of a mechanical system, in his *Les méthodes nouvelles de la mécanique céleste* [1892, 1893, 1899], probably the greatest advance in celestial mechanics since Newton. Poincaré's topological ideas not only breathed new life into complex analysis and mechanics; they amounted to the creation of a major new field, algebraic topology. In papers between 1892 and 1904, Poincaré built up an arsenal of techniques and concepts that were to keep topologists going for the next 30 years. It was not until Hurewicz discovered higher dimensional analogues of the fundamental group in 1933 that a significant new weapon was added to Poincaré's arsenal. Recently, as mentioned in Section 19.1, there has been a return to geometric methods in topology. It would be fitting if these methods finally succeed in resolving the main unsolved problem left by Poincaré, the so-called *Poincaré conjecture*. The conjecture states that the 3-sphere (the space obtained by completing $\mathbb{R}^3$ by a point at infinity, as the plane was completed to an ordinary sphere in Section 14.2) is the only closed three-dimensional space whose fundamental group is trivial.

Poincaré was perhaps the last mathematician to have a general grasp of all branches of mathematics. Like Euler, he wrote fluently and copiously on all parts of mathematics, and in fact he surpassed Euler in his popular writing. He wrote many volumes on science and its philosophy, which were best-sellers in the early part of this century. Poincaré would perhaps have been as prolific as Euler if ill health had not overtaken him in his fifties. In 1911 he took the unusual step of published an unfinished paper, on periodic solutions of the three-body problem, believing he might not live to complete the proof. "Poincaré's last theorem" was indeed still open when he died in 1912, but the proof was completed in 1913 by the American mathematician G. D. Birkhoff.

CHAPTER 20

Sets, Logic, and Computation

20.1. An Explanation

In any survey of the history of mathematics it is hard to ignore the twentieth century. At the very least, we have to admit that our conception of classical mathematics is influenced by the mathematical ideas that are fashionable today. Indeed, it can be argued that most of classical mathematics becomes clearer when presented in modern terms, and that this is the best way to make it accessible to mathematicians who are not professional historians. As will be evident by now, this is the point of view I have adopted in this book.

At the same time, twentieth-century mathematics is far more than a mere instrument for viewing the mathematics of the past. It includes many results that are themselves classical and hence eligible for inclusion in our survey. The problem is that twentieth-century mathematics is so vast that no one can grasp it all, and even some single results are based on theories too large to be explained in a book of this size. Under these circumstances, no chapter on twentieth-century mathematics is likely to be representative, and the reader is entitled to an explanation of the author's choice of topics.

I believe that the choice of sets, logic, and computation is appropriate for the following reasons:

(i) The three topics are linked, historically and logically.
(ii) They are distinctively of the twentieth-century (or in the case of set theory, post-1870), having developed virtually from scratch in recent times.
(iii) Because of this they are accessible to anyone with a smattering of mathematics—certainly more accessible than any other modern topic of comparable importance.
(iv) They throw completely new light on the question "What is mathematics?" In fact they arise from the first serious attempts to answer this question.

20.2. Sets

Sets became established in mathematics in the late nineteenth century as a result of attempts to answer certain questions about the real numbers. First, what *is* a real number? Several equivalent answers were given around 1870, all involving infinite sets or sequences. The simplest was that of Dedekind [1872], who defined a real number to be a partition (or "cut") of the rational numbers into two sets, L and U, such that each member of L is less than all members of U. If one has a preconceived notion of real number, such as a point x on a line, then L and U are uniquely determined by x as the sets of rational points to left and right of it, respectively. Thus if x is preconceived, then L and U are no more than auxiliary concepts which enable x to be handled in terms of rationals, as Eudoxus did (Section 4.2). Dedekind's break-through was to realize that no preconceived x was necessary: x could be *defined* as the pair (L, U). Thus the concept of sets of rationals was a basis for the concept of real number.

Dedekind cuts gave a precise model for the continuous number line $\mathbb{R}$, since they filled all the gaps in the rationals. Indeed, wherever there is a gap in the rationals, the real number that fills it is essentially the gap itself: the pair of sets L, U to left and right of it. Other formulations of this *completeness* property of $\mathbb{R}$ are also easy consequences of Dedekind's definition. For example, each bounded set of reals (L_i, U_i) has a least upper bound (L, U). L is simply the union of the sets L_i.

Dedekind seemed to have settled the ancient problem of explaining the continuous in terms of the discrete, but in penetrating as far as he did, he also uncovered deeper problems. The central problem is that the completeness of $\mathbb{R}$ entails its *uncountability*, a phenomenon discovered by Cantor [1874]. The *countable* sets are those that can be put in one-to-one correspondence with $\mathbb{N} = \{0, 1, 2, \ldots\}$, and they include the set of rationals and the set of algebraic numbers, as Cantor also discovered. But if $\mathbb{R}$ is countable, this means that all reals can be included in a sequence $x_0, x_1, x_2, \ldots$. Cantor [1874] showed that this is impossible by constructing for each sequence $\{x_m\}$ of distinct reals a sequence $a_0, b_0, a_1, b_1, a_2, b_2, \ldots$, such that

$$a_0 < a_1 < a_2 < \cdots < b_2 < b_1 < b_0$$

and with the property that each x_m is outside one of the nested intervals $(a_0, b_0) \supset (a_1, b_1) \supset (a_2, b_2) \supset \cdots$. It follows that any common element of all the (a_n, b_n) is a real $x \neq$ each x_m. A common element obviously exists if the sequence of intervals is finite, and if the sequence is infinite, it exists by completeness, as the least upper bound of the a_n.

The uncountability of $\mathbb{R}$ has been a great challenge to set theorists and logicians ever since its discovery. The most successful response to this chal-lenge has been the theory of *ordinal numbers*, which grew out of Cantor's [1872] investigation of Fourier series (cf. Section 12.4). The existence of a Fourier series for a function f depends largely on the structure of the set of discontinuities of f, and thus leads to the problem of analyzing the complexity of point sets. Cantor measured complexity by the number of iterations of the prime operation (') of taking the limit points of a set. For example, if $S = \{0, 1/2, 3/4, 7/8, \ldots, 1\}$, then the prime can be applied once, and $S' = \{1\}$. It can happen that S' itself has limit points, so that S'' also exists. In fact, one can find a set S for which $S', S'', \ldots, S^{(n)}, \ldots$, exist for all finite n, so one can envisage iterating the prime operation an infinite number of times. In the case where all the $S^{(n)}$ exist, Cantor [1880] defined

$$S^{(\infty)} = \bigcap_{n=1,2,3,\ldots} S^{(n)}$$

He viewed ∞ as the first infinite number. To avoid confusion with higher infinite numbers soon to appear, I shall use the modern notation ω for the first infinite ordinal.

Having made the leap to ω, it is easy to go further. $(S^{(\omega)})' = S^{(\omega+1)}$, $(S^{(\omega+1)})' = S^{(\omega+2)}, \ldots$, and this new infinite sequence can be intersected to give $S^{(\omega \cdot 2)}$, where $\omega \cdot 2$ is the first infinite number after $\omega, \omega + 1, \omega + 2, \ldots$. After $\omega \cdot 2$, one has $\omega \cdot 2 + 1, \omega \cdot 2 + 2, \ldots, \omega \cdot 3, \ldots, \omega \cdot 4, \ldots, \ldots, \omega \cdot \omega, \ldots$. All these can actually be realized as numbers of iterations of the prime on sets of reals. But of course it is natural to investigate the ordinal numbers independently of this realization, as a generalization of the concept of natural number.

Cantor [1883] viewed the ordinals as being generated by two operations:

(i) Successor, which for each ordinal α gives the next ordinal, $\alpha + 1$.
(ii) Least upper bound, which for each set $\{\alpha_i\}$ of ordinals gives the least ordinal $\geq$ each α_i.

The most elegant formalization of these notions was given by von Neumann [1923]. The empty set $\varnothing$ (not considered by Cantor) is taken to be the ordinal 0, the successor of α is $\alpha \cup \{\alpha\}$, and the least upper bound of $\{\alpha_i\}$ is simply the union of the α_i. Thus

$$0 = \varnothing$$
$$1 = \{0\}$$
$$2 = \{0, 1\}$$
$$\ldots\ldots$$
$$\omega = \{0, 1, 2, \ldots, n, \ldots\}$$
$$\omega + 1 = \{0, 1, 2, \ldots, \omega\}$$

and so on. The natural ordering of the ordinals is then given by set membership, $\in$, and in particular the members of an ordinal α are all ordinals smaller than α.

Cantor's principle (ii) generates ordinals of breathtaking size, since it gives the power to transcend any set of ordinals already defined. In particular, an ordinal of uncountable size is on the horizon as soon as one thinks of the concept of countable ordinal, as Cantor did [1883]. He defined an ordinal α to be countable (or, as he later put it, *of cardinality* $\aleph_0$) if α could be put in one-to-one correspondence with $\mathbb{N}$. For example, $\omega \cdot 2 = \{0, 1, 2, \ldots, \omega, \omega + 1, \omega + 2, \ldots\}$ is countable because of the obvious correspondence with $\mathbb{N} = \{0, 2, 4, \ldots, 1, 3, 5, \ldots\}$. The least upper bound of the countable ordinals is the least *uncountable* ordinal, ω_1. Sets in one-to-one correspondence with ω_1 define the next cardinality, $\aleph_1$. Ordinals of cardinality $\aleph_1$ have a least upper bound ω_2 of cardinality $\aleph_2$, and so on.

Having found this orderly way of generating successive uncountable cardinals, Cantor reconsidered the uncountable set $\mathbb{R}$. Although no method of generating members of $\mathbb{R}$ in the manner of ordinals was apparent, Cantor conjectured that the cardinality of $\mathbb{R}$ was $\aleph_1$. This conjecture has since become known as the *continuum hypothesis*. By 1900 it was recognized as the outstanding open problem of set theory, and Hilbert [1900] made it number one on the famous list of problems he presented to the mathematical community. There have been two outstanding results on the continuum problem since 1900, but they seem to make it less likely that we will ever know whether the continuum hypothesis is correct. Gödel [1938] showed that the continuum hypothesis is *consistent* with standard axioms for set theory, but Cohen [1963] showed that its negation is also consistent. Thus the continuum hypothesis is independent of standard set theory in much the same way that the parallel postulate is independent of Euclid's other postulates. Whether this means that the notion of "set" is open to different natural interpretations, like the notion of "straight line," is not yet clear.

EXERCISES

1. Show that for any sequence x_0, x_1, x_2, $\ldots$, of distinct reals there is a sequence $a_0, b_0, a_1, b_1, a_2, b_2, \ldots$, of the type described above.

2. Give a rule for continuing the sequence

$$1/2, 1/3, 2/3, 1/4, 3/4, 1/5, 2/5, 3/5, 4/5, \ldots$$

   that will include all rationals in $(0, 1)$.

3. Construct a linear fractional mapping of $(0, 1)$ onto the positive reals. Use this or another method to conclude that the set of positive rationals is countable. How can one then conclude that the set of all rationals is countable?

4. The words on a fixed finite alphabet can be enumerated by listing first the one-letter words, then the two-letter words, and so on. Use this observation to show that the set of polynomial equations with integer coefficients is countable and hence that the set of algebraic numbers is countable.

20.3. Measure

The reason for the investigation of sets of discontinuities in connection with Fourier series was the discovery of Fourier [1822] that these series depend on integrals. Assuming that

$$f(x) = \tfrac{1}{2}a_0 + \sum_{n=1}^{\infty} (a_n \cos n\pi x + b_n \sin n\pi x)$$

Fourier derived the formulas

$$a_n = \int_{-1}^{1} f(x) \cos n\pi x \, dx, \qquad b_n = \int_{-1}^{1} f(x) \sin n\pi x \, dx$$

Thus the existence of the series depends on the existence of the integrals for
a_n and b_n, and this in turn depends on how discontinuous f is. It was known
(though not rigorously proved) that every continuous function had an integral,
so the next question was how the integral should, or could, be defined for
discontinuous functions. The first precise answer was the Riemann [1854']
integral concept, familiar to all calculus students, based on approximating
the integrand by step functions. Any function with a finite number of discontinuities has a Riemann integral, and indeed so have certain functions with
infinitely many discontinuities, but not all. The classic function for which the
Riemann integral does not exist is the function of Dirichlet [1829]:

$$f(x) = \begin{cases} 1 & \text{if } x \text{ is rational} \\ 0 & \text{if } x \text{ is irrational} \end{cases}$$

Eventually a more general integral, the Lebesgue integral, was introduced
to cope with such functions, but not until the focus of attention had shifted
from the problem of integration to the more fundamental problem of *measure*.
Measure generalizes the concept of length (on the line $\mathbb{R}$), area (in the plane
$\mathbb{R}^2$), and so on, to quite general point sets. Since an integral can be viewed as
the area under a graph, its dependence on the concept of measure is clear,
though it was not realized at first that the measure of sets on the line had to
be clarified first. The need for clarification arose from the discovery of Harnack
[1885] that any countable subset $\{x_0, x_1, x_2, \ldots\}$ of $\mathbb{R}$ could be covered by
a collection of intervals of arbitrarily small total length (cover x_0 by an interval
of length $\varepsilon/2$, x_1 by an interval of length $\varepsilon/4$, x_2 by an interval of length
$\varepsilon/8$, ..., so that the total length of intervals used is $\leqslant \varepsilon$). This seemed to
show that countable sets were "small" (of *measure zero*, as we now say), but
mathematicians were reluctant to say this of dense countable sets, like the
rationals. The first response was to define measure analogously to the Riemann
integral, using finite unions of intervals to approximate subsets of $\mathbb{R}$ (Jordan
[1892]). Under this definition, "sparse" countable sets like $\{0, 1/2, 3/4, 7/8, \ldots\}$
did have measure zero, but dense sets like the rationals were not measurable
at all.

The first to take the hint from Harnack's result that countable unions of
intervals should be used to measure subsets of $\mathbb{R}$ was Borel [1898]. He defined
the measure of any countable union of intervals to be its total length and
extended measurability to more and more complicated sets by *complementation*
and *countable disjoint unions*. That is, if a set S contained in an interval I has
measure $\mu(S)$, then $\mu(I - S) = \mu(I) - \mu(S)$, and if S is a disjoint union of sets
S_n with measures $\mu(S_n)$, then $\mu(S) = \sum_{n=1}^{\infty} \mu(S_n)$. The sets that can be formed

from intervals by complementation and countable unions are now called *Borel sets*. Borel's idea was pushed to its logical conclusion by Lebesgue [1902], who assigned measure zero to any subset of a Borel set of measure zero. Since not all such sets are Borel, Lebesgue extended measurability to a larger class of sets: those that differ from Borel's by sets of measure zero. Whether these are *all* subsets of $\mathbb{R}$ is an interesting question to which we shall return shortly.

The distinctive property of Borel–Lebesgue measure is *countable additivity*: if $S_0, S_1, S_2, \ldots$, are disjoint measurable sets, then

$$\mu(S_0 \cup S_1 \cup S_2 \cup \cdots) = \mu(S_0) + \mu(S_1) + \mu(S_2) + \cdots$$

Lebesgue showed that this gives a concept of integral which is better behaved with respect to limits than the Riemann integral. For example, if $f_n \to f$ as $n \to \infty$, then $\int f_n \, dx \to \int f \, dx$ for the Lebesgue integral, whereas this is not generally true for the Riemann integral (see exercise 1). Another motivation for countable additivity pointed out by Borel was the theory of probability. If an "event" E is formalized as a set S of points ("favorable outcomes"), then the probability of E can be defined as the measure of S. Some quite natural events turn out to be countable unions, hence it is necessary for probability measure to be countably additive. In informal probability theory, countable additivity was assumed as far back as James Bernoulli [1690], who answered the following question he had posed in 1685:

> *A* and *B* play with a die, the one that throws an ace first being declared the winner. *A* throws once, then *B* throws once also. *A* then throws twice, and *B* does the same, and so on, until one wins. What is the ratio of their chances of success?

To solve this problem James Bernoulli decomposed the event of a win for *A* (or *B*) into the subevents of a win at *A*'s (*B*'s) first, second, third, …, turn, and summed the probabilities of these countably many subevents. Formal probability theory, which was created by Kolmogorov [1933], bases all such arguments on the theory of countably additive measures.

It could be said that set theory paved the way for measure theory by showing the uncountability of $\mathbb{R}$, thus enabling countable subsets of $\mathbb{R}$ to be regarded as "small." On the other hand, measure theory itself shows the uncountability of $\mathbb{R}$ (look again at Harnack's result), and in fact measure theory's assessment of the "smallness" of countable sets greatly influenced the later development of set theory. "Measure theoretically desirable" axioms, such as the measurablity of all subsets of $\mathbb{R}$, turned out to conflict with "set theoretically desirable" axioms such as the continuum hypothesis, and efforts to resolve the conflict brought more fundamental questions about sets to light. These questions do not reduce to clear-cut alternatives—the way geometric questions reduce to alternative parallel axioms, for example—but they do seem to gravitate toward the so-called *choice* and *large cardinal* axioms.

The usual axiom of choice states that for any set S (of sets) there is a *choice function* f such that $f(x) \in x$ for each $x \in S$. (Thus f "chooses" an element from each set x in S.) The axiom seems so plausible that early set theorists used it almost unconsciously, and it first attracted attention in Zermelo's [1904] proof that any set S could be *well ordered* (i.e., put in one-to-one correspondence with an ordinal). This looked like progress toward the continuum hypothesis. But Zermelo's proof gave no more than the existence of a well-ordering of S, given the existence of a choice function for the set of subsets of S. There was still no sign of an explicit well-ordering of $\mathbb{R}$. And of course if one doubted the existence of a well-ordering of $\mathbb{R}$, this threw doubt on the axiom of choice. Further doubts were raised when the axiom of choice was found to have incredible consequences in measure theory.

The first of these, discovered by Vitali [1905], was that the circle can be decomposed into countably many disjoint congruent sets. Since congruent sets have the same Lebesgue measure, it easily follows that the sets in question are not Lebesgue measurable (by countable additivity; see exercises below). Even more paradoxical decompositions were given by Hausdorff [1914] (for the sphere) and Banach and Tarski [1924] (for the ball). The Banach–Tarski theorem states that the unit ball can be decomposed into finitely many sets which, when rigidly moved in space, form *two* unit balls! This shows that not all subsets of the ball are measurable, even if one asks only for finite, rather than countable, additivity. For an excellent discussion of the paradoxical decompositions and their connections with other parts of mathematics, see Wagon [1985].

The measure-theoretic consequences of the paradoxical decompositions follow from the geometrically natural assumption that congruent sets have the same measure. If one drops this assumption and asks only for countable additivity and nontriviality (i.e., not all subsets have measure zero), then the conflict with the axiom of choice seems to disappear. No contradiction has yet been derived from these assumptions, but Ulam [1930] showed that any set possessing such a measure must be extraordinarily large—large enough, in fact, to be a model for set theory itself, and in particular larger than $\aleph_1, \aleph_2, \ldots, \aleph_\omega, \ldots$. Thus if $\mathbb{R}$ has a nontrivial countably additive measure, then $\mathbb{R}$ must be far larger than $\aleph_1$, and we still have a conflict with the continuum hypothesis. (For more on the "largeness" of models see Section 20.6.)

An even more desirable axiom than measurability would be Lebesgue measurability of all subsets of $\mathbb{R}$. This conflicts with the axiom of choice, by Vitali's theorem, and requires the existence of large cardinals, by Shelah [1984], but Solovay [1970] showed that the axiom of Lebesgue measurability was nevertheless consistent, assuming the existence of a large cardinal. Thus measurability of all subsets of $\mathbb{R}$ is intimately connected with the existence of sets large enough to model the whole of set theory. This mind-boggling concept seems to be the answer to many fundamental questions. We shall find ourselves drawn to it again in the next sections when we explore the influence of set theory on logic. Meanwhile, for those who would like a more detailed account of the development of set theory, and the contentious axioms in particular, we refer to van Dalen [1972].

EXERCISES

1. Show that a function f_n which is zero at all but n points has Riemann integral zero over any interval, and that the non–Riemann integrable function of Dirichlet is a limit as $n \to \infty$ of such f_n.

2. Call f *sequentially continuous* at $x = a$ if, for any sequence $\{a_n\}$ such that $a_n \to a$ we have $f(a_n) \to f(a)$. Show, assuming the axiom of choice, that f is continuous at a if f is sequentially continuous at a. (It is a consequence of Cohen [1963] that this result *cannot* be proved without the axiom of choice.)

3. For each θ between 0 and 2π let $S(\theta)$ be the set of points on the unit circle whose angle differs from θ by a rational multiple of 2π. Thus $S(\theta) = S(\phi)$ if $\theta - \phi = 2\pi \times a$ rational, and $S(\theta) \cap S(\phi) = \varnothing$ otherwise. Let S be a set (existing by virtue of the axiom of choice) which contains exactly one element from each distinct $S(\theta)$ and let

$$S + r = \{\theta + r | \theta \in S\} \qquad \text{for each rational } r$$

Show that the circle is a countable union of sets $S + r$.

4. Show that both assumptions $\mu(S) = 0$ and $\mu(S) > 0$ lead to contradictions, and hence conclude that S is nonmeasurable.

20.4. The Diagonal Argument

The uncountability of $\mathbb{R}$ was shown again in a strikingly simple way by Cantor [1891]. His argument applies most directly to the set $2^{\mathbb{N}}$ of all subsets of $\mathbb{N}$, but there are variants that work similarly on the set $\mathbb{N}^{\mathbb{N}}$ of integer functions and on $\mathbb{R}$ (which can be identified with a set of integer functions in various ways). To show that there are uncountably many subsets of $\mathbb{N}$ one shows that any countable collection $S_0, S_1, S_2, \ldots,$ of sets $S_n \subseteq \mathbb{N}$ is incomplete, by constructing a new set S different from each S_n. S is the so-called *diagonal set* $\{n | n \notin S_n\}$, which obviously differs from S_n with respect to the number n. Q.E.D.

The "diagonal" nature of S can be seen by visualizing a table of 0's and 1's in which

$$m\text{th entry in } n\text{th row} = \begin{cases} 0 & \text{if } m \notin S_n \\ 1 & \text{if } m \in S_n \end{cases}$$

In other words, the nth row consists of the values of the characteristic function of S_n. The characteristic function of S is simply the diagonal of the table, with all values reversed. A sequence $x_0, x_1, x_2, \ldots,$ of real numbers can be diagonalized similarly by forming the table whose nth row consists of the decimal digits of x_n. A suitable way to "reverse" the digits on the diagonal is to change any 1 to a 2 and any other digit to a 1. (The resulting sequence of 1's and 2's, after a decimal point, then defines a real number x whose decimal expansion is unique. Hence x is not just different from each x_n in its decimal expansion but is definitely a different number.) More generally, for any table of integers, that is, sequence of integer functions f_n, one can construct an integer function f unequal to each f_n by changing the values along the diagonal of the table. The diagonal argument was in fact first given in this context, by duBois-Reymond [1875], in order to construct an f with a greater rate of growth than all functions in a sequence $f_0, f_1, f_2, \ldots$ (exercise 1). With hindsight, one can even see a diagonal construction in Cantor's first [1874] argument for the uncountability of $\mathbb{R}$ (exercise 2).

The diagonal argument is important in set theory because it readily generalizes to show that any set has more subsets than elements (exercise 3), and hence that there is no largest set. What was not noticed at first is that the diagonal argument also has consequences at a more concrete level. This is because the diagonal of a table is *computable* if the table as a whole is computable. Hence the argument does not merely show how to add a new function f to a list $f_0, f_1, f_2, \ldots$—it shows how to add a new computable function to a computable list. In other words, *it is impossible to compute a list of all computable functions.* And of course the same goes for lists of computable real numbers. This remarkable result went unnoticed in the early days of the diagonal argument because computability was not then regarded as an interesting concept, or indeed a mathematical concept at all. The controversies over the axiom of choice, however, helped to sharpen awareness of the difference between constructive and nonconstructive functions. In the 1920s logicians began to investigate the concept of computability more seriously and by a "kind of miracle," as Gödel [1946] later expressed it, computability turned out to be a mathematically precise notion.

The notion was first formalized by Turing [1936] and Post [1936], who arrived independently at a definition of computing machine, now called a *Turing machine*. A Turing machine M is given by finite sets $\{q_0, q_1, \ldots, q_m\}$, of *internal states*, and $\{s_0, s_1, \ldots, s_n\}$, of *symbols*, and *transition function* T which formalizes the behavior of M for pairs (q_i, s_j). M is visualized as having an infinite tape, divided into squares, each of which can carry one of the symbols s_j. (For most purposes, M is assumed to start on a tape which has all but finitely many squares blank; s_0 is taken to denote the blank symbol.) Depending on its internal state q_i, M will make a *transition*: changing s_j to s_k, then moving one square right or left and going into a new state q_l. Thus the transition function is given by finitely many equations

$$T(q_i, s_j) = (m, s_k, q_l)$$

where $m = \pm 1$ indicates a move to right or left.

To use M to compute a function $f: \mathbb{N} \to \mathbb{N}$, some convention must be adopted for inputs (arguments of f) and outputs (values of f). The simplest is seen in Figure 20.1. M starts in state q_0 on the leftmost 1 of a block of n 1's, on an otherwise blank tape, and halts on the leftmost 1 of a block of $f(n)$ 1's, on an otherwise blank tape. M halts by virtue of entering a *halting state*, that is, a state q_h for which M has no transition from the pair $(q_h, 1)$. A

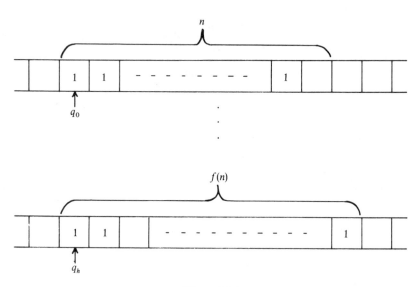

Figure 20.1

computable function f is one that can be represented in this way by a Turing machine M.

It follows that there are only countably many computable functions $f: \mathbb{N} \to \mathbb{N}$ since there are only countably many Turing machines. In fact we can compute a list of all (nonisomorphic) Turing machines by first listing the finitely many machines with one transition, then those with two transitions, and so forth. This may seem to contradict the previous discovery that a list of all computable functions cannot be computed but, as Turing [1936] realized, it does not. The catch is that not all machines define functions, and *it is impossible to pick out those that do.* Of course it is possible to rule out any machine that halts in a situation unlike that in Figure 20.1; the difficulty is in knowing whether halting is going to occur. It is precisely this difficulty that prevents computation of the diagonal function. If it could be decided, for each machine M and each input, whether M eventually halted, then we could pick out in succession the first machine to halt on input 1, the next after that to halt on input 2, the next after that to halt on input 3, and so on. But changing the corresponding outputs according to some rule (say, adding 1 if the output is a number, and taking the value 1 otherwise), we could compute a function different from each computable function.

This contradiction shows that the problem of deciding, given a machine and an input, whether halting eventually occurs, is *unsolvable.* This problem is called the *halting problem* and unsolvability means precisely that no Turing machine can solve it. That is, if the questions "Does M on input n eventually halt?" are formulated in some fixed finite alphabet, then there is no machine which, given these questions as inputs, will give their answers as outputs.

The significance of this result is that, as far as we know, all possible rules or algorithms for the solution of problems can be realized by Turing machines. This is the "kind of miracle" referred to by Gödel [1946].

Now that computers are everywhere, it is taken for granted that "computability" has a precise, absolute meaning—synonymous with Turing machine computability. It is even a familiar fact that all computations can be done on a single, sufficiently powerful machine; this corresponds to the discovery of Turing [1936] of a *universal Turing machine*. However, these claims were surprising in the 1930s, particularly to Gödel, who had shown [1931] that the related notion of "provability" is *not* absolute. This will be discussed further in the next section. Briefly, the reason for the difference is that new computable functions cannot be created by diagonalization, whereas new theorems can.

The halting problem was of no obvious mathematical significance in 1936, but it certainly seemed no more difficult than other unsolved algorithmic problems in mathematics. Thus for the first time it was reasonable to suspect that such problems were unsolvable. Moreover, if it could be shown that a solution of a particular problem P implied a solution of the halting problem, then the unsolvability of P would be rigorously established. This method was used to demonstrate the unsolvability of some problems in formal logic by Turing [1936] and Church [1936]; Church [1938] also put forward a strong candidate for unsolvability in ordinary mathematics: the word problem for groups.

This is the problem of deciding, given a finite set of defining relations for a group F (cf. Section 18.6) and a word w, whether $w = 1$ in G. There is more than a superficial analogy between the word problem and the halting problem. G corresponds to a machine M, words in G correspond to expressions on M's tape, and $w = 1$ corresponds to halting. The defining relations of G roughly correspond to the transition function of M, but unfortunately there is no machine equivalent of the cancellation of inverses in G. This creates fierce technical difficulties, but they were overcome by Novikov [1955]. He succeeded in establishing the validity of the analogy and hence the unsolvability of the word problem. This led to unsolvability results for a host of significant mathematical problems, among them the homeomorphism problem mentioned in Section 19.7.

A profound reworking of Novikov's ideas, due to Higman [1961], shows that computability is a "mathematically natural" concept in the context of groups. Higman showed that a finitely generated group H has a computable set of defining relations if and only if H is a subgroup of a finitely generated group F with a finite set of defining relations.

EXERCISES

1. Given integer functions f_0, f_1, f_2, ..., define an integer function f such that $f(m)/f_n(m) \to \infty$ as $m \to \infty$ for each n. *Hint*: Arrange that $f(m) \geq nf_n(m)$ for all $m \geq n$.

2. Show that if $a_0 < a_1 < a_2 < \cdots$ is a bounded sequence of real numbers, then $a =$ least upper bound of $\{a_0, a_1, a_2, \ldots\}$ is a "diagonal number" of the sequence in the following sense. There are integers $k_0 < k_1 < k_2 < \cdots$ such that the decimal digits of a exceed those of a_n after the k_nth place.

3. Let I be any set, and let $\{S_i\}$ be a collection of subsets of I in one-to-one correspondence with the elements i of I. Show that the natural "diagonal" set S of this collection is a subset of I unequal to each S_i.

20.5. Logic and Gödel's Theorem

Since the time of Leibniz, and perhaps earlier, attempts have been made to mechanize mathematical reasoning. Little success was achieved until the late nineteenth century, when the subject matter of mathematics was clarified by defining all mathematical objects in terms of sets. The reduction of the many concepts of number, space, function, and the like, to the single concept of set brought with it a corresponding reduction in the number of axioms that seemed to be necessary for mathematics. At about the same time, investigation of the principles of logic by Boole [1847], and particularly Frege [1879], led to a system of rules by which all logical consequences of a given set of axioms could be inferred. These two lines of investigation together offered the possibility of a complete, rigorous, and, in principle, *mechanical* system for the derivation of all mathematics.

The most thorough attempt to realize this possibility was the massive *Principia Mathematica* of Whitehead and Russell [1910, 1912, 1913]. *Principia* used axioms of set theory, together with a small collection of rules of inference, to derive a substantial part of ordinary mathematics in a completely formal language. The purpose of the formal language was to avoid the vagueness and ambiguity of natural language, so that proofs could be checked mechanically. Mechanical proof-checking was not then regarded as a goal in itself but rather as a guarantee of rigor. When Whitehead and Russell began writing their *Principia* in 1900, they believed they were about to reach the nineteenth-century goal of a complete and absolutely rigorous mathematical system. They did not realize that the rigor of their system—the possibility of checking proofs mechanically—was in fact *incompatible* with completeness. Gödel [1931] showed that there are true sentences which can be expressed in the language of *Principia Mathematica* but which do not follow from its axioms. (Unless *Principia* is inconsistent, in which case all sentences follow from its axioms. The assumption of consistency is actually a weighty one, as we shall see by the end of this section.)

Gödel's theorem created a sensation when it first appeared. Not only did it shatter previous conceptions of mathematics and logic, but its proof was of a new and bewildering kind. Gödel exploited the mechanical nature of proof in *Principia* to define the relation "the nth sentence of *Principia* is provable" within the language of *Principia* itself. Using this, he was able to concoct a sentence that says, in effect, "This sentence is not provable." The Gödel sentence, if true, is therefore not provable. And if false, it is provable, and so *Principia* proves a false sentence. Either way, provability in *Principia* is not the same as truth.

Gödel's proof was very difficult for his contemporaries to understand. Combined with the novelty of treating symbols and sentences as mathematical objects was the near inconsistency of a sentence which expresses its own unprovability (a sentence which says "This sentence is not true" *is* inconsistent). Post [1944] presented Gödel's theorem in a less paradoxical way by deriving it from the classical diagonal argument. The key to Post's approach is the concept of a *recursively enumerable set*. A set W is called recursively enumerable if a list of its members can be computed, say by a Turing machine which prints them on its tape. (Of course if W is infinite, the computation lasts forever.) The paradigm of a recursively enumerable set is the set of theorems of a formal system, such as *Principia Mathematica*. For such a system one can compute a list of all sentences, a list of all finite sequences of sentences, and, by picking out those sequences that are proofs, a list of all theorems—since a theorem is simply the last line of a proof.

Post's idea was to look at the theorems about recursively enumerable sets proved in a given system Σ and to compute a "diagonal sentence" from them. Since recursively enumerable sets are associated with Turing machines, it is possible to enumerate the recursively enumerable subsets of $\mathbb{N}$ as W_0, W_1, $W_2, \ldots$, by letting W_n be the set of numbers output by the nth machine, under some reasonable convention. (Incidentally, there is no problem of picking out suitable machines, like there is for computable functions, since we do not mind if W_n is empty.) The diagonal set $D = \{n \mid n \notin W_n\}$, being unequal to each W_n, is of course not recursively enumerable, but the following set is:

$$\mathrm{Pr}(D) = \{n \mid \Sigma \text{ proves } ``n \notin W_n"\}$$

This "provable part" of D is recursively enumerable because we can enumerate the theorems of Σ and pick out those of the form "$n \notin W_n$." $\mathrm{Pr}(D) \subseteq D$, assuming Σ proves only correct sentences, but $\mathrm{Pr}(D) \neq D$ since $\mathrm{Pr}(D)$ is recursively enumerable and D is not. This shows immediately that there is an n_0 in D which is not in $\mathrm{Pr}(D)$, that is, an $n_0 \notin W_{n_0}$ for which "$n_0 \notin W_{n_0}$" is not provable.

Better still, a specific n_0 with this property is the index of the recursively enumerable set $\mathrm{Pr}(D)$. If $W_{n_0} = \mathrm{Pr}(D)$, then $n_0 \in W_{n_0}$ is equivalent to $n_0 \in \mathrm{Pr}(D)$, which means that "$n_0 \notin W_{n_0}$" is provable. But then it is true that $n_0 \notin W_{n_0}$, assuming Σ proves only correct sentences, and we have a contradiction. Thus $n_0 \notin W_{n_0}$. This in turn is equivalent to $n_0 \notin \mathrm{Pr}(D)$, which means "$n_0 \notin W_{n_0}$" is not provable. (Notice, incidentally, that the last part of this argument reveals "$n_0 \notin W_{n_0}$" to be a sentence which expresses its own unprovability.)

It seems that Post was aware of this approach to Gödel's theorem in the 1920s, before Gödel's own proof appeared. However, Post's more general view of incompleteness as a property of arbitrary recursively enumerable systems held him up until he was satisfied that computability was a mathematically definable concept. In December 1925 Post formulated a plan for proving *Principia Mathematica* incomplete but, as he later wrote, "The plan, however, included prior calisthenics at other mathematical and logical work, and did not count on the appearance of a Gödel!" (Post [1941], p. 418).

Gödel's theorem comes from reflection on the nature of proofs in ordinary mathematics. An even more devastating theorem, known as Gödel's second theorem, comes from reflection on the proof of Gödel's theorem itself. The latter proof, unusual though it is, can in fact be expressed in ordinary mathematical language. We described Post's proof of Gödel's theorem in an informal language of Turing machines but, with some effort, it can be expressed in a small language for number theory called *Peano Arithmetic* (PA). PA is a language of addition and multiplication on $\mathbb{N}$, with basic logic and mathematical induction as the proof machinery. Turing machines can be discussed in PA by interpreting sequences of symbols on the tape as numerals, so that the changes they undergo in the course of a computation become operations on numbers. Under this interpretation, "$n_0 \notin W_{n_0}$" and "Σ does not prove '$n_0 \notin W_0$'" become sentences of PA.

At this point it is important to remember the hypothesis about Σ that was used in the proof of Gödel's theorem; Σ proves only correct sentences. This assumption cannot be dropped (since one incorrect theorem usually allows *all* sentences to be derived), but it can be weakened to the consistency assumption that Σ does not prove the sentence "$0 = 1$." Since the latter assumption says that a certain element (the number of the sentence "$0 = 1$") does not belong to a certain recursively enumerable set (the set of theorems of Σ), it can be expressed as a sentence of PA, Con(Σ). In particular, PA can express its own consistency by the sentence Con(PA). After these modifications, Gödel's theorem becomes the following sentence of PA:

$$\text{Con(PA)} \Rightarrow \text{"}n_0 \notin W_{n_0}\text{" is not provable}$$

An equivalent sentence is simply

$$\text{Con(PA)} \Rightarrow n_0 \notin W_{n_0}$$

since, as we have seen, the truth of "$n_0 \notin W_{n_0}$" is equivalent to its unprovability.

Now Gödel noticed that his proof could be carried out in PA, which was verified in detail by Hilbert and Bernays [1936]. Consequently, if Con(PA) can be proved in PA, then so can "$n_0 \notin W_{n_0}$," by basic logic. But if PA is consistent, "$n_0 \notin W_{n_0}$" *cannot* be proved in it, by Gödel's theorem, hence neither can Con(PA). (Gödel of course had a different unprovable sentence, but it was similarly implied by Con(PA).)

Thus the assertion Con(PA) that the axioms of PA are consistent is in some way stronger than the axioms themselves. Similarly, if Σ is any system that includes PA (such as *Principia Mathematica* and other systems of set theory), then Con(Σ) cannot be proved in Σ, if Σ is consistent. This is Gödel's second theorem.

20.6. Provability and Truth

In the previous section it was stressed that Gödel's theorem is a statement of alternatives: a formal system Σ either fails to prove a true sentence or else proves a false one. The second theorem identifies a sentence, Con(Σ), which is either true and unprovable or false and provable, but the proof does not say which alternative actually holds for particular Σ, such as PA or *Principia Mathematica*. How could it, without violating Gödel's theorem itself? Unless Σ is actually *inconsistent*, there can be no formal proof that Con(Σ) is true!

Nevertheless, Gödel's theorem tells us we have nothing to lose by adding Con(Σ) to the system Σ. If Σ is inconsistent, then it is already worthless, and we are no worse off for having added Con(Σ). And if Σ is consistent, we actually gain, because Con(Σ) is a new mathematical truth not provable from Σ alone. In this sense, Gödel's theorem gives a means of transcending any given formal system. Knowing that Con(Σ) is beyond the scope of Σ (if Σ is consistent) is of practical value to mathematicians, as it means there is no point in trying to prove any sentence which implies Con(Σ). If one wants to use such a sentence, it should be taken as a new axiom.

Sentences of mathematical interest actually arise in this way, most simply in set theory, where consistency is implied by the existence of a "large set." The usual axioms of set theory (called ZF) say roughly that

(i) $\mathbb{N}$ is a set.
(ii) Further sets result from certain operations, the most important of which are *power* (taking all subsets of a set) and *replacement* (taking the range of a function whose domain is a set).

Because of this, the axioms of ZF can be modeled by any set that contains $\mathbb{N}$ and is closed under power and replacement. Such a set has to be very large—larger than any set whose existence can be proved in ZF—but if it exists, then ZF must be consistent, since two contradictory sentences cannot be true of an actually existing object. Thus the existence of a set which is large in the above sense implies Con(ZF).

If ZF is consistent, then ZF + Con(ZF) is also consistent, but an even larger set is required to satisfy the enlarged axiom system. These large-set existence axioms are called *axioms of infinity*. Since they imply Con(ZF), they cannot be proved in ZF. In particular, it follows that one cannot prove the existence of a nontrivial measure on all subsets of $\mathbb{R}$ since, as mentioned in Section 20.3, this implies the existence of a large set. In fact, the existence of a nontrivial measure on $\mathbb{R}$ is an axiom of infinity far stronger than those previously mentioned. Gödel [1946] made the interesting speculation that any true but unprovable proposition is a consequence of some axiom of infinity.

Quite recently, some "largeness" properties in number theory have been found which imply Con(PA). The first of these was found by Paris and Harrington [1977], using a modification of a combinatorial theorem of Ramsey [1929]. Paris and Harrington found a sentence σ which says that for each $n \in \mathbb{N}$ there is an m such that sets of size $\geq m$ have a certain combinatorial property $C(n)$. They showed that σ follows from Ramsey's theorem on infinite sets, but that the function

$$f(n) = \text{least } m \text{ such that sets of size } m \text{ have property } C(n)$$

grows faster than any computable function whose existence can be proved in PA. Thus σ in some sense asserts the existence of a "large" function. The property $C(n)$ is such that one can decide whether a finite set has it or not, hence σ implies that f is computable. This shows immediately that σ cannot be proved in PA, but Paris and Harrington in fact showed the stronger result that σ implies Con(PA).

Gödel's theorem shows that something is missing in the purely formal view of mathematics, and the axioms of infinity show that the missing elements may be mathematically interesting and important. Despite this, the official view still seems to be that mathematics consists in the formal deduction of theorems from fixed axioms. As early as 1941 Post protested against this view:

> It is to the writer's continuing amazement that ten years after Gödel's remarkable achievement current views on the nature of mathematics are thereby affected only to the point of seeing the need of many formal systems, instead of a universal one. Rather has it seemed to us to be inevitable that these developments will result in a reversal of the entire axiomatic trend of the late nineteenth and early twentieth centuries, with a return to meaning and truth. (Post [1941], p. 345)

I believe that what Post was saying was this. Before Gödel, the goal of mathematical logic had been to distill all mathematics into a set of axioms. It was expected that all of number theory, for example, could be recovered by formal deduction from PA, that is, *by forgetting that the axioms of PA had any meaning*. Gödel showed that this was not so, and in particular that the sentence Con(PA), which expresses consistency, could not be so recovered. But it is precisely by knowing the *meaning* of the PA axioms that one knows they are consistent: contradictory sentences cannot hold in the actual structure of $\mathbb{N}$ with $+$ and $\times$. Thus it is the ability to see meaning in PA that enables us to see the truth of Con(PA) and hence to transcend the power of formal proof.

20.7. Biographical Notes: Gödel

Kurt Gödel (Fig. 20.2) was born in 1906 in Brünn, Moravia (now Brno, Czechoslovakia) and died in Princeton in 1978. He was the second son of Rudolf Gödel, the manager of a textile firm, and Marianne Handschuh. Both his parents were members of the substantial German-speaking minority of the region, and his mother had received some of her education at the French school in Brünn. Her influence seems to have been dominant in Kurt's upbringing, at least in the matter of church and school. He attended Lutheran institutions and was unsympathetic to the Catholic church, to which his father nominally belonged.

Figure 20.2. Kurt Gödel.

Gödel had a generally happy childhood and was noted for his curiosity, being known to his family as *Herr Warum* (Mr. Why). The family was fortunate that Brünn was relatively untouched by World War I, and even after the war the absorption of Moravia into the new nation of Czechoslovakia had little effect on the Gödel family. The most disturbing event of Gödel's childhood was an attack of rheumatic fever at the age of six or seven, followed by his learning, at the age of eight, that rheumatic fever can damage the heart. To the end of his life he was convinced that he had a weak heart and, when doctors found no evidence of this, he developed a distrust of the medical profession as well. This led to a brush with death in the 1940s, when he left a duodenal ulcer untreated, and he became obsessively cautious and prone to depression.

After completing secondary school, Gödel moved to Vienna (his father's birthplace) to enter university. He was at first undecided between mathematics and physics but opted for mathematics after hearing a brilliant cycle of lectures by the number theorist Fürtwangler. He was introduced to logic and set theory by Hans Hahn, who was interested in point set problems in the theory of real functions. Hahn got Gödel involved in the famous Vienna Circle of philosophers in 1926–1928 and later became his thesis supervisor. The Vienna Circle aimed to put science and philosophy on a rigorous basis by means of formal logic and no doubt had a strong influence on Gödel's work. However, his incompleteness theorem was obviously a blow to the Vienna Circle, just as it was to formalists in mathematics. In fact Gödel began to drift away from the Vienna Circle long before he discovered his theorem, as his philosophical position tended toward the diametric opposite of theirs. The Vienna Circle based their philosophy on strictly material data, whereas Gödel was metaphysical to the point of being interested in ghosts and demons (see, e.g., Kreisel [1980], p. 155).

In 1927 Gödel met his future wife, Adele Nimbursky, a dancer at a nightclub in Vienna. His parents objected to her, on the grounds that she was six years older than Gödel and had been married before, and the couple did not marry until 1938. The marriage endured, and friends noted how much warmer Gödel became in her company. They had no children, and Adele was probably the only person in Gödel's life who could bring him down to earth occasionally.

Gödel became an Austrian citizen in 1929 and rapidly rose to fame after the publication of the incompleteness theorem in 1931. He was invited to the United States and made three visits to the Institute of Advanced Study in Princeton. In between, however, he suffered bouts of depression and spent some time in mental hospitals. In 1938 Hitler annexed Austria and the atmosphere became increasingly oppressive, though Gödel does not seem to have been perceptive about the menace of Nazism. He blamed the situation on Austrian "sloppiness" and decided to leave only when he was judged fit for military service—an obviously incompetent judgment in his opinion.

During this tense period of his life (1937–1940), Gödel made his important discoveries on the consistency of the axiom of choice and the continuum hypothesis. Thus he arrived at Princeton in 1940 on a second wave of fame. He settled into a position at the Institute of Advanced Study, where he was to stay for the rest of his life. In the early 1940s he continued to work hard on set theory. In 1942 he found a proof of the independence of the axiom of choice but left his work unpublished when he found he was unable to do the same for the continuum hypothesis (i.e., to show that if set theory is consistent, one can consistently assume that the axiom of choice is true but the continuum hypothesis is false). These are the results, of course, that were eventually obtained by Cohen [1963].

From 1943 Gödel devoted himself mainly to philosophy. Indeed, Kreisel [1980], p. 150, argues that *all* of Gödel's discoveries stemmed from his philosophical acuteness—allied with the appropriate, but generally elementary mathematical techniques. The incompleteness theorem, for example, comes from observing the difference between provability and truth. Gödel [1949] made an unexpected foray into another area of mathematics of philosophical interest, the theory of relativity. He showed that there are solutions of Einstein's equations which contain closed timelike lines, theoretically allowing the possibility of time travel. Gödel later calculated that the amount of energy required to travel into one's own past was prohibitively large, but the feasibility of signals to and from the past remained open. Indeed, he seems to have believed that this was a possible basis for the existence of ghosts (Kreisel [1980], p. 155).

Gödel was understandably reticent about expressing such opinions publicly. Even in the case of the incompleteness theorem, whose implications for the question of minds versus machines were widely debated, he did not publish his opinions. His private view, that the mind is more powerful than a machine, may, however, have been important in enabling him to foresee the incompleteness theorem in the first place. Indeed, it may not be too much to say that Gödel's receptiveness to scientifically unconventional ideas paved the way for his unconventional theorems. We shall probably learn more about the ideas that influenced his work when his unpublished notes appear in forthcoming volumes of his collected works (see Feferman [1986]).

References

The books and papers in this reference section include all items mentioned in name [date] format in the text. In addition, some are suitable as general introductions to the history of mathematics or as introductions to the history of particular topics. I would like to recommend the following:

Introductory Works

General

Boyer [1968], Burton [1985], Kline [1972].

Source Books (English translations of classic papers)

Struik [1969], Fauvel and Gray [1987].

Ancient Mathematics

Heath [1921], van der Waerden [1954, 1983].

Algebra

Edwards [1984], van der Waerden [1985].

Geometry

Gray [1980], Hildebrandt and Tromba [1985].

Algebraic Geometry

Boyer [1956], Brieskorn and Knörrer [1981], Dieudonné [1985].

Calculus

Baron [1969], Boyer [1949], Edwards [1979].

Number Theory

Weil [1984].

Mechanics

Dugas [1957, 1958].

Group Theory

Chandler and Magnus [1982], Wussing [1969].

Topology

Stillwell [1980].

References Cited

N. H. Abel
[1826]   Démonstration de l'impossibilité de la résolution algébrique des
                équations générales qui passent le quatrième degré.
                J. reine angew. Math. **1**, 65–84.
                Oeuvres Complètes 1: 66–87.
[1827]   Recherches sur les fonctions elliptiques.
                J. reine angew. Math. **2**, 101–181; **3**, 160–190.
                Oeuvres Complètes 1: 263–388.

S. I. Adian
[1957]   The unsolvability of certain algorithmic problems in the theory
                  of groups. (Russian)
                  Tr. Moskov. Mater. Obsc. **6**, 231–298.

L. B. Alberti
[1436]   *Trattato della pittura.*

J. Le R. D'Alembert
[1746]   Recherches sur le calcul integral.
                  Hist. Acad. Sci. Berlin **2**, 182–224.
[1747]   Recherches sur la courbe que forme une corde tendue mise en
                  vibration.
                  Hist. Acad. Sci. Berlin **3**, 214–219.
[1752]   *Essai d'une nouvelle théorie de la résistance des fluides.*
                  Paris.

Al-Khowârizmi
[830]   *Al-jabr w'al muqâbala.*

A. Amthor
[1880]   Das Problema bovinum des Archimedes.
                  *Z. Math. Phys. (hist.-lit. Abt.)***25**, 153–171.

R. Apéry
[1981]   Interpolation de fractions continues et irrationalité de certains
                  constantes.
                  Mathematics, pp. 37–53, CTHS: Bull. Sec. Sci.; III, Bib. Nat.
                  Paris.

J. R. Argand
[1806]   Essai sur une manière de représenter les quantités imaginaires
                  dans les constructions géométriques.
                  Paris.

R. Ayoub
[1984]   The lemniscate and Fagnano's contributions to elliptic integrals.
                  Arch. Hist. Ex. Sci. **29**, 131–149.

C. Bachet de Meziriac
[1621]   *Arithmetica* (Edition of Diophantus).

A. Baillet
[1691]   *Vie de Monsieur Descartes.*
                  Paris.

W. W. R. Ball
[1890]   Newton's classification of cubic curves.
                  Proc. London Math. Soc. **22**, 104–143.

J. Baltrušaitis
[1977]   *Anamorphic Art.*
                  Harry N. Abrams, New York.

S. Banach and A. Tarski
[1924]   Sur la décomposition des ensembles de points en parties respec-
                  tivement congruents.
                  Fund. Math. **6**, 244–277.

J. Banville
[1981]   *Kepler: A Novel.*
                  Secker and Warburg, London.

M. E. Baron
[1969]   *The Origins of the Infinitesimal Calculus.*
                  Pergamon Press, Elmsford, N.Y.

I. G. Bashmakova
[1981]   Arithmetic of algebraic curves from Diophantus to Poincaré.
                  Hist. Math. **8**, 393–416.

I. Beeckman
[1615]   *Journal tenu par Isaac Beeckman de* 1604 *à* 1634.
                    Ed. C. de Waard, 4 vols.
                    Nijhoff, The Hague.
[1628]   Journal.
                    Quoted in *Oeuvres de Descartes* 10: 344–346.
E. Beltrami
[1865]   Risoluzione del problema: "Riportare i punti di una superficie
                    sopra un piano in modo che le linee geodetiche vengano rap-
                    presentate da linee rette."
                    Ann. Mat. pura appl., ser. 1, **7**, 185–204.
                    Opere Matematiche. 1: 262–280.
[1868]   Saggio di interpretazione della geometria non-euclidea.
                    Giorn. Mat. **6**, 284–312.
                    Opere Matematiche. 1: 374–405. Partial English translation in
                    Fauvel and Gray [1987].
[1868']   Teoria fondamentale degli spazii di curvatura costante.
                    Ann. Mat. pura appl., ser. 2, **2**, 232–255.
                    Opere Matematiche. 1: 406–429.
D. Bernoulli
[1738]   *Hydrodynamica.*
[1743]   Letter to Euler, 4 September 1743.
                    In Eneström [1906].
[1753]   Réflexions et éclaircissemens sur les nouvelles vibrations des
                    cordes exposées dans les mémoires de l'academie de 1747 &
                    1748.
                    Hist. Acad. Sci. Berlin **9**, 147–172.
James Bernoulli
[1690]   Quæstiones nonnullae de usuris cum solutione problematis de
                    sorte alearum propositi in Ephem. Gall. A. 1685.
                    Acta Erud. **9**, 219–233.
[1692]   Lineae cycloidales, evolutae, ant-evolute .... Spira mirabilis.
                    Acta Erud. **11**, 207–213.
[1694]   Curvatura laminae elasticae.
                    Acta Erud. **13**, 262–276.
[1697]   Solutio problematum fraternorum.
                    Acta Erud. **16**, 211–217.
                    Opera 1: 768–778.
[1713]   Ars Conjectandi.
                    Opera 3: 107–286.
James and John Bernoulli
[1689–1704]   *Über unendliche Reihen.*
                    Ostwald's Klassiker, vol. 171. Engelmann, Leipzig, 1909.
John Bernoulli
[1691]   Solutio problematis funicularii.
                    Acta Erud. **10**, 274–276.
                    Opera Omnia 1: 48–51.
[1691/2]   *Lectiones de Calculo Differentialum.*
                    Ed. P. Schafheitlein. Basel, 1922.
[1691/2']   Lectiones mathematicae de methodo integralium aliisque.
                    Opera Omnia 3: 385–558.
[1696]   Problema novum ad cujus solutionem Mathematici invitantur.
                    Acta Erud. **15**, 270.
                    Opera Omnia 1: 161.

[1697]   Curvatura radii in diaphanis non uniformibus.
                  Acta Erud. **16**, 206–211.
                  Opera Omnia 1: 187–193.
[1697']   Principia calculi exponentialum.
                  Acta Erud. **16**, 125–133.
                  Opera Omnia 1: 179–187.
[1699]   *Disputatio medico-physica de nutritione.*
                  Groningen.
                  Opera Omnia 1: 275–306.
[1702]   Solution d'un problème concernant le calcul intégral, avec quel-
                  ques abregés par raport à ce calcul.
                  Mém. Acad. Roy. Sci. Paris, pp. 289–297.
                  Opera Omnia 1: 393–400.
[1712]   Angulorum arcuumque sectio indefinita.
                  Acta Erud. **31**, 274–277.
                  Opera Omnia 1: 511–514.

J. Bertrand
[1860]   Remarque à l'occasion de la Note précédente.
                  Comp. rend. **50**, 781–782.

E. Bézout
[1779]   *Théorie générale des équations algébriques.*
                  Paris.

G. Birkhoff (Ed.)
[1973]   *A Source Book in Classical Analysis.*
                  Harvard University Press, Cambridge.

P. du Bois-Reymond
[1875]   Über asymptotischer Werte, infinitäre Approximationen und
                  infinitäre Auflösung von Gleichungen.
                  Math. Ann. **8**, 363–414.

V. G. Boltianskii
[1978]   *Hilbert's Third Problem.*
                  V. H. Winston, Silver Spring, Md.

J. Bolyai
[1832]   Scientiam spatii absolute veram exhibens: A vertitate aut falsit-
                  ate Axiomatis XI Euclidei (a priori hand unquam decidanda)
                  independentem.
                  Appendix to W. Bolyai [1832].
                  English translation in Bonola [1912].

W. Bolyai
[1832]   *Tentamen. Juventutem studiosam in elementa matheseos purae,*
                  elementaris ac sublimioris, methodo intuitiva, evidentiaque huic
                  propria, introducendi.
                  Marosvásárhely.

B. Bolzano
[1817]   *Rein analytischer Beweis des Lehrsatzes dass zwischen je zwey*
                  Werthen, die ein entgegengesetzes Resultat gewähren, wenigstens
                  eine reelle Wurzel der Gleichung liege.
                  Ostwald's Klassiker, vol. 153. Engelmann, Leipzig, 1905.

R. Bombelli
[1572]   *L'Algebra.*
                  Feltrinelli Editore, Milan, 1966.

O. Bonnet
[1848]   Mémoire sur la théorie générale des surfaces.
                  J. Éc. Polytech. **19**, 1–146.

R. Bonola
[1912]   *Noneuclidean Geometry.*
                                    Dover, New York, 1955.
G. Boole
[1847]   *Mathematical Analysis of Logic.*
                                    Basil Blackwell, London, 1948.
E. Borel
[1898]   *Leçons sur la théorie des fonctions.*
                                    Gauthier-Villars, Paris.
H. J. M. Bos
[1981]   On the representation of curves in Descartes' *Géometrie.*
                                    Arch. Hist. Ex. Sci. **24**, 295–338.
[1984]   Arguments on motivation in the rise and decline of a mathematical
                                    theory: The "Construction of Equations," 1637–c. 1750.
                                    Arch. Hist. Ex. Sci. **30**, 331–380.
A. Bosse
[1648]   *Manière universelle de Mr. Desargues.*
R. Bourgne and J. P. Azra (Eds.)
[1962]   *Ecrits et Mémoires Mathematiques d'Evariste Galois: Edition
                                    Critique Intégrale de ses Manuscrits et Publications.*
                                    Gauthier-Villars, Paris.
C. B. Boyer
[1949]   *The History of the Calculus and Its Conceptual Development.*
                                    Hafner Publishing Company, Inc.; Dover, New York, 1959.
[1956]   *History of Analytic Geometry.*
                                    Scripta Mathematica, New York.
[1968]   *A History of Mathematics.*
                                    Wiley, New York.
H. R. Brahana
[1921]   Systems of circuits on 2-dimensional manifolds.
                                    Ann. Math. **23**, 144–168.
Brahmagupta
[628]   *Brâhma-sphuta-siddhânta.*
                                    Partial English translation in Colebrooke [1817].
E. Brieskorn and H. Knörrer
[1981]   *Ebene algebraischer Kurven.*
                                    Birkhäuser, Boston.
                                    English translation: *Plane Algebraic Curves.*
                                    Birkhäuser, Basel, 1986.
H. Briggs
[1624]   *Arithmetica logarithmica, sive logarithmorum chiliades triginta.*
                                    London.
E. S. Bring
[1786]   Meletemata quaedam mathematica circa transformationem aequ-
                                    ationum algebraicarum.
                                    Lund University. Promotionschrift.
W. K. Bühler
[1981]   *Gauss. A Biographical Study.*
                                    Springer-Verlag, New York.
D. M. Burton
[1985]   *The History of Mathematics. An Introduction.*
                                    Allyn and Bacon, Newton, Mass.
F. Cajori
[1913]   History of the exponential and logarithmic concepts.
                                    Am. Math. Mon. **20**, 5–14, 35–47, 75–84, 107–117, 148–151,
                                    173–182, 205–210.

G. Cantor
[1872]   Über die Ausdehnung eines Satzes aus der Theorie der trigono-
                metrischen Reihen.
                Math. Ann. **5**, 123–132.
                Gesammelte Abhandlungen, pp. 92–102.
[1874]   Über eine Eigenschaft des Inbegriffes aller reellen algebraischen
                Zahlen.
                J. reine angew. Math. **77**, 258–262.
                Gesammelte Abhandlungen, pp. 115–118.
[1880]   Über unendlich linear Punktmannigfaltigkeiten, 2.
                Math. Ann. **17**(1880) 355–358.
                Gesammelte Abhandlungen, pp. 145–148.
[1883]   *Grundlagen einer allgemeinen Mannigfaltigkeitslehre.*
                Leipzig.
                Gesammelte Abhandlungen, pp. 165–204.
[1891]   Über eine elementare Frage der Mannigfaltigkeitslehre.
                Jahrb. Dtsch. Math. Ver. **1**, 75–78.
                Ges. Abh., pp. 278–280.
G. Cardano
[1545]   *Ars Magna.*
                English translation: *The Great Art.*
                M.I.T. Press, Cambridge, 1968.
[1575]   *De Vita Propria Liber.*
                English translation: *The Book of My Life.*
                Dutton, New York, 1930; Dover, New York, 1962.
A.-L. Cauchy
[1813]   Démonstration du théorème général de Fermat sur les nombres
                polygones.
                Mém. Sci. Math. Phys. Inst. France, ser. 1, **14**, 177–220.
                Oeuvres, ser. 2, 6: 320–353.
[1815]   Mémoire sur le nombre des valeurs qu'une fonction peut acquerir,
                lorsqu'on y permute de toutes les manières possibles les quantités
                qu'elle renferme.
                J. Éc. Polytech. **18**, 10: 1–28.
                Oeuvres, ser. 2, 1: 62–90.
[1825]   *Mémoire sur les intégrales définies prises entre des limites imag-
                inaires.*
                Paris.
[1837]   Letter to Coriolis, 29 January 1837.
                Comp. rend. **4**, 214–218.
                Oeuvres, ser. 1, 4: 38–42.
[1844]   Mémoire sur les arrangements que l'on peut former avec des
                lettres données, et sur les permutations ou substitutions à l'aide
                desquelles on passe d'un arrangement à un autre.
                Ex. anal. phys. math. **3**, 151–252.
                Oeuvres, ser. 2, 13: 171–282.
[1846]   Sur les intégrales qui s'étendent à tous les points d'une courbe
                fermée.
                Comp. rend. **23**, 251–255.
                Oeuvres, ser. 1, 10: 70–74.
B. Cavalieri
[1635]   Geometria indivisibilibus continuorum nova quadam ratione
                promota.
A. Cayley
[1854]   On the theory of groups, as depending on the symbolic equation
                $\theta^n = 1$.

Phil. Mag. **7**, 40–47.
Collected Mathematical Papers 2: 123–130.

[1859] A sixth memoir upon quantics.
Phil. Trans. Roy. Soc. **149**, 61–90.
Collected Mathematical Papers 2: 561–592.

[1878] The theory of groups.
Am. J. Math. **1**, 50–52.
Collected Mathematical Papers 10: 401–403.

[1878'] The theory of groups: Graphical representation.
Am. J. Math. **1**, 174–176.
Collected Mathematical Papers 10: 403–405.

B. Chandler and W. Magnus
[1982] The History of Combinatorial Group Theory.
Springer-Verlag, New York.

A. Church
[1936] An unsolvable problem of elementary number theory.
Am. J. Math. **58**, 345–363.

[1938] Review.
J. Symb. Logic **3**, 46.

M. Clagett
[1959] The Science of Mechanics in the Middle Ages.
University of Wisconsin Press, Madison.

[1968] Nicole Oresme and the Medieval Geometry of Qualities and Motions.
University of Wisconsin Press, Madison.

A.-C. Clairaut
[1731] Sur les courbes que l'on forme en coupant un surface courbe quelconque par un plan donnée de position.
Mém. Acad. Sci. Paris, pp. 490–493.

[1740] Sur l'integration ou la construction des équations différentielles du premier ordre.
Mém. Acad. Sci. Paris, p. 294.

[1743] Théorie de la figure de la Terre tirée des principes de l'hydro-dynamique.
Durand, Paris.

A. Clebsch
[1864] Über einen Satz von Steiner und einige Punkte der Theorie der Curven dritter Ordnung.
J. reine angew. Math. **63**, 94–121.

M. R. Cohen and I. E. Drabkin
[1958] Source Book in Greek Science.
Harvard University Press, Cambridge.

P. Cohen
[1963] The independence of the continuum hypothesis I, II.
Proc. Nat. Acad. Sci. **50**, 1143–1148; **51**, 105–110.

H. T. Colebrooke
[1817] Algebra, with Arithmetic and Mensuration, from the Sanscrit of Brahmegupta and Bháscara.
John Murray, London.
Martin Sandig, Wiesbaden, 1973.

R. Connelly
[1978] A counterexample to the rigidity conjecture for polyhedra.
Inst. Hautes Études Sci., Publ. Math. **47**, 333–338.

J. L. Coolidge
[1945] A History of the Conic Sections and Quadric Surfaces.
Clarendon Press, Oxford.

N. Copernicus
[1543]   *De revolutionibus orbium coelestium.*
R. Cotes
[1714]   Logometria.
                Phil. Trans. **29**, 5–45.
[1722]   *Harmonia Mensurarum.*
                Cambridge.
D. A. Cox
[1984]   The arithmetic-geometric mean of Gauss.
                Enseign. Math. **30**, 275–330.
H. S. M. Coxeter and W. O. J. Moser
[1980]   *Generators and Relations for Discrete Groups*, 4th ed.
                Springer-Verlag, New York.
G. Cramer
[1750]   *Introduction à l'analyse des lignes courbes algebriques.*
                Geneva.
J. N. Crossley
[1988]   *The Emergence of Number*, 2nd ed.
                World Scientific, Singapore.
D. van Dalen
[1972]   Set theory from Cantor to Cohen.
                In *Sets and Integration* (D. van Dalen and A. F. Monna, eds.).
                Wolters-Noordhoff, Groningen, pp. 1–74.
J. H. Davenport
[1981]   *On the Integration of Algebraic Functions.*
                Lecture Notes in Computer Science. Springer-Verlag, New
                York.
F. David
[1962]   *Games, Gods and Gambling.*
                Charles Griffin, London.
M. Davis
[1965]   *The Undecidable.*
                Raven Press, New York.
[1973]   Hilbert's tenth problem is unsolvable.
                Am. Math. Mon. **80**, 233–269.
R. Dedekind
[1872]   *Stetigkeit und die Irrationalzahlen.*
                English translation in *Essays on the Theory of Numbers.*
                Open Court, Chicago, 1901.
[1876]   Bernhard Riemann's Lebenslauf.
                In Riemann, *Werke* 2nd ed., pp. 539–558.
P. Dedron and J. Itard
[1973]   *Mathematics and Mathematicians*, vol. 1.
                Open University Press, Milton Keynes.
M. Dehn
[1900]   Über raumgleiche Polyeder.
                Göttinger Nachrichten, pp. 345–354.
[1910]   Über die Topologie des dreidimensionalen Raumes.
                Math. Ann. **69**, 137–168.
                English translation in Dehn [1987], pp. 92–126.
[1912]   Über unendliche diskontinuierliche Gruppen.
                Math. Ann. **71**, 116–144.
                English translation in Dehn [1987], pp. 133–178.
[1987]   *Papers on Group Theory and Topology.*
                Springer-Verlag, New York.

M. Dehn and P. Heegaard
[1907]   Analysis situs.
                          Enzyklopädie Mathematische Wissenschaft, vol. IIIAB3, pp. 153–
                          220.
                          Leipzig.

G. Desargues
[1639]   Brouillon projet d'une atteinte aux événemens des recontres du
                          cône avec un plan.
                          In Taton [1951], pp. 99–180.
                          English translation in Field and Gray [1987].

R. Descartes
[1637]   *La Géométrie.*
                          English translation: *The Geometry.*
                          Dover, New York, 1954.
[1638]   Letter to Mersenne, 18 January 1638.
                          Oeuvres 1: 490.

L. E. Dickson
[1903]   *Introduction to the Theory of Algebraic Equations.*
                          Wiley, New York.
[1919, 1920, 1923]   *History of the Theory of Numbers*, 3 vols.
                          Carnegie Institute of Washington.
                          Chelsea, New York, 1951.

J. Dieudonné
[1985]   *History of Algebraic Geometry.*
                          Wadsworth, Belmont, Calif.

P. G. L. Dirichlet
[1829]   Sur la convergence des séries trigonometriques qui servent à
                          représenter une fonction arbitraire entre des limites données.
                          J. reine angew. Math., **4**, 157–169.
                          Werke 1: 117–132.
[1837]   Beweis des Satzes, dass jede unbegrenzte arithmetische Progres-
                          sion, deren erstes Glied und Differenz ganze Zahlen ohne gemein-
                          schaftlichen Factor sind, unendliche viele Primzahlen enthält.
                          Abh. Akad. Wiss. Berlin, pp. 45–81.
                          Werke 1: 315–342.

P. Dombrowski
[1979]   150 Years after Gauss' "Disquisitiones generales circa superficies
                          curveas."
                          Astérisque **62**.

S. K. Donaldson
[1983]   An application of gauge theory to four dimensional topology.
                          J. Diff. Geom. **18**, 279–315.

S. Dostrovsky
[1975]   Early vibration theory: Physics and music in the seventeenth
                          century.
                          Arch. Hist. Ex. Sci. **14**, 169–218.

R. Dugas
[1957]   *A History of Mechanics.*
                          Routledge and Kegan Paul, London.
[1958]   *Mechanics in the Seventeenth Century.*
                          Éditions du Griffon, Neuchâtel.

A. Dürer
[1525]   *Underweysung der Messung, mit dem Zirckel und Richtscheyt.*
                          Verlag Walter Uhl, Unterscheidheim, 1972.

W. Dyck
[1882]   Gruppentheoretischen Studien.
                Math. Ann. **20**, 1–44.
[1883]   Gruppentheoretischen Studien II.
                Math. Ann. **22**, 70–108.
C. H. Edwards
[1979]   *The Historical Development of the Calculus.*
                Springer-Verlag, New York.
H. M. Edwards
[1974]   *Riemann's Zeta Function.*
                Academic Press, New York.
[1977]   *Fermat's Last Theorem.*
                Springer-Verlag, New York.
[1984]   *Galois Theory.*
                Springer-Verlag, New York.
G. Eisenstein
[1847]   Beiträge zur Theorie der elliptischen Functionen.
                J. reine angew. Math. **35**, 137–274.
G. Eneström
[1906]   Der Briefwechsel zwischen Leonhard Euler und Daniel
                Bernoulli.
                Bibl. Math., ser. 3, **7**, 126–156.
S. B. Engelsman
[1984]   *Families of Curves and the Origins of Partial Differentiation.*
                North-Holland, Amsterdam.
L. Euler
[1728]   Letter to John Bernoulli, 10 December 1728.
                Bibl. Math., ser. 3, **4**, 352–354.
[1728']   De linea brevissima in superficie quacunque duo quaelibet puncta
                iungente.
                Comm. Acad. Sci. Petrop. **3**, 110–124.
                Opera Omnia, ser. 1, 25: 1–12.
[1734/5]   De summis serierum reciprocarum.
                Comm. Acad. Sci. Petrop. **7**, 123–134.
                Opera Omnia, ser. 1, 14: 73–86.
[1736]   Theorematum quorundam ad numeros primos spectantium de-
                monstratio.
                Comm. Acad. Sci. Petrop. **8**, 141–146.
                Opera Omnia, ser. 1, 2: 33–37.
[1743]   *Additamentum I de curvis elasticus.*
                Lausanne and Geneva.
                Opera Omnia, ser. 1, 24: 231–297.
                English translation: *Isis* **20** (1933), 72–160.
[1746]   Letter to Goldbach, 14 June 1746.
                Briefwechsel *Opera Omnia*, ser. quarta A, 1, 152.
[1748]   *Introductio in Analysin Infinitorum I.*
                Opera Omnia, ser. 1, 8.
[1748']   *Introductio in Analysin Infinitorum II.*
                Opera Omnia, ser. 1, 9.
[1750]   Letter to Goldbach, 9 June 1750.
                Corresp. Math. Phys. **I**, (1843), pp. 521–524.
[1752]   Elementa doctrinae solidorum.
                Novi Comm. Acad. Sci. Petrop. **4**, 109–140.
                Opera Omnia, ser. 1, 26: 71–93.

[1758]   Theoremata arithmetica nova methodo demonstrata.
                Novi Comm. Acad. Sci. Petrop. **8**, 74–104.
                Opera Omnia, ser. 1, 2: 531–555.
[1760]   Recherches sur la courbure des surfaces.
                Mém. Acad. Sci. Berlin **16**, 119–143.
                Opera Omnia, ser. 1, 28: 1–22.
[1768]   *Institutiones Calculi Integralis.*
                Opera Omnia, ser. 1, 11.
[1770]   *Vollständige Einleitung zur Algebra.*
                English translation: *Elements of Algebra.*
                Longman, Orme and Co., London, 1840.
                Springer-Verlag, New York, 1984.
[1777]   De repraesentatione superficiei sphaericae super plano.
                Acta Acad. Sci. Imper. Petrop. **1**, 107–132.
[1849]   De numeris amicabilibus.
                Comm. Arith. **2**, 627–636.
                Opera Omnia, ser. 1, 5: 353–365.

G. C. de'T. di Fagnano
[1718]   Metodo per misurare la lemniscata.
                Giorn. lett. d'Italia **29**.
                Opere Matematiche. 2: 293–313.

G. Faltings
[1983]   Endlichkeitssätze für abelsche Varietäten über Zahlkörpern.
                Inv. Math. **73**, 349–366.

J. Fauvel and J. Gray
[1987]   *The History of Mathematics. A Reader.*
                Macmillan, New York.

P. J. Federico
[1982]   *Descartes on Polyhedra.*
                Springer-Verlag, New York.

S. Feferman
[1986]   Preface to *Kurt Gödel Collected Works*, vol. 1.
                Oxford University Press, Oxford.

P. Fermat
[1629]   Ad locos planos et solidos isagoge.
                Oeuvres 1: 92–103.
                English translation in D. E. Smith, *A Source Book in Mathematics*
                2: 389–396.
                Dover, New York, 1959.
[1640]   Letter to Frenicle, 18 October 1640.
                Oeuvres 2: 209.
[1670]   Observations sur Diophante.
                Oeuvres 3: 241–276.

Fibonacci (Leonardo of Pisa)
[1202]   *Liber Abaci.*
[1225]   *Flos Leonardi Bigolli Pisani super solutionibus quarundam quaes-*
                tionum ad numerum et ad geometriam pertinentium.

J. Field and J. Gray
[1987]   *The Geometrical Work of Girard Desargues.*
                Springer-Verlag, New York.

J. Fourier
[1822]   *La théorie analytique de la chaleur.*
                Didot, Paris.
                English translation: *The Analytical Theory of Heat.*

Cambridge University Press, Cambridge, 1878. (Dover, New York, 1955.)

D. H. Fowler
[1980]   Book II of Euclid's Elements and a pre-Eudoxan theory of ratio.
                Arch. Hist. Ex. Sci. **22**, 5–36.
[1982]   Book II of Euclid's Elements and a pre-Eudoxan theory of ratio.
                Part 2: Sides and diameters.
                Arch. Hist. Ex. Sci. **26**, 193–209.

G. Frege
[1879]   *Begriffschrift.*
                English translation in van Heijenoort [1967], pp. 5–82

R. Fritsch
[1984]   The transcendence of π has been known for about a century—but who was the man who discovered it?
                Results Math. **7**, 164–183.

Galileo Galilei
[1604]   Letter to Paolo Sarpi, 16 October 1604.
                Works of Galileo 10: 115.
[1638]   *Dialogues Concerning Two New Sciences.*
                Dover, New York, 1952.

E. Galois
[1831]   Mémoire sur les conditions de résolubilité des équations par radicaux.
                Écrits Mém. Math., pp. 43–71.
[1831′]   Analyse d'un mémoire sur la résolution algebrique des équations.
                Écrits Mém. Math., pp. 163–165.

C. F. Gauss
[1799]   Demonstratio nova theorematis omnem functionem algebraicum rationalem integram unius variabilis in factores reales primi vel secundi gradus resolvi posse.
                Helmstedt dissertation.
                Werke 3: 1–30.
[1801]   *Disquisitiones Arithmeticae.*
                English translation, Yale University Press, New Haven, Conn., 1966.
[1811]   Letter to Bessel, 18 December 1811.
                Briefwechsel mit F. W. Bessel.
                Georg Olms Verlag, Hildesheim, 1975, pp. 155–160.
[1816]   Demonstratio nova altera theorematis omnem functionem algebraicum rationalem integram unius variabilis in factores reales primi vel secundi gradus resolvi posse.
                Comm. Recentiores (Gottingae) **3**, 107–142.
                Werke 3: 31–56.
[1818]   Determinatio attractionis quam in punctum quodvis positionis datae exerceret planeta si eius massa per totam orbitam ratione temporis quo singulae partes describuntur uniformiter esset dispertita. *Comm. Soc. Reg. Sci. Gottingensis Rec.* vol. 4.
                Werke 3: 331–355.
[c. 1819]   Die Kugel.
                Werke 8: 351–356.
[1822]   Allgemeine Auflösung der Aufgabe; Die Theile einer gegebenen Fläche so abzubilden, dass die Abbildung dem Abgebildeten in

den kleinstein Theilen ähnlich wird.
Astr. Abh. **3**, 1–30.
Werke 4: 189–216.
English translation: *Phil. Mag.*, new ser., **4** (1828), 104–113, 206–215.

[1825] Die Seitenkrümmung.
Werke 8: 386–395.

[1827] Disquisitiones generales circa superficies curvas.
König. Ges. Wiss. Göttingen.
English translation in Dombrowski [1979].

[1828] Letter to Bessel, 30 March 1828.
Briefwechsel mit F. W. Bessel.
Georg Olms Verlag, Hildesheim, 1975, pp. 477–478.

[1831] Letter to Schumacher, 12 July 1831.
Werke 8: 215–218.

[1832] Letter to W. Bolyai, 6 March 1832.
Briefwechsel zwischen C. F. Gauss und Wolfgang Bolyai.
Ed. F. Schmidt and P. Stäckel. Leipzig, 1899.
Werke 8: 220–224.

[1832′] Cubirung der Tetraeder.
Werke 8: 228–229.

[1846] Letter to Gerling, 2 October 1846.
Briefwechsel mit Chr. L. Gerling.
Georg Olms Verlag, Hildesheim, 1975, pp. 738–741.

[1846′] Letter to Schumacher, 28 November 1846.
Excerpt translated in Bühler [1981], p. 150.

A. O. Gelfond
[1960] *The Solution of Equations in Integers.*
P. Noordhoff, Groningen.

K. Gödel
[1931] Über formal unentscheidbare Sätze der Principia Mathematica und verwandte Systeme I.
Monat. Math. Phys. **38**, 173–198.
English translation in van Heijenoort [1967], 596–616.

[1938] The consistency of the axiom of choice and the generalized continuum hypothesis.
Proc. Nat. Acad. Sci. **25**, 220–224.

[1946] Remarks before the Princeton bicentennial conference on problems in mathematics.
In Davis [1965], pp. 84–88.

H. H. Goldstine
[1977] *A History of Numerical Analysis from the 16th through the 19th Century.*
Springer-Verlag, New York.

F. Gomes Teixeira
[1908, 1909, 1915] *Traité des Courbes*, 3 vols.
Coimbra.
New York, 1971.

E. Goursat
[1900] Sur la définition générale des fonctions analytiques, d'après Cauchy.
Trans. Am. Math. Soc. **1**, 14–16.

G. Grandi
[1723] Florum geometricorum manipulus.
Phil. Trans. **32**, 355–371.

J. Gray
[1980]   *Ideas of Space.*
                       Oxford University Press, Oxford.

G. Green
[1828]   An essay on the application of mathematical analysis to the
                       theories of electricity and magnetism.
                       Papers, 1–115.

J. Gregory
[1667]   *Vera Circuli et Hyperbolae Quadratura.*
[1668]   *Geometriae Pars Universalis.*
[1670]   Letter to Collins, 23 November 1670.
                       In Turnbull [1939], pp. 118–133.
[1671]   Letter to Gideon Shaw, 29 January 1671.
                       In Turnbull [1939], pp. 350–357.

B. Grünbaum
[1985]   Geometry strikes again.
                       Math. Mag. **58**, 12–18.

M. Hall
[1967]   *Combinatorial Theory.*
                       Wiley, New York.

W. R. Hamilton
[1856]   Memorandum respecting a new system of roots of unity.
                       Phil. Mag. **12**, 496.
                       Mathematical Papers. 3: 610.

A. Harnack
[1885]   Über den Inhalt von Punktmengen.
                       Math. Ann. **25**, 241–250.

F. Hausdorff
[1914]   *Grundzüge der Mengenlehre.*
                       Leipzig.

T. L. Heath
[1897]   *The Works of Archimedes.*
                       Cambridge University Press, Cambridge.
[1910]   *Diophantus of Alexandria.*
                       Cambridge University Press, Cambridge.
[1921]   *A History of Greek Mathematics*, 2 vols.
                       Clarendon Press, Oxford.
                       Dover, New York, 1981.
[1925]   *Translation of Euclid's Elements*, 3 vols.
                       Cambridge University Press, Cambridge.
                       Dover, New York, 1956.

J. van Heijenoort
[1967]   *From Frege to Gödel. A Source Book in Mathematical Logic
                       1879–1931.*
                       Harvard University Press, Cambridge.

C. Hermite
[1858]   Sur la résolution de l'équation du cinquième degré.
                       Comp. Rend. **46**, 508–515.
                       Oeuvres 2: 5–12.
[1873]   Sur la fonction exponentielle.
                       Comp. Rend. **77**, 18–24, 74–79, 226–233, 285–293.
                       Oeuvres 3: 150–181.

G. Higman
[1961]   Subgroups of finitely presented groups.
                       Proc. Roy. Soc. Lond., ser. A, **262**, 455–475.

D. Hilbert
[1899]   *Grundlagen der Geometrie.*
                  Teubner, Leipzig.
                  English translation: *Foundations of Geometry.*
                  Open Court, Chicago, 1971.
[1900]   Mathematische Probleme.
                  Göttinger Nachrichten, pp. 253–297.
                  Gesammelte Abhandlungen 3: 290–329.
[1901]   Über Flächen von constanter Gaussscher Krümmung.
                  Trans. Am. Math. Soc. **2**, 87–99.
                  Gesammelte Abhandlungen 2: 437–448.
D. Hilbert and P. Bernays
[1936]   *Grundlagen der Mathematik*, vol. 1.
                  Springer, Berlin.
D. Hilbert and S. Cohn-Vossen
[1932]   *Anschauliche Geometrie.*
                  Springer, Berlin.
                  English translation: *Geometry and the Imagination.*
                  Chelsea, New York, 1952.
S. Hildebrandt and A. Tromba
[1985]   *Mathematics and Optimal Form.*
                  Scientific American Library, New York.
T. Hobbes
[1656]   Six lessons to the professors of mathematics.
                  The English Works of Thomas Hobbes 7: 181–356.
                  Scientia Aalen, Aalen, West Germany, 1962.
[1672]   Considerations upon the answer of Doctor Wallis.
                  The English Works of Thomas Hobbes 7: 443–448.
                  Scientia Aalen, Aalen, West Germany, 1962.
O. Hölder
[1896]   Über den Casus Irreducibilis bei der Gleichung dritten Grades.
                  Math. Ann. **38**, 307–312.
J. E. Hofmann
[1974]   *Leibniz in Paris.*
                  Cambridge University Press, Cambridge.
R. Hooke
[1675]   A description of helioscopes, and some other instruments.
                  In R. T. Gunther, *Early Science in Oxford*, vol. 8, Oxford, 1931,
                  pp. 119–152.
G. F. A. de L'Hôpital
[1696]   *Analyse des infiniment petits.*
                  Paris.
[1697]   Solutio problematis de linea celerrimi descensus.
                  Acta Erud. **16**, 217–220.
C. Huygens
[1646]   Letters to Mersenne, November 1646.
                  Oeuvres Complètes 1: 34–40.
[1659]   Fourth part of a treatise on quadrature.
                  Oeuvres Complètes 14: 337.
[1659']   Recherches sur la théorie des développées.
                  Oeuvres Complètes 14: 387–405.
[1659'']   Piece on the cycloid, 1 December 1659.
                  Oeuvres Complètes 16: 392–413.
[1671]   Letter to Lodewijk Huygens, 29 October 1671.
                  Oeuvres Complètes 7: 112–113.

[1673]	*Horologium oscillatorium.* Paris. *Oeuvres Complètes* 18: 69–368. English translation: *The Pendulum Clock.* Iowa State University Press, Ames, Iowa, 1986.
[1691]	Christianii Hugenii, dynastae in Zülechem, solutio ejusdem problematis. *Acta Erud.* **10**, 281–282. *Oeuvres Complètes* 10: 95–98.
[1692]	Letter to the Marquis de l'Hôpital, 29 December 1692. *Oeuvres Complètes* 10: 348–355.
[1693]	Letter to H. Basnage de Beauval, February 1693. *Oeuvres Complètes* 10: 407–417.
[1693']	Appendix to Huygens [1693]. *Oeuvres Complètes* 10: 418–422.

C. G. J. Jacobi

[1829]	*Fundamenta Nova Theoriae Functionum Ellipticarum.* Königsberg.
[1834]	De usu theoriae integralium ellipticorum et integralium abelianorum in analysi diophantea. *J. reine angew. Math.* **13**, 353–355. *Werke* 2: 53–55.

C. Jordan

[1866]	Sur la déformation des surfaces. *J. Math.*, ser. 2, **11**, 105–109.
[1870]	*Traité des Substitutions.* Gauthier-Villars, Paris.
[1892]	Remarques sur les intégrales défines. *J. Math.*, ser. 4, **8**, 69–99.

M. Kac

[1984]	How I became a mathematician. *Am. Sci.* **72**, 498–499.

A. G. Kaestner

[1761]	*Anfangsgründe der Analysis der Unendlichen—Der mathematisches Anfangsgründe*, 3. Teil, 2 Abteilung. Göttingen.

D. Kahn

[1967]	*The Codebreakers.* Weidenfeld and Nicholson, London.

J. Kepler

[1596]	*Mysterium Cosmographicum.*
[1604]	*Ad Vitellionem Paraliopmena, quibus Astronomiae pars Optica Traditur.*
[1609]	*Astronomia Nova.*

B. M. Kiernan

[1971]	The development of Galois theory from Lagrange to Artin. *Arch. Hist. Ex. Sci.* **8**, 40–154.

F. Klein

[1871]	Über die sogenannte Nicht-Euklidische Geometrie. *Math. Ann.* **4**, 573–625. *Gesammelte Mathematische Abhandlungen.* 1: 254–305.
[1872]	*Vergleichende Betrachtungen über neuere geometrische Forschungen (Erlanger Programm).* Akademische Verlagsgesellschaft, Leipzig, 1974. *Gesammelte Mathematische Abhandlungen.* 1: 460–497.

[1874]   Bemerkungen über den Zusammenhang der Flächen.
                Math. Ann. **7**, 549–557.
                Gesammelte Mathematische Abhandlungen. 2: 63–77.
[1876]   Über binäre Formen mit linearem Transformation in sich selbst.
                Math. Ann. **9**, 183–208.
                Gesammelte Mathematische Abhandlungen. 2: 275–301.
[1882]   Neue Beiträge zur Riemannschen Funktionentheorie.
                Math. Ann. **21**, 141–218.
                Gesammelte Mathematische Abhandlungen. 3: 630–710.
[1882']   Letter to Poincaré, 14 May 1882.
                Gesammelte Mathematische Abhandlungen. 3: 615–616.
[1884]   *Lectures on the Icosahedron.*
                Leipzig.
                Dover, New York, 1956.
[1924]   *Elementary Mathematics from an Advanced Standpoint: Arithmetic,
                Algebra, Analysis*, 3rd ed.
                Springer, Berlin.
                Dover, New York.
[1925]   *Elementary Mathematics from an Advanced Standpoint: Geometry,*
                3rd ed.
                Springer, Berlin.
                Dover, New York.
[1928]   *Vorlesungen über Nicht-Euklidische Geometrie.*
                Springer, Berlin.
M. Kline
[1972]   *Mathematical Thought from Ancient to Modern Times.*
                Oxford University Press, Oxford.
N. Koblitz
[1985]   *Introduction to Elliptic Curves and Modular Forms.*
                Springer-Verlag, New York.
P. Koebe
[1907]   Zur Uniformisierung der beliebiger analytischer Kurven.
                Göttinger Nachrichten, pp. 191–210.
A. Koestler
[1959]   *The Sleepwalkers.*
                Hutchinson, London.
                Penguin, New York, 1964.
A. N. Kolmogorov
[1933]   *Grundbegriffe der Wahrscheinlicheitsrechnung.*
                Springer, Berlin.
                English translation: *Foundations of the Theory of Probability.*
                Chelsea, New York, 1956.
G. Kreisel
[1980]   Kurt Gödel.
                Biog. Mem. Fellows Roy. Soc. **26**, 149–224.
L. Kronecker
[1881]   Zur Theorie der Elimination einer Variablen aus zwei algebra-
                ischen Gleichungen.
                Monatsber. König. Preuss. Akad. Wiss. Berlin, pp. 535–600.
                Werke 2: 113–192.
J. L. Lagrange
[1768]   Solution d'un problème d'arithmétique.
                Miscellanea Taurinensia. **4**, 19 ff.
                Oeuvres 1: 671–731.

[1770]   Demonstration d'un théorème d'arithmétique.
         Nouv. Mém. Acad. Berlin 1, 123–133.
         Oeuvres 3: 189–201.
[1771]   Réflexions sur la résolution algébrique des équations.
         Nouv. Mém. Acad. Berlin.
         Oeuvres 3: 205–421.
[1772]   Recherches sur la manière de former des tables des planetes
         d'après les seules observations.
         Mém. Acad. Roy. Sci. Paris.
         Oeuvres 6: 507–627.
[1773]   Solutions analytiques de quelques problèmes sur les pyramides
         triangulaires.
         Nouv. Mém. Acad. Berlin.
         Oeuvres 3: 658–692.
[1779]   Sur la construction des cartes géographiques.
         Nouv. Mém. Acad. Berlin.
         Oeuvres 4: 637–692.
[1785]   Sur une nouvelle méthode de calcul intégral.
         Mém. Acad. Roy. Sci. Turin 2.
         Oeuvres 2: 253–312.

P. de la Hire
[1673]   *Nouvelle Méthode en Géométrie.*
         Paris.

J. H. Lambert
[1766]   Die Theorie der Parallelinien.
         Mag. reine angew. Math. (1786), 137–164, 325–358.
[1772]   *Anmerkungen und Zusätze zur Entwerfung der Land- und Him-*
         melscharten.
         English translation by Waldo R. Tobler,
         Michigan Geographical Publication No. 8.
         Department of Geography, University of Michigan, 1972.

P. S. Laplace
[1787]   Mémoire sur les inégalités séculaires des planètes et des
         satellites.
         Mém. Acad. Roy. Sci. Paris, pp. 1–50.
         Oeuvres Complètes 11: 49–92.

P.-A. Laurent
[1843]   Extension du théorème de M. Cauchy relatif à la convergence du
         développement d'une fonction suivant les puissances ascendantes
         de la variable.
         Comp. Rend. 17, 348–349.
         J. Éc. Polytech. 23 (1863), 75–204.

H. Lebesgue
[1902]   Intégrale, longeur, aire.
         Ann. Mat., ser. 3, 7, 231–359.

A.-M. Legendre
[1794]   *Élements de géometrie.*
         Paris.
[1825]   *Traité des fonctions elliptiques.*
         Paris.

G. W. Leibniz
[1666]   *Dissertatio de Arte Combinatoria.*
         Leipzig.
         Mathematische Schriften 5: 7–79.

[1675]   De Bisectione Laterum.
                    (See Schneider [1968], p. 224.)
[1684]   Nova methodis pro maximis et minimis.
                    Acta Erud. **3**, 467–473.
                    English translation in Struik [1969].
                    Mathematische Schriften 5: 220–226.
[1686]   De geometria recondita et analysi indivisibilium atque infinitorum.
                    Acta Erud. **5**, 292–300.
                    Mathematische Schriften 5: 226–233.
[1691]   De linea in quam flexile se pondere proprio curvat, ejusque
                    usu insigni ad inveniendas quotcunque medias proportionales et
                    logarithmos.
                    Acta Erud. **10**, 277–281.
                    Mathematische Schriften 5: 243–247.
[1697]   Communicatio suae pariter duarumque alienarum ad edendum
                    sibi primum a Dn. Joh. Bernoullio.
                    Acta Erud. **16**, 205–210.
                    Mathematische Schriften 5: 331–336.
[1702]   Specimen novum analyseos pro scientia infiniti circa summas et
                    quadraturas.
                    Acta Erud. **21**, 210–219.
                    Mathematische Schriften 5: 350–361.

Levi ben Gershon
[1321]   *Maaser Hoshev.*
                    German translation by Gerson Lange: *Sefer Maassei Choscheb.*
                    Frankfurt, 1909.

Li Yan and Du Shiran
[1987]   *Chinese Mathematics: A Concise History.*
                    English translation by J. N. Crossley and A. W.-C. Lun.
                    Oxford University Press, Oxford.

U. Libbrecht
[1973]   *Chinese Mathematics in the Thirteenth Century.*
                    M.I.T. Press, Cambridge.

F. Lindemann
[1882]   Über die Zahl π.
                    Math. Ann. **20**, 213–225.

J. Liouville
[1833]   Mémoire sur les transcendantes elliptiques de première et de
                    seconde espèce considérées comme fonctions de leur amplitude.
                    J. Éc. Polytech. **23**, 37–83.
[1850]   Note IV to Monge's *Application de l'analyse à la géometrie*, 5th ed.
                    Bachelier, Paris.

N. I. Lobachevsky
[1829]   *On the Foundations of Geometry.* (Russian)
                    Kazanskij Věstnik.
[1836]   Application of imaginary geometry to some integrals. (Russian)
                    Kazan Zap. Univ. **1**, 3–166.

J. A. Lohne
[1965]   Thomas Harriot als Mathematiker.
                    Centaurus **11**, 19–45.
[1979]   Essays on Thomas Harriot.
                    Arch. Hist. Ex. Sci. **20**, 189–312.

L. A. Lyusternik
[1966]   *Convex Figures and Polyhedra.*
                    D. C. Heath, Boston.

C. Maclaurin
[1720]   *Geometrica organiza sive descriptio linearum curvarum universalis.*
                        London.

W. Magnus
[1930]   Über diskontinuierliche Gruppen mit einer definierenden Rela-
                        tion (Der Freiheitssatz).
                        J. reine angew. Math. **163**, 141–165.

M. S. Mahoney
[1973]   *The Mathematical Career of Pierre de Fermat.*
                        Princeton University Press, Princeton, N.J.

A. A. Markov
[1958]   Unsolvability of the problem of homeomorphy. (Russian)
                        Proc. Int. Cong. Math., pp. 300–306.

A. Masotti
[1974]   *Cartelli di Sfida Matematica.*
                        Ateneo di Brescia, Brescia.

J. V. Matiasevich
[1970]   Enumerable sets are diophantine. (Russian)
                        Dokl. Akad. Nauk SSSR **191**, 279–282.
                        English translation: *Soviet Math. Dokl.* **11**, 354–357.

Z. A. Melzak
[1976]   *Mathematical Ideas, Modeling and Applications.*
                        Wiley, New York.

P. Mengoli
[1650]   *Novae quadraturae arithmeticae.*
                        Bologna.

N. Mercator
[1668]   *Logarithmotechnica.*
M. Mersenne
[1625]   *Vérité des Sciences.*
                        Paris.

[1636]   *Harmonie Universelle.*
                        Paris.

F. Minding
[1839]   Wie sich entscheiden lässt, ob zwei gegebene krumme Flächen
                        auf einander abwickelbar sind oder nicht; nebst Bemerkungen
                        über die Flächen von unveränderlichen Krümmungensmasse.
                        J. reine angew. Math. **19**, 370–387.

[1840]   Beiträge zur Theorie der kürzesten Linien auf krummen
                        Flächen.
                        J. reine angew. Math. **20**, 323–327.

A. F. Möbius
[1827]   Der barycentrische Calcul.
                        Werke 1: 1–388.

[1863]   Theorie der elementaren Verwandtschaft.
                        Werke 2: 433–471.

E. E. Moise
[1963]   *Elementary Geometry from an Advanced Standpoint.*
                        Addison-Wesley, Reading, Mass.

A. de Moivre
[1698]   A method of extracting the root of an infinite equation.
                        Phil. Trans. **20**, 190–193.

[1707]   Æquationum quarundum potestatis tertiae, quintae, septimae,
                        nonae & superiorum, ad infinitum usque pergendo, in terminus
                        finitis, ad instar regularum pro cubicus que vocantur Cardani,

resolutio analytica.
Phil. Trans. **25**, 2368–2371.

[1730]   *Miscellanea Analytica.*
                London.

L. J. Mordell
[1922]   On the rational solutions of the indeterminate equations of the
                third and fourth degrees.
                Proc. Camb. Phil. Soc. **21**, 179–192.

M. B. Nathanson
[1987]   A short proof of Cauchy's polygonal number theorem.
                Proc. Am. Math. Soc. **99**, 22–24.

O. Neugebauer and A. Sachs
[1945]   *Mathematical Cuneiform Texts.*
                Yale University Press, New Haven, Conn.

C. Neumann
[1865]   *Vorlesungen über Riemann's Theorie der Abel'schen Integral.*
                Teubner, Leipzig.

[1870]   Zur Theorie des Logarithmischen und des Newtonschen Poten-
                tiales, Zweite Mitteilung.
                Ber. König. Sächs. Ges. Wiss., math.-phys. Cl, 264–321.

J. von Neumann
[1923]   Zur Einführung der transfiniten Zahlen.
                Acta lit. ac sci. Reg. U. Hungar. Fran. Jos. Sec. Sci. **1**, 199–208.
                English translation in van Heijenoort [1967], pp. 347–354.

I. Newton
[1665]   Annotations on Wallis.
                *The Mathematical Papers of Isaac Newton, Cambridge University
                Press, Cambridge*, 1967–1981. vol. 1: 96–111.

[1665′]   The geometrical construction of equations.
                *The Mathematical Papers of Isaac Newton, Cambridge University
                Press, Cambridge*, 1967–1981. vol. 1: 492–516.

[1665″]   Normals, curvature and the resolution of the general problem of
                tangents.
                *The Mathematical Papers of Isaac Newton, Cambridge University
                Press, Cambridge*, 1967–1981. vol. 1, pp. 245–297.

[1667]   Enumeratio curvarum trium dimensionum.
                *The Mathematical Papers of Isaac Newton, Cambridge University
                Press, Cambridge*, 1967–1981. vol. 12, pp. 10–89.

[1669]   De analysi.
                *The Mathematical Papers of Isaac Newton, Cambridge University
                Press, Cambridge*, 1967–1981. vol. 2, pp. 206–247.

[1671]   De methodis serierum et fluxionum.
                *The Mathematical Papers of Isaac Newton, Cambridge University
                Press, Cambridge*, 1967–1981. vol. 3, 32–353.

[1676]   Letter to Oldenburg, 13 June 1676.
                In Turnbull [1960], pp. 20–47.

[1676′]   Letter to Oldenburg, 24 October 1676.
                In Turnbull [1960]. pp. 110–149.

[late 1670s]   De resolutione quaestionum circa numeros.
                Math. Papers 4: 110–115.

[1687]   *Philosophiae naturalis principia mathematica.*
                English translation: *Mathematical Principles of Natural Philo-
                sophy.*
                University of California Press, Berkeley, 1934.

[1695]   Enumeratio linearum tertii ordinis.
                   Math. Papers 7: 588–645.
[1697]   The twin problems of Johann Bernoulli's "Programma" solved.
                   Phil. Trans. **17**, 388–389.
                   Math. Papers 8: 72–79.

J.-F. Niceron
[1638]   *La perspective curieuse.*

J. Nielsen
[1927]   Untersuchungen zur Topologie der geschlossen zweiseitigen
                   Flächen.
                   Acta Math. **50**, 189–358.

P. S. Novikov
[1955]   On the algorithmic unsolvability of the word problem in group
                   theory. (Russian)
                   Dokl. Akad. Nauk SSSR Mat. Inst. Tr. **44**.
                   English translation: *Am. Math. Soc. Transl.*, ser. 2, **9**, 1–122.

O. Ore
[1953]   *Cardano the Gambling Scholar.*
                   Princeton University Press, Princeton, N.J.
                   Dover, New York, 1965.
[1957]   *Niels Henrik Abel.*
                   Chelsea, New York.

N. Oresme
[c. 1350]   *Tractatus de configurationibus qualitatum et motuum.*
                   English translation in Clagett [1968].
[c. 1350′]   *Questiones super geometriam Euclidis.*
                   Ed. H. L. L. Busard. Brill, Leiden, 1961.

M. Ostrogradsky
[1828]   Démonstration d'un théorème du calcul integral.
                   Mém. Acad. Sci. St. Petersburg, ser. 6, **1** (1831), 39–53.

A. Ostrowski
[1920]   Über den ersten und vierten Gausschen Beweis des Fundamental-
                   satzes der Algebra.
                   Gauss Werke, 10, Part 2, 1–18.

L. Pacioli
[1509]   *De Divina Proportione.*

J. Paris and L. Harrington
[1977]   A mathematical incompleteness in Peano arithmetic.
                   In *Handbook of Mathematical Logic* (J. Barwise, ed.).
                   North-Holland, Amsterdam, pp. 1133–1142.

B. Pascal
[1640]   *Essay pour les coniques.*
                   Paris.
[1654]   Traité du triangle arithmétique, avec quelques autres petits traités
                   sur la même manière.
                   English translation in *Great Books of the Western World*:
                   Pascal. Encyclopedia Brittanica, London, 1952, pp. 447–
                   473.

K. Pearson
[1978]   *The History of Statistics in the 17th and 18th Centuries.*
                   Charles Griffin and Co., London.

J. Pierpont
[1895]   Zur Geschichte der Gleichung des V. Grades.
                   Monats. Math. Phys. **6**, 15–68.

J. Plücker
[1830]   Über ein neues Coordinatensystem.
                        J. reine angew. Math. **5**, 1–36.
                        Gesammelte Mathematische Abhandlungen, pp. 124–158.
[1847]   Nôte sur le théorème de Pascal.
                        J. reine angew. Math. **34**, 337–340.
                        Gesammelte Mathematische Abhandlungen, pp. 413–416.
H. Poincaré
[1882]   Théorie des groupes fuchsiens.
                        Acta Math. **1**, 1–62.
                        Oeuvres 2: 108–168.
                        English translation in Poincaré [1985], pp. 55–127.
[1883]   Mémoire sur les groupes Kleinéens.
                        Acta Math. **3**, 49–92.
                        Oeuvres 2: 258–299.
                        English translation in Poincaré [1985], pp. 255–304.
[1892, 1893, 1899]   *Les Méthodes Nouvelles de la Mécanique Céleste*, 3
                        vols.
                        Gauthier-Villars, Paris.
                        Dover, New York, 1957.
[1895]   Analysis situs.
                        J. Éc. Polytech., ser. 2, **1**, 1–123.
                        Oeuvres 6: 193–288.
[1901]   Sur les propriétés arithmetiques des courbes algébriques.
                        J. Math., ser. 5, **7**, 161–233.
                        Oeuvres 5: 483–548.
[1904]   Cinquième complément à l'analysis situs.
                        Palermo Rend. **18**, 45–110.
                        Oeuvres 6: 435–498.
[1907]   Sur l'uniformisation des fonctions analytiques.
                        Acta Math. **31**, 1–63.
                        Oeuvres 4: 70–139.
[1918]   *Science et Méthode.*
                        Flammarion, Paris.
                        English translation in *The Foundations of Science*,
                        Science Press, New York, 1929, pp. 357–553.
[1955]   *Le Livre du Centenaire de la Naissance de Henri Poincaré.*
                        Gauthier-Villars, Paris.
                        Oeuvres 11.
[1985]   *Papers on Fuchsian Functions.*
                        Springer-Verlag, New York.
G. Polya
[1954]   *Induction and Analogy in Mathematics.*
                        Princeton University Press, Princeton, N.J.
J. V. Poncelet
[1822]   *Traité des propriétés projectives des figures.*
E. Post
[1936]   Finite combinatory processes. Formulation 1.
                        J. Symb. Logic **1**, 103–105.
[1941]   Absolutely unsolvable problems and relatively undecidable prop-
                        ositions. Account of an anticipation.
                        In Davis [1965], pp. 340–433.

[1944]   Recursively enumerable sets of integers and their decision problems.
Bull. Am. Math. Soc. **50**, 284–316.

E. Prouhet
[1860]   Remarques sur un passage des oeuvres inédits de Descartes.
Comp. Rend. **50**, 779–781.

V.-A. Puiseux
[1850]   Recherches sur les fonctions algébriques.
J. Math. **15**, 365–480.

N. L. Rabinovitch
[1969]   Rabbi Levi ben Gershon and the origins of mathematical induction.
Arch. Hist. Ex. Sci. **6**, 237–248.

C. T. Rajagopal and M. S. Rangachari
[1977]   On an untapped source of medieval Keralese mathematics.
Arch. Hist. Ex. Sci. **18**, 89–102.

[1986]   On medieval Kerala mathematics.
Arch. Hist. Ex. Sci. **35**, 91–99.

F. P. Ramsey
[1929]   On a problem of formal logic.
Proc. Lond. Math. Soc. **30**, 291–310.

F. V. Raspail
[1839]   *Lettres sur les Prisons de Paris*, vol. 2.
Paris.

G. F. B. Riemann
[1851]   Grundlagen für eine allgemeine Theorie der Functionen einer veränderlichen complexen Grösse.
Werke, 2nd ed., pp. 3–48.
Dover, New York, 1953.

[1854]   Über die Hypothesen, welche der Geometrie zu Grunde liegen.
Werke, 2nd ed., pp. 272–287.

[1854']   Über die Darstellbarkeit einer Function durch eine trigonometrische Reihe.
Werke, 2nd ed., pp. 227–264.

[1857]   Theorie der Abel'schen Funktionen.
J. reine angew. Math. **54**, 115–155.
Werke, 2nd ed., pp. 82–142.

[1858]   *Elliptische Funktionen.*
Ed. H. Stahl. Leipzig, 1899.

[1858/9]   Vorlesungen über die hypergeometrische Reihe.
Werke Nachträge (M. Noether and W. Wirtinger, eds.), 1902, pp. 69–93.
In *Werke*, Dover, New York, 1953.

[1859]   Über die Anzahl der Primzahlen unter einer gegebenen Grösse.
Werke, 2nd ed., 145–153.
English translation in Edwards [1974], pp. 299–305.

A. Robert
[1973]   *Elliptic Curves.*
Springer-Verlag, New York.

A. Robinson
[1966]   *Non-standard Analysis.*
North-Holland, Amsterdam.

P. L. Rose
[1975]   *The Italian Renaissance of Mathematics.*
                          Libraire Droz, Geneva.

M. Rosen
[1981]   Abel's theorem on the lemniscate.
                          Am. Math. Mon. **88**, 387–395.

T. Rothman
[1982]   Genius and biographers: The fictionalization of Evariste Galois.
                          Am. Math. Mon. **89**, 84–106.

P. Ruffini
[1799]   *Teoria generale delle equazioni in cui si dimostra impossibile la
                          soluzione algebraica delle equazioni generali di grade superiore al
                          quarto.*
                          Bologna.

G. Saccheri
[1733]   *Euclides ab omni naevo vindicatus.*
                          Milan.
                          English translation: Open Court, Chicago, 1920.

G. Salmon
[1851]   Théorèmes sur les courbes de troisième degré.
                          J. reine angew. Math. **42**, 274–276.

I. Schneider
[1968]   Der Mathematiker Abraham de Moivre (1667–1754).
                          Arch. Hist. Ex. Sci. **5**, 177–317.

F. van Schooten
[1659]   *Geometria a Renato Des Cartes.*
                          Amsterdam.

H. A. Schwarz
[1870]   Über einen Grenzübergang durch alternirendes Verfahren.
                          Vierteljahrssch. Natur. Ges. Zürich **15**, 272–286.
                          Mathematische Abhandlungen 2: 133–143.

[1872]   Über diejenigen Fälle, in welchem die Gaussische hypergeometri-
                          sche Reihe eine algebraische Function ihres vierten Elementes
                          darstellt.
                          J. reine angew. Math. **75**, 292–335.
                          Mathematische Abhandlungen 2: 211–259.

G. P. Scott
[1983]   The geometries of 3-manifolds.
                          Bull. Lond. Math. Soc. **15**, 401–487.

J. F. Scott
[1952]   *The Scientific Work of René Descartes.*
                          Taylor and Francis, London.

H. Seifert and W. Threlfall
[1934]   *Lehrbuch der Topologie.*
                          Teubner, Leipzig.
                          English translation: *A Textbook of Topology.*
                          Academic Press, New York, 1980.

S. Shelah
[1984]   Can you take Solovay's inaccessible away?
                          Israel J. Math. **48**, 1–47.

J. W. Shirley
[1983]   *Thomas Harriot, A Biography.*
                          Oxford University Press, Oxford.

C. L. Siegel
[1969]   *Topics in Complex Function Theory*, vols. 1, 2, 3.
                          Wiley-Interscience, New York.

R. F. de Sluse
[1673]   A method of drawing tangents to all geometrical curves.
                          Phil. Trans. **7**, 5143–5147.

D. E. Smith
[1929]   *A Source Book in Mathematics.*
                          Dover, New York, 1959.

R. Solovay
[1970]   A model of set theory in which every set of reals is Lebesgue
                          measurable.
                          Ann. Math. **92**, 1–56.

P. Stäckel
[1901]   Die Entdeckung der nichteuklidischen Geometrie durch Johann
                          Bolyai.
                          Math.-natur. Ber. Ungarn. Budapest **17**, 1–19.
[1913]   *Wolfgang und Johann Bolyai. Geometrische Untersuchungen.*
                          Teubner, Leipzig.

S. Sternberg
[1969]   *Celestial Mechanics*, Part 1.
                          W.A. Benjamin, Menlo Park, Calif.

S. Stevin
[1586]   *Statics.*
J. C. Stillwell
[1980]   *Classical Topology and Combinatorial Group Theory.*
                          Springer-Verlag, New York.
[1982]   The word problem and the isomorphism problem for groups.
                          Bull. Am. Math. Soc., new ser., **6**, 33–56.

J. Stirling
[1717]   *Lineae Tertii Ordinis Neutonianae.*
K. Strubecker
[1964]   *Differentialgeometrie* I, II, III.
                          Walter de Gruyter, Berlin.

D. J. Struik
[1969]   *A Source Book in Mathematics* 1200–1800.
                          Harvard University Press, Cambridge.

I. Szabó
[1977]   *Geschichte der mechanischen Prinzipien.*
                          Birkhäuser, Basel.

N. Tartaglia
[1546]   *Quesiti et Inventioni Diverse.*
                          Facsimile of 1554 edition. Ed. A. Masotti.
                          Ateneo di Brescia, Brescia.

R. Taton
[1951]   *L'Oeuvre Mathématique de G. Desargues.*
                          Presses Universitaires de France, Paris.

F. A. Taurinus
[1826]   *Geometriae prima elementa.*
                          Cologne.

B. Taylor
[1713]   De motu nervi tensi.
                          Phil. Trans. **28**, 26–32.

[1715]   *Methodos incrementorum directa et inversa.*
                          London.
W. P. Thurston
[1982]   *The Geometry and Topology of 3-manifolds.*
                          Princeton Mathematics Department.
H. Tietze
[1908]   Über die topologischen Invarianten mehrdimensionaler Mannig-
                          faltigkeiten.
                          Monatsh. Math. Phys. **19**, 1–118.
E. Torricelli
[1643]   *De solido hyperbolico acuto.*
                          Partial English translation in Struik [1969], pp. 227–231.
[1644]   *De dimensione parabolae.*
[1645]   *De infinitis spiralibus.*
                          Ed. E. Carruccio, Pisa, 1955.
[1646]   *De infinitis hyperbolis.*
C. A. Truesdell
[1954]   Rational Fluid Mechanics 1687–1765.
                          Euleri Opera Omnia, ser. 2, 12: IV–CXXV.
[1960]   The Rational Mechanics of Flexible or Elastic Bodies 1638–
                          1788.
                          Euleri Opera Omnia, ser. 2, 11, section 2.
A. Turing
[1936]   On computable numbers, with an application to the Entscheid-
                          ungsproblem.
                          Proc. Lond. Math. Soc., ser. 2, **42**, 230–265.
H. W. Turnbull (Ed.)
[1939]   *James Gregory Tercentenary Memorial Volume.*
                          G. Bell and Sons. London.
[1960]   *The Correspondence of Isaac Newton*, vol. 2.
                          Cambridge University Press, Cambridge.
S. Ulam
[1930]   Zur Masstheorie in der allgemeinen Mengenlehre.
                          Fund. Math. **15**, 140–150.
A.-T. Vandermonde
[1771]   Mémoire sur la résolution des équations.
                          Hist. Acad. Roy. Sci.
F. Viète
[1579]   *Universalium inspectionium ad canonem mathematicum liber
                          singularis.*
[1591]   De aequationum recognitione et emendatione.
                          Opera, pp. 82–162.
                          English translation in Viète [1983], pp. 159–310.
[1593]   Variorum de rebus mathematicus responsorum libri octo.
                          Opera, pp. 347–435.
[1615]   Ad angularium sectionum analyticen theoremata.
                          Opera, pp. 287–304.
                          English translation in Viète [1983], pp. 418–450.
[1983]   *The Analytic Art*, translated by R. T. Witmer.
                          Kent State University Press, Kent, Ohio.
G. Vitali
[1905]   *Sul problema della misura dei gruppi di punti di una retta.*
                          Bologna.

J. R. Vrooman
[1970]   *René Descartes. A Biography.*
                  Putnam, New York.
B. L. van der Waerden
[1949]   *Modern Algebra.*
                  Frederick Ungar, New York.
[1954]   *Science Awakening.*
                  P. Noordhoff, Groningen.
[1976]   Pell's equation in Greek and Hindu mathematics.
                  Russ. Math. Surv. **31**, 210–225.
[1983]   *Geometry and Algebra in Ancient Civilizations.*
                  Springer-Verlag, New York.
[1985]   *A History of Algebra.*
                  Springer-Verlag, New York.
S. Wagon
[1985]   *The Banach–Tarski Paradox.*
                  Cambridge University Press, Cambridge.
J. Wallis
[1655]   De sectionibus conicis.
                  Opera 1: 291–354.
[1655′]   Arithmetica infinitorum.
                  Opera 1: 355–478.
[1657]   Mathesis universalis.
                  Opera 1: 11–228.
[1659]   Tractatus duo. Prior, de cycloide. Posterior, de cissoide.
                  Opera 1: 489–569.
[1663]   De Postulato quinto; et definitione quinta Lib. 6 Euclidis.
                  Opera 2: 669–678.
[1673]   On Imaginary Numbers.
                  English translation from his *Algebra*, vol. 2,
                  In Smith [1929], 1: 46–54.
[1696]   Autobiography.
                  Notes and Records, Roy. Soc. Lond. **25** (1970), 17–46.
P. L. Wantzel
[1837]   Recherches sur les moyens de reconnaitre si un problème de
                  géométrie peut se resoudre avec la règle et le compas.
                  J. Math. **2**, 366–372.
K. Weierstrass
[1863–1875]   Vorlesungen über die Theorie der elliptischen Funktionen.
                  Mathematische Werke 5.
[1874]   Einleitung in die Theorie der analytischen Funktionen.
                  Summer Semester 1874. Notes by G. Hettner.
                  Mathematische Institut der Universität Göttingen.
A. Weil
[1973]   Review of "The mathematical career of Pierre de Fermat, by
                  M. S. Mahoney."
                  Bull. Am. Math. Soc. **79**, 1138–1149.
[1976]   *Elliptic Functions According to Eisenstein and Kronecker.*
                  Springer-Verlag, New York.
[1984]   *Number Theory. An Approach Through History.*
                  Birkhäuser, Basel.
C. Wessel
[1797]   Om diretioneus analytiske Betegning, et Forsög anwendt for-

nemelig til plane og sphaeriske Polygoners Oplössing.
Danske Selsk. Skr. N. Samml. **5**.
English translation in Smith [1929], 1: 55–66.

R. S. Westfall
[1980]   *Never at Rest.*
                           Cambridge University Press, Cambridge.

A. N. Whitehead and B. Russell
[1910, 1912, 1913]   *Principia Mathematica*, 3 vols.
                           Cambridge University Press, Cambridge.

D. T. Whiteside
[1961]   Patterns of mathematical thought in the later seventeenth century.
                           Arch. Hist. Ex. Sci. **1**, 179–388.
[1964]   Introduction.
                           The Mathematical Works of Isaac Newton.
                           Johnson Reprint Corp., New York.
[1966]   Newton's marvellous year: 1666 and all that.
                           Notes and Records, Roy. Soc. Lond. **21**, 32–41.

L. Wright
[1983]   *Perspective in Perspective.*
                           Routledge and Kegan Paul, London.

H. Wussing
[1969]   *Die Genesis des Abstrakten Gruppenbegriffes.*
                           VEB Deutscher Verlag der Wissenschaften, Berlin.
                           English translation: *The Genesis of the Abstract Group Concept.*
                           MIT Press, Cambridge, 1984.

Yáng Huí
[1261]   Compendium of analyzed mathematical methods in the "Nine Chapters."

H. G. Zeuthen
[1903]   *Geschichte der Mathematik im 16. und 17. Jahrhundert.*
                           Teubner, Leipzig.
                           Johnson Reprint Corp., New York, 1977.

E. Zermelo
[1904]   Beweis das jede Menge wohlgeordnet werden kann.
                           Math. Ann. **59**, 514–516.
                           English translation in van Heijenoort [1967], pp. 139–141.

Zhū Shìjié
[1303]   *Siyuan yujian* (Jade mirror of four unknowns).

Index

Undergraduate Texts in Mathematics

(continued)

Ross: Elementary Analysis: The Theory of Calculus.
Scharlau/Opolka: From Fermat to Minkowski.
Sigler: Algebra.
Simmonds: A Brief on Tensor Analysis.
Singer/Thorpe: Lecture Notes on Elementary Topology and Geometry.
Smith: Linear Algebra. Second edition.
Smith: Primer of Modern Analysis.
Stanton/White: Constructive Combinatorics.
Stillwell: Mathematics and Its History.
Strayer: Linear Programming and Its Applications.
Thorpe: Elementary Topics in Differential Geometry.
Troutman: Variational Calculus with Elementary Convexity.
Wilson: Much Ado About Calculus.